AF739064

Current Topics in Behavioral Neurosciences

Volume 42

Series Editors
Mark A. Geyer, La Jolla, CA, USA
Bart A. Ellenbroek, Wellington, New Zealand
Charles A. Marsden, Nottingham, UK
Thomas R.E. Barnes, London, UK
Susan L. Andersen, Belmont, MA, USA

Current Topics in Behavioral Neurosciences provides critical and comprehensive discussions of the most significant areas of behavioral neuroscience research, written by leading international authorities. Each volume in the series represents the most informative and contemporary account of its subject available, making it an unrivalled reference source. Each volume will be made available in both print and electronic form.

With the development of new methodologies for brain imaging, genetic and genomic analyses, molecular engineering of mutant animals, novel routes for drug delivery, and sophisticated cross-species behavioral assessments, it is now possible to study behavior relevant to psychiatric and neurological diseases and disorders on the physiological level. The Behavioral Neurosciences series focuses on translational medicine and cutting-edge technologies. Preclinical and clinical trials for the development of new diagnostics and therapeutics as well as prevention efforts are covered whenever possible. Special attention is also drawn on epigenetical aspects, especially in psychiatric disorders.

CTBN series is indexed in PubMed and Scopus.

More information about this series at http://www.springer.com/series/7854

Elisabeth B. Binder • Torsten Klengel
Editors

Behavioral Neurogenomics

Springer

Editors
Elisabeth B. Binder
Department of Translational
Research in Psychiatry
Max Planck Institute of Psychiatry
Munich, Germany

Torsten Klengel
McLean Hospital, Harvard Medical School
Belmont, MA, USA

ISSN 1866-3370 ISSN 1866-3389 (electronic)
Current Topics in Behavioral Neurosciences
ISBN 978-3-030-31264-0 ISBN 978-3-030-31265-7 (eBook)
https://doi.org/10.1007/978-3-030-31265-7

© Springer Nature Switzerland AG 2019
Chapter "Role of MicroRNAs in Anxiety and Anxiety-Related Disorders" is licensed under the terms of the Creative Commons Attribution 4.0 International License (http://creativecommons.org/licenses/by/4.0/). For further details see licence information in the chapter.
This work is subject to copyright. All rights are reserved by the Publisher, whether the whole or part of the material is concerned, specifically the rights of translation, reprinting, reuse of illustrations, recitation, broadcasting, reproduction on microfilms or in any other physical way, and transmission or information storage and retrieval, electronic adaptation, computer software, or by similar or dissimilar methodology now known or hereafter developed.
The use of general descriptive names, registered names, trademarks, service marks, etc. in this publication does not imply, even in the absence of a specific statement, that such names are exempt from the relevant protective laws and regulations and therefore free for general use.
The publisher, the authors, and the editors are safe to assume that the advice and information in this book are believed to be true and accurate at the date of publication. Neither the publisher nor the authors or the editors give a warranty, expressed or implied, with respect to the material contained herein or for any errors or omissions that may have been made. The publisher remains neutral with regard to jurisdictional claims in published maps and institutional affiliations.

This Springer imprint is published by the registered company Springer Nature Switzerland AG.
The registered company address is: Gewerbestrasse 11, 6330 Cham, Switzerland

Preface

Understanding the interaction of genotype, environment, and phenotype is a central challenge in decoding normative and disease-associated behavior. We are currently at a critical intersection of increasingly larger, population-wide human studies and rapid technological advances, particularly in the fields of neurobiology, genomics, and digital phenotyping. Efforts combining high-throughput genomic technologies with global collaborative clinical studies have been rewarded, for example, by a growing number of robust, replicable genetic associations from genome-wide association studies of psychiatric disorders. Moreover, human and animal model studies provide converging evidence for the profound effects of environmental factors on long-lasting behavioral changes, which are likely mediated by epigenetic factors. Recent technological development in single-cell technologies, human in vitro cellular models, and functional genomics now provide the opportunity to understand how genetic and epigenetic factors influence gene and cell function and by this shape behavior and disease risk. Combining these new technologies with robust clinical studies carries the promise of a better understanding of the etiology, prevention, and treatment of psychiatric disorders.

This book illustrates how these new developments have recently furthered our understanding in the area of *Behavioral Neurogenomics*. In the first chapter, Divya Mehta and Darina Czamara examine the progress in genome-wide association studies, which arguably provide unprecedented insight into the complex genetic architecture of behavioral traits and psychiatric disorders. At the same time, these studies fall short in providing mechanistic insight into genetic moderation of certain behaviors or disease risk. In part this is due to the limited understanding of genome organization and function. John Fullard, Samir Rahman, and Panos Roussos focus in the second chapter on cis-regulatory elements in our genome and how genetic variation identified through large-scale studies may influence non-coding, regulatory genomic features. Fullard and colleagues provide an example of how association study findings may be utilized to investigate the molecular mechanisms of genetically driven phenotypes. However, besides behavioral variance/variation as a function of genotype, environmental factors can have long-lasting effects on behavior and disease risk. Two chapters, one by Rustad, Papale, and Alisch and the other by Merrill, Gladish, and Kobor, focus on mechanisms of how environmental factors may change molecular trajectories that eventually have an impact on behavior. While Rustad and colleagues focus on the dynamic function of DNA methylation and hydroxymethylation in the brain, Merill, Gladish, and Kobor focus on how the social environment, in particular in early life, shapes epigenetic modifications with long-lasting effects across the life span. The chapter by Fischer expands the view on epigenetic modifications and behavior specifically focusing on the role of histone modifications in learning and memory formation. Importantly, Fischer discusses the enzymatic machinery involved in histone biology and points to potential therapeutic opportunities to treat brain diseases. In the next chapter, Kaindl and Winner highlight the opportunities that arise with human *in vitro* cellular models as animal models currently cannot model the complex genetic heterogeneity of human behavior. Murphy and Singewald then return to molecular mechanisms underlying behavior. Their focus on microRNAs, a species of non-coding RNAs strongly expressed in the brain, highlights the importance of regulatory RNAs expressed in central hubs of the anxiety neurocircuitry modulating behavior in animal models. Furthermore, they link microRNA expression with known pathways regulating neurotransmitter release and signaling, synaptic plasticity, and stress-hormone axis function. Guffanti, Bartlett, DeCrescenzo, Macciardi, and Hunter explore the far less known role of transposable elements. The depiction of transposable elements as junk DNA seems shortsighted and data reviewed by Guffanti and colleagues provide interesting insights into the possible function of these elements in the brain influencing behavior and risk for disease. Finally, Vassoler, Toorie, and Byrnes review evidence for inter- and transgenerational effects of environmental cues. Behavioral phenotypes associated with inter-, multi-, and transgenerational effects following a variety of parental exposures are intriguing and may have an important role in shaping disease risk. Yet, the molecular substrate linked to the transmission of these cues remains elusive with multiple studies pointing towards DNA methylation or small RNA signaling mechanisms.

The chapters in this volume of *Current Topics in Behavioral Neurogenomics* thus provide a broad overview of the dynamic field of *Behavioral Neurogenomics*. They are written for both basic scientists and research-oriented clinicians and provide an in-depth introduction into the various areas of research. We believe that additional technological developments in basic and clinical neurogenetics will further accelerate our understanding of how genetics, environmental factors, and the resulting phenotypes interact to shape human behavior in health and disease. Knowledge of the underlying machinery driving human behavior will eventually influence preventive, diagnostic, and therapeutic strategies for complex psychiatric disorders.

Munich, Germany
Belmont, MA, USA
September 2019

Elisabeth B. Binder
Torsten Klengel

Contents

GWAS of Behavioral Traits

Divya Mehta and Darina Czamara

Contents

Abstract Over the past decade, genome-wide association studies (GWAS) have evolved into a powerful tool to investigate genetic risk factors for human diseases via a hypothesis-free scan of the genome. The success of GWAS for psychiatric disorders and behavioral traits have been somewhat mixed, partly owing to the complexity and heterogeneity of these traits. Significant progress has been made in the last few years in the development and implementation of complex statistical

D. Mehta (✉)
School of Psychology and Counselling, Faculty of Health, Institute of Health and Biomedical Innovation, Queensland University of Technology, Kelvin Grove, QLD, Australia
e-mail: divya.mehta@qut.edu.au

D. Czamara
Department of Translational Psychiatry, Max Planck Institute of Psychiatry, Munich, Germany

© Springer Nature Switzerland AG 2019

Curr Topics Behav Neurosci (2019) 42: 1–34
DOI 10.1007/7854_2019_105
Published Online: 13 August 2019

methods and algorithms incorporating GWAS. Such advanced statistical methods applied to GWAS hits in combination with incorporation of different layers of genomics data have catapulted the search for novel genes for behavioral traits and improved our understanding of the complex polygenic architecture of these traits.

This chapter will give a brief overview on GWAS and statistical methods currently used in GWAS. The chapter will focus on reviewing the current literature and highlight some of the most important GWAS on psychiatric and other behavioral traits and will conclude with a discussion on future directions.

Keywords Genome-wide association studies · Heritability · Psychiatric disorders

1 Introduction to Genome-Wide Association Studies (GWAS)

In genome-wide association studies (GWAS), association analysis between a categorical or quantitative trait and genome-wide single-nucleotide polymorphisms (SNPs) is performed. The number of single-nucleotide polymorphisms (SNPs) involved in disease susceptibility discovered in GWAS has increased considerably throughout the past decades (Visscher et al. 2012). Two main reasons drove this high increase in GWAS: the fast progressing development of new technologies and the fact that SNP genotyping on whole-genome level became not only available but also cheaper and quicker. Today, millions of SNPs can be genotyped using high-throughput methods. Furthermore, Risch and Merikangas (1996) showed that association analysis has greater power to detect susceptibility loci as compared to linkage analysis if the loci exert only a small effect on the disease. In the 1990s, linkage studies, which are familial studies assessing if markers are inherited to affected offspring more often than would be expected by chance, were commonly performed. However, the linkage method performs best if rare mutations with large effects are present, but as the genetic basis of psychiatric and behavioral disorders is rather complex (Smoller 2014), association studies are an ideal tool. GWAS in particular are suited for this purpose because they are hypothesis-free as SNPs throughout the whole genome are analyzed. This is in contrast to candidate gene studies that focus on specific genetic regions identified in previous studies or of interest because of their biological function.

In the first GWAS, susceptibility genes for several complex traits including type I diabetes (Wellcome Trust Case Control C 2007), type II diabetes (Sladek et al. 2007), and breast cancer (Easton et al. 2007) were identified. Nevertheless, for other complex diseases, this approach was not very successful, with only few genome-wide significant loci, or borderline associations were observed.

Above a certain "inflection point" (a critical sample size), the number of observed significant associations increases linearly with the sample size (Levinson et al. 2014). This was described, for example, for lipids and blood pressure which present with linear or even more rapid increase in findings for sample sizes above this critical threshold (Panagiotou et al. 2013). Based on this, it is very likely that complex diseases are predicted by a polygenic model where a high number of SNPs, each one only exhibiting a small effect size, are involved in disease susceptibility. A consequence of this is that large sample sizes of several thousand individuals are necessary to detect such small effects and indeed, the larger the sample sizes got, the more association hits were found (Visscher et al. 2012; Levinson et al. 2014).

To reach sample sizes of thousands of individuals, a large number of national and international consortia have been established, which combine datasets of individual studies in meta- or mega-analyses. One of the first was the Wellcome Trust Case Control Consortium (WTCCC) in 2005 which studied diabetes, bipolar disorder, and Crohn's disease, among other disorders (Wellcome Trust Case Control C 2007). The Psychiatric Genomics Consortium (PGC) is one of the largest consortia today including over 900,000 samples across several psychiatric disorders such as schizophrenia, bipolar disorder, major depression, or ADHD (Sullivan 2010).

2 Methods for GWAS

2.1 *Simple Analysis of Single Markers*

In a genome-wide association study, the association between a phenotype of interest and SNPs spread across the whole genome is tested. Different genotyping arrays covering up to 2,300,000 SNPs are available. After array processing, genotypes for each sample are called based on their signal intensities. This step is followed by quality control which removes SNPs as well as individuals with low performance. Usually, SNPs with low call rates, with low minor allele frequency, and with significant deviation from Hardy-Weinberg equilibrium as well as individuals with low call rates and large deviations in mean heterozygosity with respect to the whole sample are excluded. The remaining SNPs and samples are then used for further statistical analysis.

Not all SNPs are covered on genotyping arrays; however, based on the structure of the human genome in linkage disequilibrium (LD) blocks, missing genotypes can be predicted to a certain extent. Large samples from different ethnicities that have been sequenced such as the HapMap (International HapMap C 2003) and the 1000 Genomes projects (Genomes Project et al. 2010) are available for imputation. Based on the LD pattern in these reference samples, missing genotypes can be imputed in the study samples (Marchini et al. 2007), and the imputation algorithms work quite accurately (Howie et al. 2009). After imputation and quality control, association analysis is performed.

Very often affected individuals, for example, depression cases, are compared to healthy control subjects with regard to significant differences in allele frequencies. This comparison can be performed with logistic regression where also possible confounding variables, such as age, gender, or ethnicity, can be used as covariates

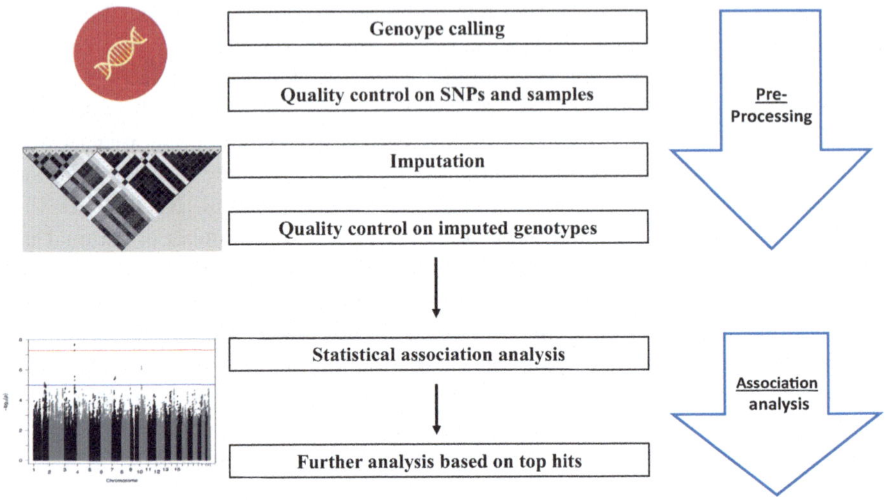

Fig. 1 GWAs workflow

in the analysis. In the GWAS summary statistics results, for each SNP the *p*-value, odds ratio (OR), standard error, and the reference allele are indicated.

If the phenotype of interest is a quantitative trait, a depression scale, for example, linear regression, is used which reports the effect size as the mean increase/decrease of the phenotype when a further copy of the reference allele is present; this analysis also allows the addition of covariates. On a genome-wide level, both logistic and linear regression can be easily performed using freely available analysis tools such as PLINK (Chang et al. 2015; Purcell et al. 2007).

Based on the assumption that one million independent SNPs are present in the European human genome, *p*-values below the threshold of 5×10^{-08} are considered as genome-wide significant (Pe'er et al. 2008). In the next steps, top hits (SNPs) are more closely investigated with regard to their genomic position, and further analyses, such as pathway analysis, are performed. Figure 1 summarizes the preprocessing and analysis pipeline of GWAS.

2.2 *Using GWAS Data to Investigate the Complex Genetic Architecture of Psychiatric Disorders: Heritability, Genetic Correlation, and Polygenic Risk Scores*

Increased risk of psychiatric disorders among relatives of affected individuals points toward a genetic contribution for these disorders, and this has led to a plethora of genome-wide association studies (GWAS) conducted for psychiatric disorders. Nevertheless, so far, GWAS have identified very few genome-wide significant

SNPs for psychiatric disorders, and the identified SNPs only explain a small proportion of variance of the phenotype. These findings suggested that newer methods were needed to account for the polygenic genetic architecture of psychiatric disorders. More recent methods demonstrated that variants associated with a particular behavioral trait can also be aggregated into a single empirical polygenic genomic risk score for each individual (Ruderfer et al. 2014). The polygenic genomic risk score method (GPRS) takes statistically independent risk alleles and their effect sizes from a GWAS (discovery or training set) and, after LD pruning, uses these to generate a combined genomic profile risk score for each individual in an independent target sample (Wray et al. 2014). A GPRS is defined as the sum of the count of risk alleles weighted by the effect size in the discovery sample (Iyegbe et al. 2014). Different inclusion thresholds for selection of SNPs based on the GWAS significance can be used to build GPRS via software tools such as PLINK (Chang et al. 2015; Purcell et al. 2007) and PRSice (Euesden et al. 2015). An advantage of this method is that the statistical power is dependent on the training set; hence, a larger training data can be used for risk prediction in smaller target samples. The GPRS method has been widely applied both within and across psychiatric disorders analyses (Cross-Disorder Group of the Psychiatric Genomics C 2013; International Schizophrenia C et al. 2009).

Genome-wide genotype data can also be used to estimate additive heritability attributable to common genetic variation, known as "SNP heritability" (h^2_{SNP}) (Lee et al. 2011a; Yang et al. 2010). Estimation of the variance explained by all SNPs (h^2_{SNP}) or the genetic similarity between unrelated individuals in the study is performed using methods called GREML (genome-based restricted maximum likelihood) (Benjamin et al. 2012). The GREML-based method GCTA (genome-wide complex trait analysis) can be used to assess the degree of genetic relatedness between individuals by building a genetic relationship matrix, based on the assumption that cases are genetically more similar to each other than to controls (Yang et al. 2011). The genetic relationship matrix is then correlated with dichotomous or quantitative phenotypes (Lee et al. 2011a; Yang et al. 2010, 2011; Speed et al. 2012). An extension of GCTA is the estimation of genetic correlation (rg_{SNP}) explained by GWAS SNPs between two disorders (Lee et al. 2012a), with a positive correlation indicating that the cases of one disorder show higher genetic similarity to the cases of the other disorder than to their own controls. This method can also be used to calculate SNP-based co-heritability across pairs of disorders. Both GPRS and GREML methods can be applied simultaneously to datasets to provide further evidence for shared polygenic relationships between traits. A limitation of these methods is that they are computationally intensive and require individual-level genotype data.

The limitations of the conventional methods sparked the development of a newer method termed cross-trait linkage disequilibrium (LD) score regression that enables estimation of genetic correlation using only summary statistics from GWAS (Bulik-Sullivan et al. 2015a). This method does not require individual genotypes, genome-wide significant SNPs, or LD pruning and is computationally fast. The LD score regression approach functions under the assumption that if a trait is genetically influenced, then variants in high LD would tag causal variants and be more significantly associated with the trait and hence have higher summary statistics as compared to variants with low LD. This method is flexible and can be adapted to estimate SNP heritability (Bulik-Sullivan et al. 2015a), partition SNP heritability by functional categories (Finucane et al. 2015), and can be used to estimate genetic correlation between different psychiatric disorders (Bulik-Sullivan et al. 2015b). Based on this method, the LD Hub web interface was built, which is a centralized database (http://ldsc.broadinstitute.org/) that allows calculation of heritability and genetic correlation via LD score regression analysis of user GWAS data against summary-level GWAS results for 173 diseases/traits (Zheng et al. 2017).

The above described methods complement each other and capitalize on GWAS data to gain further insights into the genetic architecture of psychiatric disorders.

2.3 Methods for Identification of Causal Genes from GWAS via Gene Enrichment, Network Analysis, and Other Functional Data

GWAS have been implemented for complex traits such as psychiatric disorders; however, identifying new risk loci and interpreting the results have been more challenging than anticipated. LD between SNPs forms the major backbone of GWAS, but it also makes it difficult to disentangle causal variants from nonfunctional variants co-occurring in large LD blocks. As a result, genetic variants and genes identified in GWAS are generally aggregated via pathway and network-based methods according to shared biology or function to identify likely causal genes within GWAS-implicated loci (Kao et al. 2017). Figure 2 shows different paths from identification of GWAS SNPs to unraveling risk genes for psychiatric disorders.

Gene set analysis allows to prioritize loci where multiple SNPs or genes within particular biological processes or molecular function groups show evidence of association. A widely used gene set analysis method is gene set enrichment analysis (GSEA) that evaluates whether members of a particular gene set tend to be more significantly correlated with the phenotype than expected by chance alone (Subramanian et al. 2005). Gene set analysis methods can bypass stringent multiple-testing corrections needed for analysis of individual signals and concentrate on biologically distinct sets of genes or processes, hence increasing the power of the study. Numerous public databases have pre-available gene sets that can be used for gene set and pathway analyses, including the Gene Ontology (Ashburner et al. 2000), WikiPathways (Kutmon et al. 2016), the Pathway Commons database (Cerami et al. 2011), the Kyoto Encyclopedia of Genes and Genomes (KEGG) (Kanehisa and Goto 2000), and DAVID (da Huang et al. 2009; Huang et al. 2007), among others. Other pathway-based GWAS analysis tools include INRICH (INterval enRICHment analysis) that tests for enriched association signals of

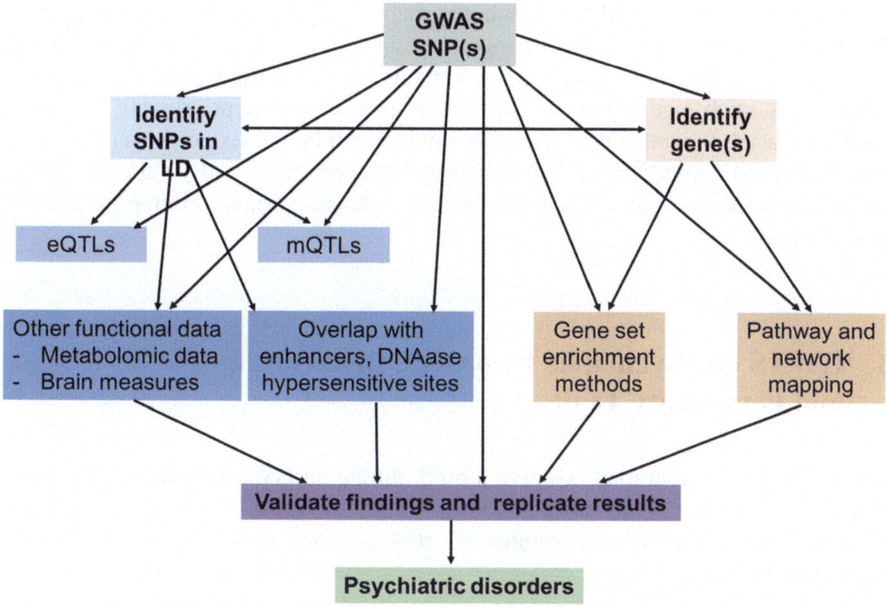

Fig. 2 From GWAS to psychiatric disorder

predefined gene sets across independent genomic intervals (Lee et al. 2012b) and Meta-Analysis Gene Set Enrichment of Variant or MAGENTA (Segre et al. 2010) that tests for enrichment of genetic associations in predefined biological processes or sets of functionally related genes. Enriched gene sets might offer deeper insight into disease mechanisms and biology of disease.

Different causal genes for the same phenotypic trait often directly interact or interact via common interaction partners, by the principle of "guilt by association"; hence, network-based approaches can also be useful to pinpoint putative causal gene(s) from LD blocks by finding genes that are nearby or related in a network to other known causal genes. Biological networks, such as protein-protein interaction (PPI) networks, can be used to define gene sets by selecting genes that interact at the protein level and plotting direct PPI neighbors and the interactions between these proteins to visualize large protein networks (Cowley et al. 2012).

Finally, using quantitative information such as gene expression or DNA methylation data can also reveal GWAS variants that are more likely to have consequences on gene activity or function. GWAS variants associated with gene expression are referred to as expression quantitative trait loci (eQTL) or expression SNPs (eSNPs). It has been demonstrated that complex trait-associated SNPs identified by GWAS are more likely to be eQTLs, confirming that genetic risk factors for complex traits will often affect phenotype by altering the amount or timing of protein production (Nicolae et al. 2010). Harnessing gene expression information can therefore enhance discovery of trait-associated SNPs and help us to gain a better understanding of the biology underlying complex traits.

Gene set enrichment and network-based methods described above are limited by the existing knowledge of the biological pathways and hence can have limited coverage and might miss interactions, reducing the sensitivity of these analyses. Despite this, such methods provide a further step toward the prioritizing candidate genes or causal variants for further experimental validation. Extensive analysis and combination of computational and experimental approaches will yield mechanistic insights into the molecular process by which a genetic variant(s) influence psychiatric disorders.

3 GWAS on Psychiatric Disorders, Behavioral, and Personality Traits

In the following paragraphs, we give a brief outline on GWAS findings for behavioral traits. In reviewing existing literature, we put the focus on the most important findings and/or the largest published studies; this review does not claim completeness.

3.1 Schizophrenia

Schizophrenia is a severe, chronic mental health disorder which is characterized by a variety of symptoms, including delusions, hallucinations, disorganized speech or behavior, and impaired cognitive abilities (Patel et al. 2014). It presents with a lifetime risk of 1% and is highly heritable with heritability estimates of up to 80% (Cardno and Gottesman 2000; Sullivan et al. 2003), stressing the importance of genetic variants in the etiology of schizophrenia. The first GWAS on schizophrenia (Mah et al. 2006; Lencz et al. 2007; Sullivan et al. 2008; O'Donovan et al. 2008; Shifman et al. 2008; Kirov et al. 2009; Need et al. 2009) were underpowered to find any genome-wide significant hits (Bergen and Petryshen 2012). The first consortia papers were published by the International Schizophrenia Consortium (ISC) (International Schizophrenia C et al. 2009), Molecular Genetics of Schizophrenia (MGS) (Shi et al. 2009), and SGENE (Stefansson et al. 2009). They reported the first genome-wide significant findings in *MHC*, *TCF4*, and *NRGN*. Apart from other GWAS with nonsignificant findings (Chen et al. 2011; Athanasiu et al. 2010), Williams et al. (2011) added *ZNF804A* to the list of genome-wide significant hits, while Steinberg et al. (2011) replicated the *MHC* and *TCF4* associations and additionally reported genome-wide significant variants in VRK2 to be associated with schizophrenia.

The sample size increased to over 9,000 cases and over 12,000 controls in the first Psychiatric Genomics Consortium (PGC) schizophrenia paper (Schizophrenia Psychiatric Genome-Wide Association Study C 2011), where 7 genome-wide significant hits including 5 new loci (*MIR137*, *PCGEM1*, *CSMD1*, *MMP16*, and *CNNM2/NT5C2*) were reported.

Interestingly, the microRNA *MIR137* and its gene targets (*TCF4*, *CACNA1C*, *CSMD1*, and *C10orf26*) were genome-wide significant in either the schizophrenia or the combined schizophrenia-bipolar disorder PGC analyses (Schizophrenia Psychiatric Genome-Wide Association Study C 2011), suggesting a shared biological pathway in schizophrenia and bipolar disorder.

Also, GWAS on non-Western populations have been published (Ikeda et al. 2011; Yue et al. 2011; Shi et al. 2011a): Yue et al. (2011) reported a novel association at 11p11.2 and could replicate associations in the *MHC* region, while Shi et al. (2011a) found two new loci at 8p12 and 1q24.2.

Combining the PGC cohorts and a Swedish sample, Ripke et al. (2013) showed that many independent, mostly common SNPs collectively accounted for at least 32% of the variance in the liability to schizophrenia, underpinning the polygenetic component of this disease. In the same study, they identified 13 new susceptibility loci for schizophrenia and replicated the *MHC* locus. Although the most replicated locus in schizophrenia is the *MHC* region, further investigations with regard to disease susceptibility have been difficult as this region is a rather complex composition of SNPs in high LD blocks across many genes involved in neurodevelopment, immunity, and other processes (Bergen and Petryshen 2012).

In the latest PCG GWAS (Schizophrenia Working Group of the Psychiatric Genomics C 2014), the sample size increased to nearly 37,000 cases and 113,000 controls, and 108 genome-wide significant loci were identified. Associated genes were enriched for genes expressed in brain regions, providing evidence for biological plausibility. It has to be noted that each of these 108 loci exhibits only a small effect, with risk-associated alleles differing mostly <2% in frequency between cases and controls.

Most of the schizophrenia-associated genes can be grouped into several categories: therapeutic targets (*DRD2* and *GRM3*), glutamatergic transmission (such as *GRIN2A* and *GRIA1*), neuronal calcium signaling (e.g., *CACNA1C* and *NRGN*), or synaptic function and plasticity (such as *MEF2C* and *CNTN4*). Interestingly, in independent studies, genes encoding calcium channels and genes involved in glutamatergic neurotransmission and synaptic plasticity have been shown to have implications in schizophrenia based on rare genetic variations (Purcell et al. 2014; Fromer et al. 2014; Kirov et al. 2012). Furthermore, GWAS hits identified in the PGC cohort significantly overlap with genes where de novo non-synonymous mutations in schizophrenia have been reported. In a combined meta-analysis of ~21,000 cases and ~20,000 controls, Marshall et al. (2017) found 8 genome-wide significant loci with copy number variants (CNVs). This suggests that both common and rare variations have implications in schizophrenia. In general, CNVs show generally greater effect sizes (odds ratios (ORs) of 4–70) on schizophrenia risk than common variants (OR < 1.3) (Birnbaum and Weinberger 2017).

Although common and rare risk loci have been identified, the causative pathophysiology of schizophrenia is still unknown. Gene expression and DNA methylation studies on postmortem brain suggest that risk factors seem to occur mainly during early brain development and not at the stage of adolescent when the diagnosis is typically made (Birnbaum and Weinberger 2017).

3.2 Bipolar Disorder

Bipolar disorder (BD) presents with high heritability estimates between 80 and 90% (Craddock et al. 2005; McGuffin et al. 2003). Main symptoms of the disease are unpredictable mood swings between mania and depression (Kerner 2014). BD is divided into two subtypes: bipolar I disorder with manic episodes and bipolar II disorder with hypomanic episodes.

The first big GWAS on bipolar disorder was published in 2007 by the WTCCC (Wellcome Trust Case Control C 2007) with 1,900 BP cases and 3,000 controls. The authors identified one genome-wide significant hit in 16p1 which includes *PALB2* (partner and localizer of *BRCA2*), *NDUFAB1*, and *DCTN5*. Baum et al. (2008) used a DNA pooling approach followed by individual genotyping of interesting genetic markers; they identified one genome-wide significant hit in *DGKH*. Sklar et al. (2008) did not find genome-wide significant hits and however reported genome-wide significance in a haplotype test in *MYO5b* and furthermore associations with *CACNA1C*, consistent with WTCCC results. The study by Ferreira et al. (2008) also supported findings in *CACNA1C* as well as in *ANK3*. Scott et al. (2009) reported variants, albeit not genome-wide significant, in *MCTP1*, *GNL3*, *NEK4*, *ITIH*, and *ANK3*. Smith et al. (2009) ran GWAS independently in Europeans and African Americans. Although no genome-wide significant SNPs were reported, they found evidence for associations in *SLITRK2* and *NTRK2*. Several other GWAS were published (Cichon et al. 2011; Djurovic et al. 2010; Hattori et al. 2009; Lee et al. 2011b; Smith et al. 2011; Yosifova et al. 2011) until the PGC reported a GWAS on BD (Psychiatric GCBDWG 2011) comparing over 10,000 bipolar cases against over 50,000 controls in the combined initial and replication sample. They confirmed genome-wide significant evidence of association for *CACNA1C* and identified a new intronic variant in *ODZ4*. Chen et al. (2013) reported novel association findings near the genes *TRANK1*, *LMAN2L*, and *PTGFR*. Mühleisen et al. (2014) ran a GWAS on ~2,200 patients with BD and ~5,000 controls and identified 56 genome-wide significant hits including the previously reported risk loci *ANK3*, *ODZ4*, and *TRANK1*, as well as two new hits in *ADCY2* and a region between *MIR2113* and *POU3F2*. Hou et al. (2016) added *ERBB2* and an intergenic region on chromosome 9p21.3 and Charney et al. (2017) *ADD3* and *XPNPEP1* to the list of possible candidates. Ikeda et al. (2017a) reported novel genome-wide significant findings at 1q12.2, a region known to contain regulatory genes for plasma lipid levels (*FADS1/2/3*), in a Japanese population. A recent study by Stahl and colleagues (2019) identified 30 genome-wide significant loci including *CACNA1C*, *GRIN2A*, *SCN2A*, *SLC4A1*, *RIMS1*, and *ANK3*.

Three of the most striking findings of BD GWAS with regard to gene functionality are possibly *ANK3* and *CACNA1* which are associated with white-matter integrity and brain volume (Gurung and Prata 2015) and *FADS1/2/3* indicating that lipid abnormality may be associated with the pathophysiology of BD (Ikeda et al. 2017b). However, the exact disease etiology remains to be elucidated.

3.3 Major Depressive Disorder (MDD)

MDD is the most common psychiatric disorder with a lifetime risk of 10–20% and heritability estimates around 40% (Major Depressive Disorder Working Group of the Psychiatric GC et al. 2013). It is characterized by depressed mood, diminished interests, and impaired cognitive function (Flint and Kendler 2014).

Up to recently, it has been proven to be difficult to identify genome-wide significant susceptibility loci for MDD. The first GWAS (Major Depressive Disorder Working Group of the Psychiatric GC et al. 2013; Sullivan et al. 2009; Wray et al. 2012; Muglia et al. 2010; Shi et al. 2011b; Shyn et al. 2011; Lewis et al. 2010; Rietschel et al. 2010) did not report genome-wide significant findings. Kohli et al. (2011) found a genome-wide significant SNP close to *SLC6A15* when they combined their initial and replication sample and used a recessive model. The CONVERGE consortium found genome-wide significant hits in the *LHPP* gene and near the *SIRT1* gene (Consortium C 2015) in a Han Chinese population. However, these identified risk alleles are very rare (low frequency) in Europeans.

The first genome-wide significant hit in Europeans was reported in a meta-analysis study combining samples from the PGC consortium with large-scale consumer genetic data from 23 and Me (Hyde et al. 2016). In this study, 15 MDD-associated loci were reported including *LFM4*, *TMEM161B–MEF2C*, *MEIS2–TMCO5A*, and *NEGR1*. Going one step further, the PGC recently published a GWAS with a combined sample size of 130,664 MDD cases and 330,470 controls. Forty-four independent genome-wide significant loci were reported (Wray et al. 2018) including 30 novel hits. The most significant hits were located in or near *OLFM4* and *NEGR*; novel associations included *RBFOX1* and *LRFN5*. Genes that were targets of antidepressant medications were strongly enriched for MDD association hits indicating the relevance of these findings for pharmacotherapy.

MDD has a modest heritability pointing to the fact that both genetic and environmental factors are involved in disease etiology. Environmental factors, such as childhood abuse, are strongly associated with the risk of developing MDD (Flint and Kendler 2014). Furthermore, twin studies reported that environment accounts for over 60% of the variance (Sullivan et al. 2000). Thus, in future, probably, gene x environment (GxE) studies will become more and more important to elucidate disease pathology of MDD.

3.4 *Attention Deficit Hyperactivity Disorder (ADHD)*

Attention deficit hyperactivity disorder (ADHD) is a common childhood disorder affecting 2–10% of school-aged children worldwide (Hawi et al. 2015). Twin studies estimate a heritability of 70–80% (Franke et al. 2012; Faraone et al. 2005). ADHD is characterized by extreme levels of motor activity, impulsivity, and inattention. In 30–60% of the affected, these symptoms occur up into adulthood and are associated with increased risk for drug abuse, for instance (Mannuzza et al. 1993; Konstenius et al. 2015). Several GWAS been conducted for ADHD (Neale et al. 2008, 2010a, b; Hinney et al. 2011; Stergiakouli et al. 2012; Yang et al. 2013; Mick et al. 2010; Lesch et al. 2008; Middeldorp et al. 2016; Zayats et al. 2015; Sanchez-Mora et al. 2015; Alemany et al. 2015), but none reported any genome-wide significant hits. However, a trend for *CDH13* was observed in some of these studies. Recently, the PGC reported a genome-wide meta-analysis including over 20,000 ADHD cases and over 35,000 controls (Demontis et al. 2019). They identified 16 genome-wide significant loci including *FOXP2*, *DUSP6*, *SEMA6D*, and *MEF2C*.

3.5 *Autism*

Autism spectrum disorder (ASD) is a severe neurodevelopmental disorder characterized by various degree of abnormal communication, social reciprocity, and restricted repetitive behaviors (Lai et al. 2014). Twin studies estimate the heritability to be more than 80% (Ronald and Hoekstra 2011) with a prevalence of about 1% (Fernell and Gillberg 2010). It is believed that rare penetrant mutations as well as common variants with multiplicative effects contribute to a complex disorder like autism (Lai et al. 2014).

The earliest report of a pathogenic mutation involved the gene *NLGN4X* (Jamain et al. 2003; Laumonnier et al. 2004). Sebat et al. reported that large de novo structural variation was present in 10% of sporadic autism families (Sebat et al. 2007). However, common genetic variations explain about half of the genetic risk in ASD variants (Gaugler et al. 2014), and thus, GWAS are the ideal tool to elucidate associated variants. While some studies could not detect genome-wide significant loci (Ma et al. 2009; Salyakina et al. 2010; Kuo et al. 2015), the Autism Genome Project Consortium (AGP) reported one genome-wide significant hit in *MACROD2* (Anney et al. 2010), and Wang et al. (2009) found genome-wide significance in the region between *CDH9* and *CDH10*. Interestingly, Kerin et al. (2012) identified the noncoding RNA *MSNP1AS* in this region to be overexpressed in postmortem cerebral cortex of individuals with ASD. Xia et al. found genome-wide significant hits in *CSDE1*, *NRAS*, and *TRIM33* (Xia et al. 2014). Connolly et al. (2013) identified genome-wide significant loci in *KCND2*, *C8ORFK32*, *NOS2A*, *NELL1*, *BIN1*, *MPN2*, *SDK1*, and *FER*.

The PGC reported no genome-wide significant association (Autism Spectrum Disorders Working Group of The Psychiatric Genomics C 2017); top-associated loci included *CUEDC2*, *MACROD2*, and *ASTN2*. Importantly, they observed significant genetic correlation between ASD and schizophrenia pointing to a shared disease etiology.

3.6 Cross Disorders: ASD, ADHD, BP, MDD, and Schizophrenia

Genetic markers for psychiatric disorders do not uniquely map to only one diagnosis. Diseases are generally separated on the basis of their symptom patterns and course of illness, but pathogenic mechanisms of psychiatric disorders are largely unknown. Therefore, although diseases present with different symptoms, they could share, at least partially, the same genetic etiology.

Most of the cross-disorder studies investigated the relationship between two disorders. Several studies suggest overlap in the genetic liability of schizophrenia and BD (Smoller and Finn 2003; Maier et al. 1993), and genetic variants associated with both schizophrenia and BD have been identified (International Schizophrenia C et al. 2009; Schizophrenia Psychiatric Genome-Wide Association Study C 2011; Psychiatric GCBDWG 2011). Ruderfer et al. (2014) compared schizophrenic and bipolar patients and created a schizophrenia versus BD genetic risk score differentiating the two diseases. Furthermore, they ran a combined GWAS on schizophrenic/bipolar cases against healthy controls and found five genome-wide significant regions (*CACNA1C*, *IFI44L*, *MHC*, *TRANK1*, *MAD1L1*, and *PIK3C2A*). Also, genetic overlap between ASD and schizophrenia (King and Lord 2011; Levinson et al. 2011), especially in the context of CNVs, has been previously reported. Furthermore, also ASD and ADHD show a certain overlap (Ronald et al. 2008; Rommelse et al. 2010; Lichtenstein et al. 2010).

Huang et al. (2010) studied cross-disorder effects between MDD, BD, and schizophrenia and reported hits, although not genome-wide significant, close to *ADM* and in *NPAS3*.

In the first cross-disorder study focusing on five diseases, the PGC investigated shared genetic effects between ASD, ADHD, BD, MDD, and schizophrenia (Cross-Disorder Group of the Psychiatric Genomics C 2013). They ran a meta-analysis including in total over 33,000 cases and 27,000 controls and conducted a pairwise-cross-disorder-polygenic risk score analysis, defining risk scores for each disorder and applying these to the other disorders. The authors found significant overlap between MDD, schizophrenia, and BD, and also ASD, schizophrenia, and BD were significantly overlapping. Interestingly, no significant overlap between MDD and ASD or between ADHD and any other disorder was found. In the overall meta-analysis combining all five disorders, four independent regions with genome-wide significant hits were identified: *ITIH3*, *CACNB2*, *AS3MT*, and *CACNA1C*.

CACNA1C had already been identified as susceptibility gene for BD (Ferreira et al. 2008; Psychiatric GCBDWG 2011), schizophrenia (Schizophrenia Psychiatric Genome-Wide Association Study C 2011), and MDD (Green et al. 2010) and also in a cross-disorder study including BD and MDD (Liu et al. 2011). *CACNB2* was reported as susceptibility gene for BD in Han Chinese (Lee et al. 2011b), while *AS3MT* was previously associated with BD and schizophrenia (Schizophrenia Psychiatric Genome-Wide Association Study C 2011; Psychiatric GCBDWG 2011). Calcium channel signaling pathways were implicated across all five disorders, and SNPs associated across the disorders were also enriched for brain eQTL markers.

Along these lines, the PGC published another paper (Cross-Disorder Group of the Psychiatric Genomics C et al. 2013) which focused on the general genetic correlation between psychiatric disorders and further highlighted the genome-wide pleiotropy of these disorders. Pairwise genetic correlation between these disorders was estimated. This correlation is positive if cases of one disorder are more genetically similar to cases of another disorder than to their own controls. The genetic correlation using common SNPs was high between schizophrenia and BD; moderate between schizophrenia and MDD, BD and MDD, and ADHD and MDD; low between schizophrenia and ASD; and nonsignificant for other pairs of disorders as well as between psychiatric disorders and the negative control of Crohn's disease.

In future, disease treatment could be focused on a nosology informed by disease cause rather than by symptoms (Cross-Disorder Group of the Psychiatric Genomics C 2013). When looking into how different diseases correlate with each other, it is also interesting to learn more about their exact relationship (Docherty et al. 2016). Crespi and colleagues (Crespi et al. 2010) surveyed four possible models to reflect how schizophrenia and ASD might be related: ASD could be a subtype of schizophrenia ("subsumed model"); ASD could be a distinct disorder ("separate model"); both diseases could be diametric, i.e., opposite conditions ("diametric model"); or they could overlap to some distinct sharing some risk factors and phenotypes but not others ("overlapping model").

Newer approaches other than the classical GWAS have recently been applied to test for cross-disorder effects, e.g., phenotype-based approaches that look more closely into which genes are involved in multiple disorders (Gonzalez-Mantilla et al. 2016). In this context, a dimensional refinement of the phenotype (Docherty et al. 2016) and concepts like the Research Domain Criteria (RDoc) (Cuthbert and Insel 2013) will probably gain importance.

3.7 Posttraumatic Stress Disorder (PTSD)

Posttraumatic stress disorder (PTSD) is a debilitating psychiatric disorder that can occur after exposure to a potentially life-threatening traumatic event. To date only a handful of PTSD GWAS studies have been performed in humans. The first GWAS study in PTSD detected a genome-wide significant association between a SNP in the retinoid-related orphan receptor alpha (*RORA*) gene and PTSD in a discovery sample (295 cases and 196 controls) of trauma exposed white non-Hispanic veterans and their intimate partners (Logue et al. 2013). The *RORA* gene plays an important role in protecting brain cells from injury, stress, and disease (Logue et al. 2013). The association of PTSD with the *RORA* gene has since then been replicated in a cohort of Florida hurricane survivors (Amstadter et al. 2013) but was not replicated in two independent replication samples (Guffanti et al. 2014).

Another GWAS was performed in an African American population (94 PTSD cases and 319 controls) of women who had been exposed to varying traumatic events (Guffanti et al. 2013). Only one SNP in a novel RNA gene *lincRNA AC068718.1* (rs10170218) reached genome-wide significance. This SNP association was replicated in a female European population (578 PTSD cases and 1,963 controls) and was found to be marginally significant. A third study performed a GWAS in 300 PTSD European Americans and 444 PTSD African Americans (Xie et al. 2013). The unknown SNP rs406001 reached genome-wide significance in the European American population, and no SNPs reached genome significance in the African American population. The genome-wide significant SNP was in a region with three other SNPs in LD, located about 630 kb downstream of the Cordon-Bleu (*COBL*) gene. The second significant locus from this study pointed toward the Tolloid-Like 1 or *TLL-1* in PTSD, a zinc-dependent metalloprotease that plays a key role in extracellular matrix remodeling (Apte and Parks 2015).

A further study employed a relatively large multiracial PTSD cohort of 2,312 Iraq and Afghanistan veterans (Ashley-Koch et al. 2015) and did not identify any individual SNPs that reached genome-wide significance. The study pointed toward several plausible candidate genes showing nominal significance including *TBC1D2*, *SDC2*, and *PCDH7*, genes previously implicated in neurologic processes.

A study using a small military cohort (Systems Biology PTSD Biomarkers Consortium, $N = 147$) identified a genome-wide significant SNP rs717947 at chromosome 4p15 associated with PTSD and replicated this finding in a larger urban community cohort (Grady Trauma Project, GTP, $N = 2{,}006$) (Almli et al. 2015). In the GTP replication sample, SNP rs717947 was associated with PTSD diagnosis in females ($N = 2{,}006$, $P = 0.005$) and was also found to be a methylation quantitative trait locus (meQTL) in the GTP replication sample ($N = 157$, $P = 0.002$). Further, the risk allele of rs717947 was also associated with decreased medial and dorsolateral cortical activation to fearful faces ($N = 53$, $P < 0.05$) in the replication sample. The peak of SNPs in LD with the genome-wide significant SNP on chromosome 4 was located in close proximity to the ncRNA *BC036345* whose function is unknown.

Another GWAS performed on samples from combat-exposed US Marines and Sailors from the Marine Resiliency Study (MRS) scheduled for deployment to Iraq and/or Afghanistan. This meta-analysis identified the phosphoribosyl transferase domain containing one gene (*PRTFDC1*) as a genome-wide significant PTSD locus (rs6482463), with a similar effect across ancestry groups (Nievergelt et al. 2015). Replication of the *PRTFDC1* signal with PTSD in an independent military cohort in the same study showed evidence for robustness of the finding.

One PTSD GWAS investigated two large US military cohorts with a total of around 10,000 samples (Stein et al. 2016). Genome-wide significance was reached for rs159572 in the *ANKRD55* gene, implicated in autoimmune and inflammatory disorders, but only in the African American samples.

One GWAS study examined symptoms defining the dissociative subtype of PTSD and found no SNPs reaching genome-wide significance but ten SNPs showing suggestive association (Wolf et al. 2014).

Given the modest GWAS results in the field of PTSD, great advances have been made via the formation of the Psychiatric Genomics Consortium-PTSD (PGC-PTSD) Workgroup in 2013 (Logue et al. 2015). To facilitate the analysis of different types of data, focus groups have been created within the PGC-PTSD Workgroup including the PGC-PTSD Epigenetics Workgroup and the PGC-PTSD Neuroimaging Genetics Workgroup among others (Logue et al. 2015). The most recent largest PTSD GWAS from the PGC-PTSD included data from 20,730 individuals across 11 multiethnic studies (Duncan et al. 2017). In addition to identifying risk loci for PTSD, the authors quantified PTSD heritability and examined shared genetic risk of PTSD with schizophrenia, BD, and MDD to identify risk loci for PTSD. The study did not reveal any significant genome-wide hits in European Americans and 13 genome-wide significant hits in African Americans (Duncan et al. 2017). SNP-based heritability estimates for European American females were 29%, similar to that for schizophrenia, and were substantially higher than that in European American males (not different from zero). Using polygenic risk score profiling, there was strong evidence of overlapping genetic risk between PTSD and schizophrenia along with more modest evidence of overlap of PTSD with BD and MDD.

In summary, a handful of scattered GWAS in PTSD have been conducted, and several candidate genes have been revealed. Further larger GWAS analyses within the PGC-PTSD consortium are currently underway and are likely to identify additional candidates for PTSD.

3.8 Obsessive-Compulsive Disorder (OCD)

Obsessive-compulsive disorder (OCD) is characterized by recurring obsessions and/or compulsions, affects up to 2% of the world's population, and is 4.5 times more common in males than females (Eaton et al. 2008). The success of GWAS in OCD has been limited, with no genome-wide significant SNPs identified so far; some of the largest OCD GWAS studies are reviewed below.

The GWAS from the International OCD Foundation Genetics Collaborative (IOCDF-GC) was performed in 1,465 cases, 5,557 ancestry-matched controls, and 400 trios (Stewart et al. 2013). The trio analysis identified the SNP rs6131295, near *BTBD3*, that exceeded the genome-wide significance threshold, but this SNP lost genome-wide significance when meta-analyzed with the case-control samples. For the top-ranked SNPs ($P < 0.01$) from the trio analysis, a significant enrichment of methylation QTLs and frontal lobe expression quantitative trait loci (eQTLs) was observed, suggesting a role in brain gene expression and OCD etiology.

The OCD Collaborative Genetics Association Study (OCGAS) GWAS was conducted in a total of 1,065 families (including 1,406 OCD patients), combined with population-based samples, comprising a total sample of 5,061 individuals (Mattheisen et al. 2015). Although this study failed to detect any genome-wide significant hits, a trend for significance was seen for a SNP near *PTPRD*, a presynaptic gene that promotes the differentiation of glutamatergic synapses.

The recent largest GWAS was a meta-analysis of the two previous IOCDF-GC and OCGAS GWAS, investigating a total of 2,688 Europeans and 7,037 genetically matched controls (International Obsessive Compulsive Disorder Foundation Genetics C, Studies OCDCGA 2017). None of the SNPs reached genome-wide significance. SNPs located within or near the genes *ASB13*, *RSPO4*, *DLGAP1*, *PTPRD*, *GRIK2*, *FAIM2*, and *CDH20*, identified in linkage peaks and the original GWASs, were among the top signals. The common SNP heritability in the combined sample was estimated to be 0.28.

3.9 Alcohol, Nicotine, and Cannabis Dependence

A large number of GWAS have been performed on alcohol and other substance dependence traits; details of these studies are available on the GWAS catalog website (MacArthur et al. 2017). Some of the largest studies are described in this section.

Alcohol Dependence

With a lifetime prevalence of 12.5% in the general American population, alcohol dependence (AD) is a common, complex psychiatric disorder characterized by excessive and compulsive use of alcohol, which often results in physical and social consequences (Hasin et al. 2007). Significant advances in the identification of risk loci for AD can be largely attributed to the advent of genome-wide association studies (GWAS), and these studies have been reviewed in-depth (Hart and Kranzler 2015; Stickel et al. 2017). The first GWAS of AD was performed in a sample of German men, comprising a discovery sample of 476 cases and 1,358 controls (Treutlein et al. 2009) and a replication sample of 1,024 AD cases and 996 controls. Although no SNPs were genome-wide significant in the discovery sample, two SNPs that mapped to the *PECR* gene reached genome-wide significance in a meta-analysis of the discovery and replication samples. A total of 15 SNPs located in or near several genes (*CAST*, *ERAP1*, *PPP2R2B*, *ESR1*, *CCD41*, *ADH1C*, *GATA4*, and *CDH13*) were genome-wide significant in the meta-analysis across the two samples.

Another GWAS (Frank et al. 2012) in a German sample of 1,333 men with severe AD and 2,168 controls identified one genome-wide significant SNP (rs1789891) located between *ADH1B* and *ADH1C*, two alcohol-metabolizing enzyme genes. Furthermore, this was the first AD GWAS to identify a robust genome-wide significant hit that was replicated in an independent sample and the first study to identify significant associations in alcohol-metabolizing enzyme genes, thereby replicating the findings from previous candidate gene studies (Bierut et al. 2012; Edenberg 2007; Olfson and Bierut 2012; Thomasson et al. 1991).

A GWAS performed by Gelernter et al. (2014) was the first study to impute the genotyped SNPs tested to include over nine million in African Americans and over six million in European Americans. The authors tested for association with a quantitative AD phenotype while controlling for cocaine, opioid, and nicotine dependence criteria in the sample and meta-analyzed the results with those from the SAGE sample (Bierut et al. 2010) and also with their discovery and replication samples. In African Americans, they identified several genome-wide hits at the alcohol dehydrogenase (*ADH*) locus on chromosome 4, with the strongest finding in *ADH1B* (rs2066702). In European Americans, the strongest finding was also in *ADH1B* (rs1229984). Both *ADH1B* SNPs were missense polymorphisms, and both associations replicated in an independent sample. Thus, as in other AD GWAS (Frank et al. 2012; Park et al. 2013), the ADH gene cluster had the greatest effect on AD risk.

To summarize, several genes including those within the ADH locus have been identified for AD.

Nicotine Dependence

Nicotine dependence is highly comorbid with several psychiatric disorders. The Tobacco and Genetics Consortium meta-analyses of several smoking phenotypes ($n = 74{,}053$) identified three loci associated with number of cigarettes smoked per day (Tobacco, Genetics C 2010). The strongest association was a SNP in the nicotinic receptor gene *CHRNA3* (rs1051730). Two other SNPs on chromosome 10 (rs1329650 and rs1028936) and one SNP in *EGLN2* (rs3733829) also exceeded genome-wide significance for cigarettes smoked per day. For smoking initiation, eight SNPs exceeded genome-wide significance, with the strongest association at a non-synonymous SNP in *BDNF* on chromosome 11 (rs6265). One SNP located near *DBH* on chromosome 9 (rs3025343) was significantly associated with smoking cessation.

In another GWAS (Gelernter et al. 2015), the chromosome 7 intergenic SNP rs13225753 was associated with nicotine dependence in Europeans Americans. In African Americans, associations were observed at several SNPs mapping to a region on chromosome 14 and two regions on chromosome 8. This study did not identify any associations at the chromosome 15 nicotinic receptor gene cluster (*CHRNA5-CHRNA3-CHRNB4*) previously associated with nicotine dependence and smoking quantity traits.

In a recently conducted largest-ever GWAS meta-analysis, the *CHRNA5-CHRNA3-CHRNB4* genes were reconfirmed to be associated with nicotine dependence, and a further novel association in the DNA methyltransferase gene *DNMT3B* was identified (Hancock et al. 2017).

Cannabis Dependence

After nicotine, cannabis is the most widely abused drug worldwide, and cannabis use is also a risk factor for psychiatric disorders including MDD (Lynskey et al. 2004) and schizophrenia (Power et al. 2014). To date, only three genome-wide studies have been conducted for cannabis-related phenotypes. The first GWAS (Sherva et al. 2016) uncovered three independent regions with genome-wide significant SNPs including rs143244591 in novel antisense transcript *RP11-206M11.7*, rs146091982 in the solute carrier family 35 member G1 gene (*SLC35G1*), and rs77378271 in the CUB and Sushi multiple domains 1 gene (*CSMD1*).

The International Cannabis Consortium meta-analysis of genome-wide association data of 13 cohorts ($N = 32{,}330$) and 4 replication samples ($N = 5{,}627$) did not find any individual SNPs reaching genome-wide significance, but gene-based tests identified four genes (*NCAM1*, *CADM2*, *SCOC*, and *KCNT2*) significantly associated with lifetime cannabis use (Stringer et al. 2016). This study demonstrated that the common SNPs explained 13–20% of the liability of lifetime cannabis use and indicated a strong genetic correlation between lifetime cannabis use and lifetime cigarette smoking ($rg = 0.83$).

Another recent meta-analysis of genome-wide association study data on 2,080 cannabis-dependent cases and 6,435 cannabis-exposed controls of European descent which identified a cluster of correlated SNPs in a novel region on chromosome 10 was genome-wide significant (Agrawal et al. 2017). The SNPs were located nearby multiple genes that are expressed in brain-derived tissues. Among the SNPs, rs1409568 was predicted to modify binding scores for several transcription factors, had active enhancer marks for addiction-relevant brain regions including the dorsolateral prefrontal cortex and the angular and cingulate gyri, and was also associated with a modest increase in right hippocampal volume (2.13%) in an independent sample.

3.10 Personality Traits: Neuroticism and Borderline Personality

Personality traits are influenced by genetic and environmental factors (Vukasovic and Bratko 2015) and have been reported to be associated with mental health (Lo et al. 2017). Genetic research of personality traits is still in its infant stages compared to other psychiatric disorders with early genetic studies performed in small samples and focused on single candidate genes. This section outlines GWAS studies in neuroticism and borderline personality traits in conjunction with other psychiatric disorders or symptoms.

3.10.1 GWAS of Neuroticism

A meta-analysis of the Genetics of Personality Consortium identified a novel locus (rs35855737) for neuroticism in *MAGI1*, a known gene that has been previously been associated with bipolar disorder and schizophrenia (Genetics of Personality C et al. 2015). Additionally, the researchers found that common genetic variants explained 15% of the variance in neuroticism and the polygenic risk score for neuroticism significantly predicted MDD in independent cohorts, providing evidence for a genetic overlap.

Another GWAS of personality traits discovered 3 genetic variants associated with subjective well-being, 2 variants associated with depressive symptoms, and 11 variants associated with neuroticism (Okbay et al. 2016). This study also found a genetic overlap across subjective well-being, depressive symptoms, and neuroticism (rg = −0.81 between subjective well-being and depressive symptoms, rg = −0.75 between subjective well-being and neuroticism, and rg = 0.75 between depressive symptoms and neuroticism).

Lo and colleagues conducted a meta-analysis of GWAS and identified six genetic loci, (five novel) significantly associated with personality traits (Lo et al. 2017). Among the traits, extraversion was associated with genetic variants in *WSCD2* and near *PCDH15*, and neuroticism was associated with genetic variants in *L3MBTL2* and on chromosome 8p23.1. The study also found high genetic correlations between neuroticism and MDD (rg = 0.54), between extraversion and attention deficit hyperactivity disorder (rg = 0.30), and between openness and schizophrenia (rg = 0.36) and BD (rg = 0.34) among others.

A recent study investigated over 329,000 individuals from the UK Biobank and reported 116 significant independent loci associated with neuroticism, 15 of which were also replicated in an unrelated cohort of over 122,000 individuals (Luciano et al. 2018). The study revealed significant genetic correlations between neuroticism and depressive symptoms (rg = 0.82), MDD (rg = 0.69), and subjective well-being (rg = −0.68,) together with other mental health traits.

3.10.2 GWAS of Borderline Personality

To the best of our knowledge, only two GWAS in borderline personality have been reported so far, one based on borderline personality features (Lubke et al. 2014) and the other based on borderline personality disorder (Witt et al. 2017). Although no genome-wide significant loci were identified, the first GWAS found seven SNPs within a promising locus corresponding to myelination-protein *SERINC5* (chromosome 5) to be associated with borderline personality features; these results were also confirmed in an independent sample (Lubke et al. 2014). The borderline personality disorder GWAS was performed in 998 patients and 1,545 controls (Witt et al. 2017). While no SNP survived multiple-testing correction for association, gene-based analysis yielded two significant genes *DPYD* and *PKP4*, both previously implicated in bipolar disorder and schizophrenia. *SERINC5*, which was the top hit of the previous GWAS of borderline personality features (Lubke et al. 2014), showed nominal significance in the study. This was the first study to demonstrate the genetic overlap of borderline personality disorder with BD (rg = 0.28), schizophrenia (rg = 0.34), and MDD (rg = 0.57).

In summary, both studies pointed toward the role of *SERINC5* among others, in borderline personality traits. The *SERINC5* protein is enriched in myelin in the brain and incorporates serine into newly forming membrane lipids (Krueger et al. 1997). This is of interest, given the previous findings that decreased myelination is associated with a reduced capacity for social interaction (Lubke et al. 2014; Liu et al. 2012).

4 Genome-Wide Association Studies in Psychiatric Disorders: Future Directions

GWAS allow the interrogation of common genetic polymorphisms in the genome without making any assumptions about the genomic location of the causal variants. Despite the large number of loci that have been identified, for most diseases, only a modest fraction of the predicted heritability has been accounted for through GWAS. Given the effects of natural selection and that genetic variants having a major effect on overall disease susceptibility would decrease reproductive fitness and would hence be maintained at a low frequency within populations has raised questions about the ability of GWAS methods to identify major risk loci for complex diseases.

The potential and limitations of GWAS have recently been reviewed in-depth (Visscher et al. 2017). GWAS methods have been limited by a variety of factors; hypothesis-free scanning of the genome, for instance, has the potential to identify many false associations, and therefore replication and/or validation in independent populations is essential. There is also no strict consensus on the statistical correction for multiple tests, and different methods and strategies including a two-step study design have been suggested (Wason and Dudbridge 2012). An additional challenge for GWAS is that it is currently based on the "common disease, common variant" (CDCV) hypothesis, which means that rare susceptibility variants are generally not detected. GWAS findings also typically point toward a region of linkage disequilibrium spanning several hundreds or thousands of base pairs harboring sets of correlated SNPs; hence, the risk SNPs implicated by GWAS are not necessarily causal but instead are more likely in linkage disequilibrium (LD) with true causal variants, and/or these intergenic and intronic risk variants represent regulatory elements. Thus, it is essential to demonstrate biological, functional, and clinical relevance for these GWAS loci via studies that confirm a functional link between a SNP and a variant in LD with the GWAS SNP with the disease (Edwards et al. 2013). Nevertheless, owing to the lack of tools and our limited knowledge, it has not been straightforward to shift from the GWAS loci to the underlying biological mechanisms nor has it been easy to identify the gene(s) involved in the susceptibility to the disease. Finally, given the complex nature of psychiatric disorders with both common and rare genetic variants as well as environmental factors contributing to risk, complex and sophisticated statistical algorithms such as those described in this chapter are required to predict susceptibility to disease.

Despite the many challenges of GWAS, these studies have uncovered valuable biological insights about genes and pathways involved in psychiatric disorders. Newer approaches in the future will advance our understanding of disease pathophysiology with a combination of deeper resequencing of loci, imputation, assessment of copy number variations, insertions/deletions, epigenetics, and gene expression in larger, well-characterized samples, to account for the missing heritability in psychiatric disorders. The growth of large population biobanks with extensive genetic and phenotype information such as the 1000 Genomes project (Genomes Project et al. 2015) and the UK biobank (Ollier et al. 2005) will also aid to identify less common genetic variants associated with psychiatric disorders and behavioral traits; such efforts are currently underway and reflect the next wave of genetic discovery. Other projects such as the ENCODE (Consortium 2012) and Roadmap Epigenomics projects (Roadmap Epigenomics et al. 2015) have already greatly facilitated our understanding of gene regulation and have provided the tools for studying gene regulation and function, to unravel the biology underlying GWAS associations. Other complementary approaches to GWAS such as phenome-wide association studies (PheWAS) incorporate methods that conduct a genotype-to-phenotype analysis have been useful in immune-related disorders (Denny et al. 2010), and such methods might add value when applied to psychiatric disorders.

GWAS findings can also have impact on personalized medicine (Smoller 2014) by identifying genetic variants associated with the disease, and hence pointing to new drug targets, by stratifying patients into different therapeutic subgroups and pharmacogenetics studies investigating if genetic variants are associated with response. While the discovery of drugs in psychiatric disorders via GWAS is still in its infant stage, there is potential for the role of genetics and novel genes identified to impact on drug discovery.

In summary, GWAS in psychiatric disorders have shed light on a number of novel genes that might play a role in the pathophysiology of the disorders. Given the heterogeneous nature of psychiatric disorders, it may be more important to identify and clearly define patients within each biological or clinical subtype before utilizing resources toward more extensive and expensive genetic studies. Systems approaches that combine different layers of genomics information and provide a greater understanding of gene function and biological etiology of psychiatric disorders will shed light on the molecular underpinnings of psychiatric disorders.

References

Agrawal A, Chou YL, Carey CE, Baranger DAA, Zhang B, Sherva R, Wetherill L, Kapoor M, Wang JC, Bertelsen S et al (2017) Genome-wide association study identifies a novel locus for cannabis dependence. Mol Psychiatry 23:1293

Alemany S, Ribases M, Vilor-Tejedor N, Bustamante M, Sanchez-Mora C, Bosch R, Richarte V, Cormand B, Casas M, Ramos-Quiroga JA et al (2015) New suggestive genetic loci and biological pathways for attention function in adult attention-deficit/hyperactivity disorder. Am J Med Genet B Neuropsychiatr Genet 168(6):459–470

Almli LM, Stevens JS, Smith AK, Kilaru V, Meng Q, Flory J, Abu-Amara D, Hammamieh R, Yang R, Mercer KB et al (2015) A genome-wide identified risk variant for PTSD is a methylation quantitative trait locus and confers decreased cortical activation to fearful faces. Am J Med Genet B Neuropsychiatr Genet 168B(5):327–336

Amstadter AB, Sumner JA, Acierno R, Ruggiero KJ, Koenen KC, Kilpatrick DG, Galea S, Gelernter J (2013) Support for association of RORA variant and post traumatic stress symptoms in a population-based study of hurricane exposed adults. Mol Psychiatry 18(11):1148–1149

Anney R, Klei L, Pinto D, Regan R, Conroy J, Magalhaes TR, Correia C, Abrahams BS, Sykes N, Pagnamenta AT et al (2010) A genome-wide scan for common alleles affecting risk for autism. Hum Mol Genet 19(20):4072–4082

Apte SS, Parks WC (2015) Metalloproteinases: a parade of functions in matrix biology and an outlook for the future. Matrix Biol 44–46:1–6

Ashburner M, Ball CA, Blake JA, Botstein D, Butler H, Cherry JM, Davis AP, Dolinski K, Dwight SS, Eppig JT et al (2000) Gene ontology: tool for the unification of biology. The gene ontology consortium. Nat Genet 25(1):25–29

Ashley-Koch AE, Garrett ME, Gibson J, Liu Y, Dennis MF, Kimbrel NA, Veterans Affairs Mid-Atlantic Mental Illness Research E, Clinical Center W, Beckham JC, Hauser MA (2015) Genome-wide association study of posttraumatic stress disorder in a cohort of Iraq-Afghanistan era veterans. J Affect Disord 184:225–234

Athanasiu L, Mattingsdal M, Kahler AK, Brown A, Gustafsson O, Agartz I, Giegling I, Muglia P, Cichon S, Rietschel M et al (2010) Gene variants associated with schizophrenia in a Norwegian genome-wide study are replicated in a large European cohort. J Psychiatr Res 44(12):748–753

Autism Spectrum Disorders Working Group of The Psychiatric Genomics C (2017) Meta-analysis of GWAS of over 16,000 individuals with autism spectrum disorder highlights a novel locus at 10q24.32 and a significant overlap with schizophrenia. Mol Autism 8:21

Baum AE, Akula N, Cabanero M, Cardona I, Corona W, Klemens B, Schulze TG, Cichon S, Rietschel M, Nothen MM et al (2008) A genome-wide association study implicates diacylglycerol kinase eta (DGKH) and several other genes in the etiology of bipolar disorder. Mol Psychiatry 13(2):197–207

Benjamin DJ, Cesarini D, van der Loos MJ, Dawes CT, Koellinger PD, Magnusson PK, Chabris CF, Conley D, Laibson D, Johannesson M et al (2012) The genetic architecture of economic and political preferences. Proc Natl Acad Sci U S A 109(21):8026–8031

Bergen SE, Petryshen TL (2012) Genome-wide association studies of schizophrenia: does bigger lead to better results? Curr Opin Psychiatry 25(2):76–82

Bierut LJ, Agrawal A, Bucholz KK, Doheny KF, Laurie C, Pugh E, Fisher S, Fox L, Howells W, Bertelsen S et al (2010) A genome-wide association study of alcohol dependence. Proc Natl Acad Sci U S A 107(11):5082–5087

Bierut LJ, Goate AM, Breslau N, Johnson EO, Bertelsen S, Fox L, Agrawal A, Bucholz KK, Grucza R, Hesselbrock V et al (2012) ADH1B is associated with alcohol dependence and alcohol consumption in populations of European and African ancestry. Mol Psychiatry 17 (4):445–450

Birnbaum R, Weinberger DR (2017) Genetic insights into the neurodevelopmental origins of schizophrenia. Nat Rev. Neurosci 18(12):727–740

Bulik-Sullivan BK, Loh PR, Finucane HK, Ripke S, Yang J, Schizophrenia Working Group of the Psychiatric Genomics C, Patterson N, Daly MJ, Price AL, Neale BM (2015a) LD score regression distinguishes confounding from polygenicity in genome-wide association studies. Nat Genet 47(3):291–295

Bulik-Sullivan B, Finucane HK, Anttila V, Gusev A, Day FR, Loh PR, ReproGen C, Psychiatric Genomics C, Genetic Consortium for Anorexia Nervosa of the Wellcome Trust Case Control C, Duncan L et al (2015b) An atlas of genetic correlations across human diseases and traits. Nat Genet 47(11):1236–1241

Cardno AG, Gottesman II (2000) Twin studies of schizophrenia: from bow-and-arrow concordances to star wars Mx and functional genomics. Am J Med Genet 97(1):12–17

Cerami EG, Gross BE, Demir E, Rodchenkov I, Babur O, Anwar N, Schultz N, Bader GD, Sander C (2011) Pathway commons, a web resource for biological pathway data. Nucleic Acids Res 39: D685–D690

Chang CC, Chow CC, Tellier LC, Vattikuti S, Purcell SM, Lee JJ (2015) Second-generation PLINK: rising to the challenge of larger and richer datasets. Gigascience 4:7

Charney AW, Ruderfer DM, Stahl EA, Moran JL, Chambert K, Belliveau RA, Forty L, Gordon-Smith K, Di Florio A, Lee PH et al (2017) Evidence for genetic heterogeneity between clinical subtypes of bipolar disorder. Transl Psychiatry 7(1):e993

Chen X, Lee G, Maher BS, Fanous AH, Chen J, Zhao Z, Guo A, van den Oord E, Sullivan PF, Shi J et al (2011) GWA study data mining and independent replication identify cardiomyopathy-associated 5 (CMYA5) as a risk gene for schizophrenia. Mol Psychiatry 16(11):1117–1129

Chen DT, Jiang X, Akula N, Shugart YY, Wendland JR, Steele CJ, Kassem L, Park JH, Chatterjee N, Jamain S et al (2013) Genome-wide association study meta-analysis of European and Asian-ancestry samples identifies three novel loci associated with bipolar disorder. Mol Psychiatry 18(2):195–205

Cichon S, Muhleisen TW, Degenhardt FA, Mattheisen M, Miro X, Strohmaier J, Steffens M, Meesters C, Herms S, Weingarten M et al (2011) Genome-wide association study identifies genetic variation in neurocan as a susceptibility factor for bipolar disorder. Am J Hum Genet 88 (3):372–381

Connolly JJ, Glessner JT, Hakonarson H (2013) A genome-wide association study of autism incorporating autism diagnostic interview-revised, autism diagnostic observation schedule, and social responsiveness scale. Child Dev 84(1):17–33

Consortium EP (2012) An integrated encyclopedia of DNA elements in the human genome. Nature 489(7414):57–74

Consortium C (2015) Sparse whole-genome sequencing identifies two loci for major depressive disorder. Nature 523(7562):588–591

Cowley MJ, Pinese M, Kassahn KS, Waddell N, Pearson JV, Grimmond SM, Biankin AV, Hautaniemi S, Wu J (2012) PINA v2.0: mining interactome modules. Nucleic Acids Res 40: D862–D865

Craddock N, O'Donovan MC, Owen MJ (2005) The genetics of schizophrenia and bipolar disorder: dissecting psychosis. J Med Genet 42(3):193–204

Crespi B, Stead P, Elliot M (2010) Evolution in health and medicine Sackler colloquium: comparative genomics of autism and schizophrenia. Proc Natl Acad Sci U S A 107(Suppl 1):1736–1741

Cross-Disorder Group of the Psychiatric Genomics C (2013) Identification of risk loci with shared effects on five major psychiatric disorders: a genome-wide analysis. Lancet 381 (9875):1371–1379

Cross-Disorder Group of the Psychiatric Genomics C, Lee SH, Ripke S, Neale BM, Faraone SV, Purcell SM, Perlis RH, Mowry BJ, Thapar A, Goddard ME et al (2013) Genetic relationship between five psychiatric disorders estimated from genome-wide SNPs. Nat Genet 45 (9):984–994

Cuthbert BN, Insel TR (2013) Toward the future of psychiatric diagnosis: the seven pillars of RDoC. BMC Med 11:126

da Huang W, Sherman BT, Lempicki RA (2009) Systematic and integrative analysis of large gene lists using DAVID bioinformatics resources. Nat Protoc 4(1):44–57

Demontis D, Walters RK, Martin J, Mattheisen M, Damm Als T, Agerbo E, Belliveau RA, Byberg-Grauholm J, Bækved-Hansen M, Cerrato F et al (2019) Discovery of the first genome-wide significant risk loci for attention deficit/hyperactivity disorder. Nat Genet 51(1):63–75

Denny JC, Ritchie MD, Basford MA, Pulley JM, Bastarache L, Brown-Gentry K, Wang D, Masys DR, Roden DM, Crawford DC (2010) PheWAS: demonstrating the feasibility of a phenome-wide scan to discover gene-disease associations. Bioinformatics 26(9):1205–1210

Djurovic S, Gustafsson O, Mattingsdal M, Athanasiu L, Bjella T, Tesli M, Agartz I, Lorentzen S, Melle I, Morken G et al (2010) A genome-wide association study of bipolar disorder in Norwegian individuals, followed by replication in Icelandic sample. J Affect Disord 126 (1–2):312–316

Docherty AR, Moscati AA, Fanous AH (2016) Cross-disorder psychiatric genomics. Curr Behav Neurosci Rep 3(3):256–263

Duncan LE, Ratanatharathorn A, Aiello AE, Almli LM, Amstadter AB, Ashley-Koch AE, Baker DG, Beckham JC, Bierut LJ, Bisson J et al (2017) Largest GWAS of PTSD (N = 20 070) yields genetic overlap with schizophrenia and sex differences in heritability. Mol Psychiatry 23:666

Easton DF, Pooley KA, Dunning AM, Pharoah PD, Thompson D, Ballinger DG, Struewing JP, Morrison J, Field H, Luben R et al (2007) Genome-wide association study identifies novel breast cancer susceptibility loci. Nature 447(7148):1087–1093

Eaton WW, Martins SS, Nestadt G, Bienvenu OJ, Clarke D, Alexandre P (2008) The burden of mental disorders. Epidemiol Rev 30:1–14

Edenberg HJ (2007) The genetics of alcohol metabolism: role of alcohol dehydrogenase and aldehyde dehydrogenase variants. Alcohol Res Health 30(1):5–13

Edwards SL, Beesley J, French JD, Dunning AM (2013) Beyond GWASs: illuminating the dark road from association to function. Am J Hum Genet 93(5):779–797

Euesden J, Lewis CM, O'Reilly PF (2015) PRSice: polygenic risk score software. Bioinformatics 31(9):1466–1468

Faraone SV, Perlis RH, Doyle AE, Smoller JW, Goralnick JJ, Holmgren MA, Sklar P (2005) Molecular genetics of attention-deficit/hyperactivity disorder. Biol Psychiatry 57(11):1313–1323

Fernell E, Gillberg C (2010) Autism spectrum disorder diagnoses in Stockholm preschoolers. Res Dev Disabil 31(3):680–685

Ferreira MA, O'Donovan MC, Meng YA, Jones IR, Ruderfer DM, Jones L, Fan J, Kirov G, Perlis RH, Green EK et al (2008) Collaborative genome-wide association analysis supports a role for ANK3 and CACNA1C in bipolar disorder. Nat Genet 40(9):1056–1058

Finucane HK, Bulik-Sullivan B, Gusev A, Trynka G, Reshef Y, Loh PR, Anttila V, Xu H, Zang C, Farh K et al (2015) Partitioning heritability by functional annotation using genome-wide association summary statistics. Nat Genet 47(11):1228–1235

Flint J, Kendler KS (2014) The genetics of major depression. Neuron 81(5):1214

Frank J, Cichon S, Treutlein J, Ridinger M, Mattheisen M, Hoffmann P, Herms S, Wodarz N, Soyka M, Zill P et al (2012) Genome-wide significant association between alcohol dependence and a variant in the ADH gene cluster. Addict Biol 17(1):171–180

Franke B, Faraone SV, Asherson P, Buitelaar J, Bau CH, Ramos-Quiroga JA, Mick E, Grevet EH, Johansson S, Haavik J et al (2012) The genetics of attention deficit/hyperactivity disorder in adults, a review. Mol Psychiatry 17(10):960–987

Fromer M, Pocklington AJ, Kavanagh DH, Williams HJ, Dwyer S, Gormley P, Georgieva L, Rees E, Palta P, Ruderfer DM et al (2014) De novo mutations in schizophrenia implicate synaptic networks. Nature 506(7487):179–184

Gaugler T, Klei L, Sanders SJ, Bodea CA, Goldberg AP, Lee AB, Mahajan M, Manaa D, Pawitan Y, Reichert J et al (2014) Most genetic risk for autism resides with common variation. Nat Genet 46(8):881–885

Gelernter J, Kranzler HR, Sherva R, Almasy L, Koesterer R, Smith AH, Anton R, Preuss UW, Ridinger M, Rujescu D et al (2014) Genome-wide association study of alcohol dependence: significant findings in African- and European-Americans including novel risk loci. Mol Psychiatry 19(1):41–49

Gelernter J, Kranzler HR, Sherva R, Almasy L, Herman AI, Koesterer R, Zhao H, Farrer LA (2015) Genome-wide association study of nicotine dependence in American populations: identification of novel risk loci in both African-Americans and European-Americans. Biol Psychiatry 77 (5):493–503

Genetics of Personality C, de Moor MH, van den Berg SM, Verweij KJ, Krueger RF, Luciano M, Arias Vasquez A, Matteson LK, Derringer J, Esko T et al (2015) Meta-analysis of genome-wide association studies for neuroticism, and the polygenic association with major depressive disorder. JAMA Psychiat 72(7):642–650

Genomes Project C, Abecasis GR, Altshuler D, Auton A, Brooks LD, Durbin RM, Gibbs RA, Hurles ME, McVean GA (2010) A map of human genome variation from population-scale sequencing. Nature 467(7319):1061–1073

Genomes Project C, Auton A, Brooks LD, Durbin RM, Garrison EP, Kang HM, Korbel JO, Marchini JL, McCarthy S, McVean GA et al (2015) A global reference for human genetic variation. Nature 526(7571):68–74

Gonzalez-Mantilla AJ, Moreno-De-Luca A, Ledbetter DH, Martin CL (2016) A cross-disorder method to identify novel candidate genes for developmental brain disorders. JAMA Psychiat 73 (3):275–283

Green EK, Grozeva D, Jones I, Jones L, Kirov G, Caesar S, Gordon-Smith K, Fraser C, Forty L, Russell E et al (2010) The bipolar disorder risk allele at CACNA1C also confers risk of recurrent major depression and of schizophrenia. Mol Psychiatry 15(10):1016–1022

Guffanti G, Galea S, Yan L, Roberts AL, Solovieff N, Aiello AE, Smoller JW, De Vivo I, Ranu H, Uddin M et al (2013) Genome-wide association study implicates a novel RNA gene, the lincRNA AC068718.1, as a risk factor for post-traumatic stress disorder in women. Psychoneuroendocrinology 38(12):3029–3038

Guffanti G, Ashley-Koch AE, Roberts AL, Garrett ME, Solovieff N, Ratanatharathorn A, de Vivo I, Dennis M, Ranu H, Smoller JW et al (2014) No association between RORA polymorphisms and PTSD in two independent samples. Mol Psychiatry 19(10):1056–1057

Gurung R, Prata DP (2015) What is the impact of genome-wide supported risk variants for schizophrenia and bipolar disorder on brain structure and function? A systematic review. Psychol Med 45(12):2461–2480

Hancock DB, Guo Y, Reginsson GW, Gaddis NC, Lutz SM, Sherva R, Loukola A, Minica CC, Markunas CA, Han Y et al (2017) Genome-wide association study across European and African American ancestries identifies a SNP in DNMT3B contributing to nicotine dependence. Mol Psychiatry 23:1911

Hart AB, Kranzler HR (2015) Alcohol dependence genetics: lessons learned from genome-wide association studies (GWAS) and post-GWAS analyses. Alcohol Clin Exp Res 39(8):1312–1327

Hasin DS, Stinson FS, Ogburn E, Grant BF (2007) Prevalence, correlates, disability, and comorbidity of DSM-IV alcohol abuse and dependence in the United States: results from the National Epidemiologic Survey on Alcohol and Related Conditions. Arch Gen Psychiatry 64(7):830–842

Hattori E, Toyota T, Ishitsuka Y, Iwayama Y, Yamada K, Ujike H, Morita Y, Kodama M, Nakata K, Minabe Y et al (2009) Preliminary genome-wide association study of bipolar disorder in the Japanese population. Am J Med Genet B Neuropsychiatr Genet 150B(8):1110–1117

Hawi Z, Cummins TD, Tong J, Johnson B, Lau R, Samarrai W, Bellgrove MA (2015) The molecular genetic architecture of attention deficit hyperactivity disorder. Mol Psychiatry 20 (3):289–297

Hinney A, Scherag A, Jarick I, Albayrak O, Putter C, Pechlivanis S, Dauvermann MR, Beck S, Weber H, Scherag S et al (2011) Genome-wide association study in German patients with attention deficit/hyperactivity disorder. Am J Med Genet B Neuropsychiatr Genet 156B (8):888–897

Hou L, Bergen SE, Akula N, Song J, Hultman CM, Landen M, Adli M, Alda M, Ardau R, Arias B et al (2016) Genome-wide association study of 40,000 individuals identifies two novel loci associated with bipolar disorder. Hum Mol Genet 25(15):3383–3394

Howie BN, Donnelly P, Marchini J (2009) A flexible and accurate genotype imputation method for the next generation of genome-wide association studies. PLoS Genet 5(6):e1000529

Huang DW, Sherman BT, Tan Q, Kir J, Liu D, Bryant D, Guo Y, Stephens R, Baseler MW, Lane HC et al (2007) DAVID bioinformatics resources: expanded annotation database and novel algorithms to better extract biology from large gene lists. Nucleic Acids Res 35:W169–W175

Huang J, Perlis RH, Lee PH, Rush AJ, Fava M, Sachs GS, Lieberman J, Hamilton SP, Sullivan P, Sklar P et al (2010) Cross-disorder genomewide analysis of schizophrenia, bipolar disorder, and depression. Am J Psychiatry 167(10):1254–1263

Hyde CL, Nagle MW, Tian C, Chen X, Paciga SA, Wendland JR, Tung JY, Hinds DA, Perlis RH, Winslow AR (2016) Identification of 15 genetic loci associated with risk of major depression in individuals of European descent. Nat Genet 48(9):1031–1036

Ikeda M, Aleksic B, Kinoshita Y, Okochi T, Kawashima K, Kushima I, Ito Y, Nakamura Y, Kishi T, Okumura T et al (2011) Genome-wide association study of schizophrenia in a Japanese population. Biol Psychiatry 69(5):472–478

Ikeda M, Takahashi A, Kamatani Y, Okahisa Y, Kunugi H, Mori N, Sasaki T, Ohmori T, Okamoto Y, Kawasaki H et al (2017a) A genome-wide association study identifies two novel susceptibility loci and trans population polygenicity associated with bipolar disorder. Mol Psychiatry 23:639

Ikeda M, Saito T, Kondo K, Iwata N (2017b) Genome-wide association studies of bipolar disorder: a systematic review of recent findings and their clinical implications. Psychiatry Clin Neurosci 72:52–63

International HapMap C (2003) The international HapMap project. Nature 426(6968):789–796

International Obsessive Compulsive Disorder Foundation Genetics C, Studies OCDCGA (2017) Revealing the complex genetic architecture of obsessive-compulsive disorder using meta-analysis. Mol Psychiatry 23:1181

International Schizophrenia C, Purcell SM, Wray NR, Stone JL, Visscher PM, O'Donovan MC, Sullivan PF, Sklar P (2009) Common polygenic variation contributes to risk of schizophrenia and bipolar disorder. Nature 460(7256):748–752

Iyegbe C, Campbell D, Butler A, Ajnakina O, Sham P (2014) The emerging molecular architecture of schizophrenia, polygenic risk scores and the clinical implications for GxE research. Soc Psychiatry Psychiatr Epidemiol 49(2):169–182

Jamain S, Quach H, Betancur C, Rastam M, Colineaux C, Gillberg IC, Soderstrom H, Giros B, Leboyer M, Gillberg C et al (2003) Mutations of the X-linked genes encoding neuroligins NLGN3 and NLGN4 are associated with autism. Nat Genet 34(1):27–29

Kanehisa M, Goto S (2000) KEGG: Kyoto encyclopedia of genes and genomes. Nucleic Acids Res 28(1):27–30

Kao PY, Leung KH, Chan LW, Yip SP, Yap MK (2017) Pathway analysis of complex diseases for GWAS, extending to consider rare variants, multi-omics and interactions. Biochim Biophys Acta 1861(2):335–353

Kerin T, Ramanathan A, Rivas K, Grepo N, Coetzee GA, Campbell DB (2012) A noncoding RNA antisense to moesin at 5p14.1 in autism. Sci Transl Med 4(128):128ra140

Kerner B (2014) Genetics of bipolar disorder. Appl Clin Genet 7:33–42

King BH, Lord C (2011) Is schizophrenia on the autism spectrum? Brain Res 1380:34–41

Kirov G, Zaharieva I, Georgieva L, Moskvina V, Nikolov I, Cichon S, Hillmer A, Toncheva D, Owen MJ, O'Donovan MC (2009) A genome-wide association study in 574 schizophrenia trios using DNA pooling. Mol Psychiatry 14(8):796–803

Kirov G, Pocklington AJ, Holmans P, Ivanov D, Ikeda M, Ruderfer D, Moran J, Chambert K, Toncheva D, Georgieva L et al (2012) De novo CNV analysis implicates specific abnormalities of postsynaptic signalling complexes in the pathogenesis of schizophrenia. Mol Psychiatry 17 (2):142–153

Kohli MA, Lucae S, Saemann PG, Schmidt MV, Demirkan A, Hek K, Czamara D, Alexander M, Salyakina D, Ripke S et al (2011) The neuronal transporter gene SLC6A15 confers risk to major depression. Neuron 70(2):252–265

Konstenius M, Larsson H, Lundholm L, Philips B, van de Glind G, Jayaram-Lindstrom N, Franck J (2015) An epidemiological study of ADHD, substance use, and comorbid problems in incarcerated women in Sweden. J Atten Disord 19(1):44–52

Krueger WH, Gonye GE, Madison DL, Murray KE, Kumar M, Spoerel N, Pfeiffer SE (1997) TPO1, a member of a novel protein family, is developmentally regulated in cultured oligodendrocytes. J Neurochem 69(4):1343–1355

Kuo PH, Chuang LC, Su MH, Chen CH, Chen CH, Wu JY, Yen CJ, Wu YY, Liu SK, Chou MC et al (2015) Genome-wide association study for autism spectrum disorder in Taiwanese Han population. PLoS One 10(9):e0138695

Kutmon M, Riutta A, Nunes N, Hanspers K, Willighagen EL, Bohler A, Melius J, Waagmeester A, Sinha SR, Miller R et al (2016) WikiPathways: capturing the full diversity of pathway knowledge. Nucleic Acids Res 44(D1):D488–D494

Lai MC, Lombardo MV, Baron-Cohen S (2014) Autism. Lancet 383(9920):896–910

Laumonnier F, Bonnet-Brilhault F, Gomot M, Blanc R, David A, Moizard MP, Raynaud M, Ronce N, Lemonnier E, Calvas P et al (2004) X-linked mental retardation and autism are associated with a mutation in the NLGN4 gene, a member of the neuroligin family. Am J Hum Genet 74(3):552–557

Lee SH, Wray NR, Goddard ME, Visscher PM (2011a) Estimating missing heritability for disease from genome-wide association studies. Am J Hum Genet 88(3):294–305

Lee MT, Chen CH, Lee CS, Chen CC, Chong MY, Ouyang WC, Chiu NY, Chuo LJ, Chen CY, Tan HK et al (2011b) Genome-wide association study of bipolar I disorder in the Han Chinese population. Mol Psychiatry 16(5):548–556

Lee SH, Yang J, Goddard ME, Visscher PM, Wray NR (2012a) Estimation of pleiotropy between complex diseases using single-nucleotide polymorphism-derived genomic relationships and restricted maximum likelihood. Bioinformatics 28(19):2540–2542

Lee PH, O'Dushlaine C, Thomas B, Purcell SM (2012b) INRICH: interval-based enrichment analysis for genome-wide association studies. Bioinformatics 28(13):1797–1799

Lencz T, Morgan TV, Athanasiou M, Dain B, Reed CR, Kane JM, Kucherlapati R, Malhotra AK (2007) Converging evidence for a pseudoautosomal cytokine receptor gene locus in schizophrenia. Mol Psychiatry 12(6):572–580

Lesch KP, Timmesfeld N, Renner TJ, Halperin R, Roser C, Nguyen TT, Craig DW, Romanos J, Heine M, Meyer J et al (2008) Molecular genetics of adult ADHD: converging evidence from genome-wide association and extended pedigree linkage studies. J Neural Transm (Vienna) 115 (11):1573–1585

Levinson DF, Duan J, Oh S, Wang K, Sanders AR, Shi J, Zhang N, Mowry BJ, Olincy A, Amin F et al (2011) Copy number variants in schizophrenia: confirmation of five previous findings and new evidence for 3q29 microdeletions and VIPR2 duplications. Am J Psychiatry 168 (3):302–316

Levinson DF, Mostafavi S, Milaneschi Y, Rivera M, Ripke S, Wray NR, Sullivan PF (2014) Genetic studies of major depressive disorder: why are there no genome-wide association study findings and what can we do about it? Biol Psychiatry 76(7):510–512

Lewis CM, Ng MY, Butler AW, Cohen-Woods S, Uher R, Pirlo K, Weale ME, Schosser A, Paredes UM, Rivera M et al (2010) Genome-wide association study of major recurrent depression in the U.K. population. Am J Psychiatry 167(8):949–957

Lichtenstein P, Carlstrom E, Rastam M, Gillberg C, Anckarsater H (2010) The genetics of autism spectrum disorders and related neuropsychiatric disorders in childhood. Am J Psychiatry 167 (11):1357–1363

Liu Y, Blackwood DH, Caesar S, de Geus EJ, Farmer A, Ferreira MA, Ferrier IN, Fraser C, Gordon-Smith K, Green EK et al (2011) Meta-analysis of genome-wide association data of bipolar disorder and major depressive disorder. Mol Psychiatry 16(1):2–4

Liu J, Dietz K, DeLoyht JM, Pedre X, Kelkar D, Kaur J, Vialou V, Lobo MK, Dietz DM, Nestler EJ et al (2012) Impaired adult myelination in the prefrontal cortex of socially isolated mice. Nat Neurosci 15(12):1621–1623

Lo MT, Hinds DA, Tung JY, Franz C, Fan CC, Wang Y, Smeland OB, Schork A, Holland D, Kauppi K et al (2017) Genome-wide analyses for personality traits identify six genomic loci and show correlations with psychiatric disorders. Nat Genet 49(1):152–156

Logue MW, Baldwin C, Guffanti G, Melista E, Wolf EJ, Reardon AF, Uddin M, Wildman D, Galea S, Koenen KC et al (2013) A genome-wide association study of post-traumatic stress disorder identifies the retinoid-related orphan receptor alpha (RORA) gene as a significant risk locus. Mol Psychiatry 18(8):937–942

Logue MW, Amstadter AB, Baker DG, Duncan L, Koenen KC, Liberzon I, Miller MW, Morey RA, Nievergelt CM, Ressler KJ et al (2015) The psychiatric genomics consortium posttraumatic stress disorder workgroup: posttraumatic stress disorder enters the age of large-scale genomic collaboration. Neuropsychopharmacology 40(10):2287–2297

Lubke GH, Laurin C, Amin N, Hottenga JJ, Willemsen G, van Grootheest G, Abdellaoui A, Karssen LC, Oostra BA, van Duijn CM et al (2014) Genome-wide analyses of borderline personality features. Mol Psychiatry 19(8):923–929

Luciano M, Hagenaars SP, Davies G, Hill WD, Clarke TK, Shirali M, Harris SE, Marioni RE, Liewald DC, Fawns-Ritchie C et al (2018) Association analysis in over 329,000 individuals identifies 116 independent variants influencing neuroticism. Nat Genet 50(1):6–11

Lynskey MT, Glowinski AL, Todorov AA, Bucholz KK, Madden PA, Nelson EC, Statham DJ, Martin NG, Heath AC (2004) Major depressive disorder, suicidal ideation, and suicide attempt in twins discordant for cannabis dependence and early-onset cannabis use. Arch Gen Psychiatry 61(10):1026–1032

Ma D, Salyakina D, Jaworski JM, Konidari I, Whitehead PL, Andersen AN, Hoffman JD, Slifer SH, Hedges DJ, Cukier HN et al (2009) A genome-wide association study of autism reveals a common novel risk locus at 5p14.1. Ann Hum Genet 73(Pt 3):263–273

MacArthur J, Bowler E, Cerezo M, Gil L, Hall P, Hastings E, Junkins H, McMahon A, Milano A, Morales J et al (2017) The new NHGRI-EBI catalog of published genome-wide association studies (GWAS catalog). Nucleic Acids Res 45(D1):D896–D901

Mah S, Nelson MR, Delisi LE, Reneland RH, Markward N, James MR, Nyholt DR, Hayward N, Handoko H, Mowry B et al (2006) Identification of the semaphorin receptor PLXNA2 as a candidate for susceptibility to schizophrenia. Mol Psychiatry 11(5):471–478

Maier W, Lichtermann D, Minges J, Hallmayer J, Heun R, Benkert O, Levinson DF (1993) Continuity and discontinuity of affective disorders and schizophrenia. Results of a controlled family study. Arch Gen Psychiatry 50(11):871–883

Major Depressive Disorder Working Group of the Psychiatric GC, Ripke S, Wray NR, Lewis CM, Hamilton SP, Weissman MM, Breen G, Byrne EM, Blackwood DH, Boomsma DI et al (2013) A mega-analysis of genome-wide association studies for major depressive disorder. Mol Psychiatry 18(4):497–511

Mannuzza S, Klein RG, Bessler A, Malloy P, LaPadula M (1993) Adult outcome of hyperactive boys. Educational achievement, occupational rank, and psychiatric status. Arch Gen Psychiatry 50(7):565–576

Marchini J, Howie B, Myers S, McVean G, Donnelly P (2007) A new multipoint method for genome-wide association studies by imputation of genotypes. Nat Genet 39(7):906–913

Marshall CR, Howrigan DP, Merico D, Thiruvahindrapuram B, Wu W, Greer DS, Antaki D, Shetty A, Holmans PA, Pinto D et al (2017) Contribution of copy number variants to schizophrenia from a genome-wide study of 41,321 subjects. Nat Genet 49(1):27–35

Mattheisen M, Samuels JF, Wang Y, Greenberg BD, Fyer AJ, McCracken JT, Geller DA, Murphy DL, Knowles JA, Grados MA et al (2015) Genome-wide association study in obsessive-compulsive disorder: results from the OCGAS. Mol Psychiatry 20(3):337–344

McGuffin P, Rijsdijk F, Andrew M, Sham P, Katz R, Cardno A (2003) The heritability of bipolar affective disorder and the genetic relationship to unipolar depression. Arch Gen Psychiatry 60(5):497–502

Mick E, Todorov A, Smalley S, Hu X, Loo S, Todd RD, Biederman J, Byrne D, Dechairo B, Guiney A et al (2010) Family-based genome-wide association scan of attention-deficit/hyperactivity disorder. J Am Acad Child Adolesc Psychiatry 49(9):898–905.e893

Middeldorp CM, Hammerschlag AR, Ouwens KG, Groen-Blokhuis MM, Pourcain BS, Greven CU, Pappa I, Tiesler CMT, Ang W, Nolte IM et al (2016) A genome-wide association meta-analysis of attention-deficit/hyperactivity disorder symptoms in population-based pediatric cohorts. J Am Acad Child Adolesc Psychiatry 55(10):896–905.e896

Muglia P, Tozzi F, Galwey NW, Francks C, Upmanyu R, Kong XQ, Antoniades A, Domenici E, Perry J, Rothen S et al (2010) Genome-wide association study of recurrent major depressive disorder in two European case-control cohorts. Mol Psychiatry 15(6):589–601

Muhleisen TW, Leber M, Schulze TG, Strohmaier J, Degenhardt F, Treutlein J, Mattheisen M, Forstner AJ, Schumacher J, Breuer R et al (2014) Genome-wide association study reveals two new risk loci for bipolar disorder. Nat Commun 5:3339

Neale BM, Lasky-Su J, Anney R, Franke B, Zhou K, Maller JB, Vasquez AA, Asherson P, Chen W, Banaschewski T et al (2008) Genome-wide association scan of attention deficit hyperactivity disorder. Am J Med Genet B Neuropsychiatr Genet 147B(8):1337–1344

Neale BM, Medland S, Ripke S, Anney RJ, Asherson P, Buitelaar J, Franke B, Gill M, Kent L, Holmans P et al (2010a) Case-control genome-wide association study of attention-deficit/hyperactivity disorder. J Am Acad Child Adolesc Psychiatry 49(9):906–920

Neale BM, Medland SE, Ripke S, Asherson P, Franke B, Lesch KP, Faraone SV, Nguyen TT, Schafer H, Holmans P et al (2010b) Meta-analysis of genome-wide association studies of attention-deficit/hyperactivity disorder. J Am Acad Child Adolesc Psychiatry 49(9):884–897

Need AC, Ge D, Weale ME, Maia J, Feng S, Heinzen EL, Shianna KV, Yoon W, Kasperaviciute D, Gennarelli M et al (2009) A genome-wide investigation of SNPs and CNVs in schizophrenia. PLoS Genet 5(2):e1000373

Nicolae DL, Gamazon E, Zhang W, Duan S, Dolan ME, Cox NJ (2010) Trait-associated SNPs are more likely to be eQTLs: annotation to enhance discovery from GWAS. PLoS Genet 6(4): e1000888

Nievergelt CM, Maihofer AX, Mustapic M, Yurgil KA, Schork NJ, Miller MW, Logue MW, Geyer MA, Risbrough VB, O'Connor DT et al (2015) Genomic predictors of combat stress vulnerability and resilience in U.S. marines: a genome-wide association study across multiple ancestries implicates PRTFDC1 as a potential PTSD gene. Psychoneuroendocrinology 51:459–471

O'Donovan MC, Craddock N, Norton N, Williams H, Peirce T, Moskvina V, Nikolov I, Hamshere M, Carroll L, Georgieva L et al (2008) Identification of loci associated with schizophrenia by genome-wide association and follow-up. Nat Genet 40(9):1053–1055

Okbay A, Baselmans BM, De Neve JE, Turley P, Nivard MG, Fontana MA, Meddens SF, Linner RK, Rietveld CA, Derringer J et al (2016) Genetic variants associated with subjective well-being, depressive symptoms, and neuroticism identified through genome-wide analyses. Nat Genet 48(6):624–633

Olfson E, Bierut LJ (2012) Convergence of genome-wide association and candidate gene studies for alcoholism. Alcohol Clin Exp Res 36(12):2086–2094

Ollier W, Sprosen T, Peakman T (2005) UK biobank: from concept to reality. Pharmacogenomics 6(6):639–646

Panagiotou OA, Willer CJ, Hirschhorn JN, Ioannidis JP (2013) The power of meta-analysis in genome-wide association studies. Annu Rev Genomics Hum Genet 14:441–465

Park BL, Kim JW, Cheong HS, Kim LH, Lee BC, Seo CH, Kang TC, Nam YW, Kim GB, Shin HD et al (2013) Extended genetic effects of ADH cluster genes on the risk of alcohol dependence: from GWAS to replication. Hum Genet 132(6):657–668

Patel KR, Cherian J, Gohil K, Atkinson D (2014) Schizophrenia: overview and treatment options. P T 39(9):638–645

Pe'er I, Yelensky R, Altshuler D, Daly MJ (2008) Estimation of the multiple testing burden for genomewide association studies of nearly all common variants. Genet Epidemiol 32 (4):381–385

Power RA, Verweij KJ, Zuhair M, Montgomery GW, Henders AK, Heath AC, Madden PA, Medland SE, Wray NR, Martin NG (2014) Genetic predisposition to schizophrenia associated with increased use of cannabis. Mol Psychiatry 19(11):1201–1204

Psychiatric GCBDWG (2011) Large-scale genome-wide association analysis of bipolar disorder identifies a new susceptibility locus near ODZ4. Nat Genet 43(10):977–983

Purcell S, Neale B, Todd-Brown K, Thomas L, Ferreira MA, Bender D, Maller J, Sklar P, de Bakker PI, Daly MJ et al (2007) PLINK: a tool set for whole-genome association and population-based linkage analyses. Am J Hum Genet 81(3):559–575

Purcell SM, Moran JL, Fromer M, Ruderfer D, Solovieff N, Roussos P, O'Dushlaine C, Chambert K, Bergen SE, Kahler A et al (2014) A polygenic burden of rare disruptive mutations in schizophrenia. Nature 506(7487):185–190

Rietschel M, Mattheisen M, Frank J, Treutlein J, Degenhardt F, Breuer R, Steffens M, Mier D, Esslinger C, Walter H et al (2010) Genome-wide association-, replication-, and neuroimaging study implicates HOMER1 in the etiology of major depression. Biol Psychiatry 68(6):578–585

Ripke S, O'Dushlaine C, Chambert K, Moran JL, Kahler AK, Akterin S, Bergen SE, Collins AL, Crowley JJ, Fromer M et al (2013) Genome-wide association analysis identifies 13 new risk loci for schizophrenia. Nat Genet 45(10):1150–1159

Risch N, Merikangas K (1996) The future of genetic studies of complex human diseases. Science 273(5281):1516–1517

Roadmap Epigenomics C, Kundaje A, Meuleman W, Ernst J, Bilenky M, Yen A, Heravi-Moussavi A, Kheradpour P, Zhang Z, Wang J et al (2015) Integrative analysis of 111 reference human epigenomes. Nature 518(7539):317–330

Rommelse NN, Franke B, Geurts HM, Hartman CA, Buitelaar JK (2010) Shared heritability of attention-deficit/hyperactivity disorder and autism spectrum disorder. Eur Child Adolesc Psychiatry 19(3):281–295

Ronald A, Hoekstra RA (2011) Autism spectrum disorders and autistic traits: a decade of new twin studies. Am J Med Genet B Neuropsychiatr Genet 156B(3):255–274

Ronald A, Simonoff E, Kuntsi J, Asherson P, Plomin R (2008) Evidence for overlapping genetic influences on autistic and ADHD behaviours in a community twin sample. J Child Psychol Psychiatry 49(5):535–542

Ruderfer DM, Fanous AH, Ripke S, McQuillin A, Amdur RL, Schizophrenia Working Group of the Psychiatric Genomics C, Bipolar Disorder Working Group of the Psychiatric Genomics C, Cross-Disorder Working Group of the Psychiatric Genomics C, Gejman PV, O'Donovan MC et al (2014) Polygenic dissection of diagnosis and clinical dimensions of bipolar disorder and schizophrenia. Mol Psychiatry 19(9):1017–1024

Salyakina D, Ma DQ, Jaworski JM, Konidari I, Whitehead PL, Henson R, Martinez D, Robinson JL, Sacharow S, Wright HH et al (2010) Variants in several genomic regions associated with Asperger disorder. Autism Res 3(6):303–310

Sanchez-Mora C, Ramos-Quiroga JA, Bosch R, Corrales M, Garcia-Martinez I, Nogueira M, Pagerols M, Palomar G, Richarte V, Vidal R et al (2015) Case-control genome-wide association study of persistent attention-deficit hyperactivity disorder identifies FBXO33 as a novel susceptibility gene for the disorder. Neuropsychopharmacology 40(4):915–926

Schizophrenia Psychiatric Genome-Wide Association Study C (2011) Genome-wide association study identifies five new schizophrenia loci. Nat Genet 43(10):969–976

Schizophrenia Working Group of the Psychiatric Genomics C (2014) Biological insights from 108 schizophrenia-associated genetic loci. Nature 511(7510):421–427

Scott LJ, Muglia P, Kong XQ, Guan W, Flickinger M, Upmanyu R, Tozzi F, Li JZ, Burmeister M, Absher D et al (2009) Genome-wide association and meta-analysis of bipolar disorder in individuals of European ancestry. Proc Natl Acad Sci U S A 106(18):7501–7506

Sebat J, Lakshmi B, Malhotra D, Troge J, Lese-Martin C, Walsh T, Yamrom B, Yoon S, Krasnitz A, Kendall J et al (2007) Strong association of de novo copy number mutations with autism. Science 316(5823):445–449
Segre AV, Consortium D, Investigators M, Groop L, Mootha VK, Daly MJ, Altshuler D (2010) Common inherited variation in mitochondrial genes is not enriched for associations with type 2 diabetes or related glycemic traits. PLoS Genet 6(8):e1001058
Sherva R, Wang Q, Kranzler H, Zhao H, Koesterer R, Herman A, Farrer LA, Gelernter J (2016) Genome-wide association study of cannabis dependence severity, novel risk variants, and shared genetic risks. JAMA Psychiat 73(5):472–480
Shi J, Levinson DF, Duan J, Sanders AR, Zheng Y, Pe'er I, Dudbridge F, Holmans PA, Whittemore AS, Mowry BJ et al (2009) Common variants on chromosome 6p22.1 are associated with schizophrenia. Nature 460(7256):753–757
Shi Y, Li Z, Xu Q, Wang T, Li T, Shen J, Zhang F, Chen J, Zhou G, Ji W et al (2011a) Common variants on 8p12 and 1q24.2 confer risk of schizophrenia. Nat Genet 43(12):1224–1227
Shi J, Potash JB, Knowles JA, Weissman MM, Coryell W, Scheftner WA, Lawson WB, DePaulo JR Jr, Gejman PV, Sanders AR et al (2011b) Genome-wide association study of recurrent early-onset major depressive disorder. Mol Psychiatry 16(2):193–201
Shifman S, Johannesson M, Bronstein M, Chen SX, Collier DA, Craddock NJ, Kendler KS, Li T, O'Donovan M, O'Neill FA et al (2008) Genome-wide association identifies a common variant in the reelin gene that increases the risk of schizophrenia only in women. PLoS Genet 4(2):e28
Shyn SI, Shi J, Kraft JB, Potash JB, Knowles JA, Weissman MM, Garriock HA, Yokoyama JS, McGrath PJ, Peters EJ et al (2011) Novel loci for major depression identified by genome-wide association study of sequenced treatment alternatives to relieve depression and meta-analysis of three studies. Mol Psychiatry 16(2):202–215
Sklar P, Smoller JW, Fan J, Ferreira MA, Perlis RH, Chambert K, Nimgaonkar VL, McQueen MB, Faraone SV, Kirby A et al (2008) Whole-genome association study of bipolar disorder. Mol Psychiatry 13(6):558–569
Sladek R, Rocheleau G, Rung J, Dina C, Shen L, Serre D, Boutin P, Vincent D, Belisle A, Hadjadj S et al (2007) A genome-wide association study identifies novel risk loci for type 2 diabetes. Nature 445(7130):881–885
Smith EN, Bloss CS, Badner JA, Barrett T, Belmonte PL, Berrettini W, Byerley W, Coryell W, Craig D, Edenberg HJ et al (2009) Genome-wide association study of bipolar disorder in European American and African American individuals. Mol Psychiatry 14(8):755–763
Smith EN, Koller DL, Panganiban C, Szelinger S, Zhang P, Badner JA, Barrett TB, Berrettini WH, Bloss CS, Byerley W et al (2011) Genome-wide association of bipolar disorder suggests an enrichment of replicable associations in regions near genes. PLoS Genet 7(6):e1002134
Smoller JW (2014) Psychiatric genetics and the future of personalized treatment. Depress Anxiety 31(11):893–898
Smoller JW, Finn CT (2003) Family, twin, and adoption studies of bipolar disorder. Am J Med Genet C Semin Med Genet 123C(1):48–58
Speed D, Hemani G, Johnson MR, Balding DJ (2012) Improved heritability estimation from genome-wide SNPs. Am J Hum Genet 91(6):1011–1021
Stahl E, Forstner AJ, McQuillin A, Ripke S, Trubetskoy V, Mattheisen M, Wang W, Wang Y, Coleman JRI, Gaspar HA et al (2019) Genome-wide association study identifies 30 loci associated with bipolar disorder. Nat Genet 51(5):793–803
Stefansson H, Ophoff RA, Steinberg S, Andreassen OA, Cichon S, Rujescu D, Werge T, Pietilainen OP, Mors O, Mortensen PB et al (2009) Common variants conferring risk of schizophrenia. Nature 460(7256):744–747
Stein MB, Chen CY, Ursano RJ, Cai T, Gelernter J, Heeringa SG, Jain S, Jensen KP, Maihofer AX, Mitchell C et al (2016) Genome-wide association studies of posttraumatic stress disorder in 2 cohorts of US army soldiers. JAMA Psychiat 73(7):695–704

Steinberg S, de Jong S, Irish Schizophrenia Genomics C, Andreassen OA, Werge T, Borglum AD, Mors O, Mortensen PB, Gustafsson O, Costas J et al (2011) Common variants at VRK2 and TCF4 conferring risk of schizophrenia. Hum Mol Genet 20(20):4076–4081

Stergiakouli E, Hamshere M, Holmans P, Langley K, Zaharieva I, deCODE Genetics; Psychiatric GWAS Consortium, Hawi Z, Kent L, Gill M et al (2012) Investigating the contribution of common genetic variants to the risk and pathogenesis of ADHD. Am J Psychiatry 169 (2):186–194

Stewart SE, Yu D, Scharf JM, Neale BM, Fagerness JA, Mathews CA, Arnold PD, Evans PD, Gamazon ER, Davis LK et al (2013) Genome-wide association study of obsessive-compulsive disorder. Mol Psychiatry 18(7):788–798

Stickel F, Moreno C, Hampe J, Morgan MY (2017) The genetics of alcohol dependence and alcohol-related liver disease. J Hepatol 66(1):195–211

Stringer S, Minica CC, Verweij KJ, Mbarek H, Bernard M, Derringer J, van Eijk KR, Isen JD, Loukola A, Maciejewski DF et al (2016) Genome-wide association study of lifetime cannabis use based on a large meta-analytic sample of 32 330 subjects from the International Cannabis Consortium. Transl Psychiatry 6:e769

Subramanian A, Tamayo P, Mootha VK, Mukherjee S, Ebert BL, Gillette MA, Paulovich A, Pomeroy SL, Golub TR, Lander ES et al (2005) Gene set enrichment analysis: a knowledge-based approach for interpreting genome-wide expression profiles. Proc Natl Acad Sci U S A 102 (43):15545–15550

Sullivan PF (2010) The psychiatric GWAS consortium: big science comes to psychiatry. Neuron 68 (2):182–186

Sullivan PF, Neale MC, Kendler KS (2000) Genetic epidemiology of major depression: review and meta-analysis. Am J Psychiatry 157(10):1552–1562

Sullivan PF, Kendler KS, Neale MC (2003) Schizophrenia as a complex trait: evidence from a meta-analysis of twin studies. Arch Gen Psychiatry 60(12):1187–1192

Sullivan PF, Lin D, Tzeng JY, van den Oord E, Perkins D, Stroup TS, Wagner M, Lee S, Wright FA, Zou F et al (2008) Genomewide association for schizophrenia in the CATIE study: results of stage 1. Mol Psychiatry 13(6):570–584

Sullivan PF, de Geus EJ, Willemsen G, James MR, Smit JH, Zandbelt T, Arolt V, Baune BT, Blackwood D, Cichon S et al (2009) Genome-wide association for major depressive disorder: a possible role for the presynaptic protein piccolo. Mol Psychiatry 14(4):359–375

Thomasson HR, Edenberg HJ, Crabb DW, Mai XL, Jerome RE, Li TK, Wang SP, Lin YT, Lu RB, Yin SJ (1991) Alcohol and aldehyde dehydrogenase genotypes and alcoholism in Chinese men. Am J Hum Genet 48(4):677–681

Tobacco, Genetics C (2010) Genome-wide meta-analyses identify multiple loci associated with smoking behavior. Nat Genet 42(5):441–447

Treutlein J, Cichon S, Ridinger M, Wodarz N, Soyka M, Zill P, Maier W, Moessner R, Gaebel W, Dahmen N et al (2009) Genome-wide association study of alcohol dependence. Arch Gen Psychiatry 66(7):773–784

Visscher PM, Brown MA, McCarthy MI, Yang J (2012) Five years of GWAS discovery. Am J Hum Genet 90(1):7–24

Visscher PM, Wray NR, Zhang Q, Sklar P, McCarthy MI, Brown MA, Yang J (2017) 10 years of GWAS discovery: biology, function, and translation. Am J Hum Genet 101(1):5–22

Vukasovic T, Bratko D (2015) Heritability of personality: a meta-analysis of behavior genetic studies. Psychol Bull 141(4):769–785

Wang K, Zhang H, Ma D, Bucan M, Glessner JT, Abrahams BS, Salyakina D, Imielinski M, Bradfield JP, Sleiman PM et al (2009) Common genetic variants on 5p14.1 associate with autism spectrum disorders. Nature 459(7246):528–533

Wason JM, Dudbridge F (2012) A general framework for two-stage analysis of genome-wide association studies and its application to case-control studies. Am J Hum Genet 90(5):760–773

Wellcome Trust Case Control C (2007) Genome-wide association study of 14,000 cases of seven common diseases and 3,000 shared controls. Nature 447(7145):661–678

Williams HJ, Norton N, Dwyer S, Moskvina V, Nikolov I, Carroll L, Georgieva L, Williams NM, Morris DW, Quinn EM et al (2011) Fine mapping of ZNF804A and genome-wide significant evidence for its involvement in schizophrenia and bipolar disorder. Mol Psychiatry 16 (4):429–441

Witt SH, Streit F, Jungkunz M, Frank J, Awasthi S, Reinbold CS, Treutlein J, Degenhardt F, Forstner AJ, Heilmann-Heimbach S et al (2017) Genome-wide association study of borderline personality disorder reveals genetic overlap with bipolar disorder, major depression and schizophrenia. Transl Psychiatry 7(6):e1155

Wolf EJ, Rasmusson AM, Mitchell KS, Logue MW, Baldwin CT, Miller MW (2014) A genome-wide association study of clinical symptoms of dissociation in a trauma-exposed sample. Depress Anxiety 31(4):352–360

Wray NR, Pergadia ML, Blackwood DH, Penninx BW, Gordon SD, Nyholt DR, Ripke S, MacIntyre DJ, McGhee KA, Maclean AW et al (2012) Genome-wide association study of major depressive disorder: new results, meta-analysis, and lessons learned. Mol Psychiatry 17 (1):36–48

Wray NR, Lee SH, Mehta D, Vinkhuyzen AA, Dudbridge F, Middeldorp CM (2014) Research review: polygenic methods and their application to psychiatric traits. J Child Psychol Psychiatry 55(10):1068–1087

Wray NR, Ripke S, Matttheisen M, Trzaskowski M, Byrne EM, Abdellaoui A, Adams M, Agerbo E, Air TM, Andlauer TFM (2018) Genome-wide association analyses identify 44 risk variants and refine the genetic architecture of major depressive disorder. Nat Genet 50 (5):668–681

Xia K, Guo H, Hu Z, Xun G, Zuo L, Peng Y, Wang K, He Y, Xiong Z, Sun L et al (2014) Common genetic variants on 1p13.2 associate with risk of autism. Mol Psychiatry 19(11):1212–1219

Xie P, Kranzler HR, Yang C, Zhao H, Farrer LA, Gelernter J (2013) Genome-wide association study identifies new susceptibility loci for posttraumatic stress disorder. Biol Psychiatry 74 (9):656–663

Yang J, Benyamin B, McEvoy BP, Gordon S, Henders AK, Nyholt DR, Madden PA, Heath AC, Martin NG, Montgomery GW et al (2010) Common SNPs explain a large proportion of the heritability for human height. Nat Genet 42(7):565–569

Yang J, Lee SH, Goddard ME, Visscher PM (2011) GCTA: a tool for genome-wide complex trait analysis. Am J Hum Genet 88(1):76–82

Yang L, Neale BM, Liu L, Lee SH, Wray NR, Ji N, Li H, Qian Q, Wang D, Li J et al (2013) Polygenic transmission and complex neuro developmental network for attention deficit hyperactivity disorder: genome-wide association study of both common and rare variants. Am J Med Genet B Neuropsychiatr Genet 162B(5):419–430

Yosifova A, Mushiroda T, Kubo M, Takahashi A, Kamatani Y, Kamatani N, Stoianov D, Vazharova R, Karachanak S, Zaharieva I et al (2011) Genome-wide association study on bipolar disorder in the Bulgarian population. Genes Brain Behav 10(7):789–797

Yue WH, Wang HF, Sun LD, Tang FL, Liu ZH, Zhang HX, Li WQ, Zhang YL, Zhang Y, Ma CC et al (2011) Genome-wide association study identifies a susceptibility locus for schizophrenia in Han Chinese at 11p11.2. Nat Genet 43(12):1228–1231

Zayats T, Athanasiu L, Sonderby I, Djurovic S, Westlye LT, Tamnes CK, Fladby T, Aase H, Zeiner P, Reichborn-Kjennerud T et al (2015) Genome-wide analysis of attention deficit hyperactivity disorder in Norway. PLoS One 10(4):e0122501

Zheng J, Erzurumluoglu AM, Elsworth BL, Kemp JP, Howe L, Haycock PC, Hemani G, Tansey K, Laurin C, Early G et al (2017) LD hub: a centralized database and web interface to perform LD score regression that maximizes the potential of summary level GWAS data for SNP heritability and genetic correlation analysis. Bioinformatics 33(2):272–279

Genetic Variation in Long-Range Enhancers

John F. Fullard, Samir Rahman, and Panos Roussos

Contents

J. F. Fullard
Department of Psychiatry, Icahn School of Medicine at Mount Sinai, New York, NY, USA

Friedman Brain Institute, Icahn School of Medicine at Mount Sinai, New York, NY, USA

Department of Genetics and Genomic Science and Institute for Multiscale Biology, Icahn School of Medicine at Mount Sinai, New York, NY, USA

S. Rahman
Friedman Brain Institute, Icahn School of Medicine at Mount Sinai, New York, NY, USA

Department of Genetics and Genomic Science and Institute for Multiscale Biology, Icahn School of Medicine at Mount Sinai, New York, NY, USA

P. Roussos (✉)
Department of Psychiatry, Icahn School of Medicine at Mount Sinai, New York, NY, USA

Friedman Brain Institute, Icahn School of Medicine at Mount Sinai, New York, NY, USA

Department of Genetics and Genomic Science and Institute for Multiscale Biology, Icahn School of Medicine at Mount Sinai, New York, NY, USA

Mental Illness Research, Education, and Clinical Center (VISN 2 South), James J. Peters VA Medical Center, Bronx, NY, USA
e-mail: Panagiotis.roussos@mssm.edu

© Springer Nature Switzerland AG 2019

Curr Topics Behav Neurosci (2019) 42: 35–50
DOI 10.1007/7854_2019_110
Published Online: 9 August 2019

Abstract *Cis*-regulatory elements (CREs), including insulators, promoters, and enhancers, play critical roles in the establishment and maintenance of normal cellular function. Within each cell, the 3D structure of chromatin is arranged in specific patterns to expose the CREs required for optimal spatiotemporal regulation of gene expression. CREs can act over large distances along the linear genome, facilitated by looping of the intervening chromatin to allow direct interaction between distal regulatory elements and their target genes. A number of pathologies are associated with dysregulation of CRE function, including developmental disorders, cancers, and neuropsychiatric disease. A majority of known neuropsychiatric disease risk loci are noncoding, and increasing evidence suggests that they contribute to disease through disruption of CREs. As such, rather than directly altering the amino acid content of proteins, these variants are instead thought to affect where, when, and to what extent a given gene is expressed. The distances over which CREs can operate often render their target genes difficult to identify. Furthermore, as many risk loci contain multiple variants in high linkage disequilibrium, identification of the causative single nucleotide polymorphism(s) therein is not straightforward. Thus, deciphering the genetic etiology of complex neuropsychiatric disorders presents a significant challenge.

Keywords Disease · Enhancers · Epigenome · Transcription

1 Introduction

The human body is made up of hundreds of different cell types, each of which shares an essentially identical genome. Human nuclei are typically between 2 and 10 μm in diameter and contain approximately 2 m of DNA. In order to fit inside the nucleus, DNA is packaged into chromatin. This packaging is not random and differs between cell types, leading to the exposure of the distinct repertoire of *cis*-regulatory elements (CREs) required to generate cell-type-specific transcriptomes (reviewed in Cremer and Cremer 2001; Bulger and Groudine 2010; Heinz et al. 2015; Romanoski et al. 2015). Thus, the spatial organization of the genome plays a key role in the regulation of gene expression, the disruption of which may lead to disease (Dekker and Mirny 2016; Dixon et al. 2016; Lupianez et al. 2016).

2 Structure of Chromatin and 3D Genome

The fundamental unit of chromatin organization is the nucleosome, an octamer of histone proteins, which accommodates approximately 150 nucleotides of DNA that wrap around the complex. Nucleosomes coalesce into chromatin fibers, which are further compacted into chromosomes. Chromosomes are highly condensed during mitosis but become de-condensed during the interphase of the cell cycle. The conformation of the genome during interphase, which comprises the major fraction of the cell cycle, during which a cell grows and expresses its genes, is not disorganized but highly structured. The three-dimensional (3D) structure of the genome is implicated in cell-type-specific regulation of gene transcription by facilitating contact between distal regulatory elements called enhancers and target promoters. In 1885, Carl Rabl proposed the "chromosome territory" model, which held that each chromosome occupies a defined volume of the nucleus and only shares nuclear space with a small number of adjacent chromosomes (Rabl 1885). This model was not validated until the 1980s, when Cremer and co-workers induced local DNA damage in the nucleus using a focused laser and, by providing radioactively labeled nucleotides to mark the chromosomal regions as the cells repaired the DNA damage, were able to observe DNA damage on only a few chromosomes (Cremer et al. 1982). Subsequent studies using fluorescent in situ hybridization confirmed that chromosomes occupy distinct regions within the nucleus during interphase (Cremer and Cremer 2001; Habermann et al. 2001; Parada et al. 2002, 2004; Tanabe et al. 2002; Meaburn and Misteli 2007).

The genome is broadly partitioned into megabase-sized compartments that consist of gene-rich regions of open chromatin, corresponding to transcriptionally active regions and, conversely, gene-poor regions of closed chromatin, corresponding to transcriptionally silent regions (Lieberman-Aiden et al. 2009). The genome is further segregated into locally self-interacting neighborhoods that are insulated from each other by boundary elements enriched for the architectural protein CTCF (Dixon et al. 2012; Rao et al. 2014). These neighborhoods, referred to as topologically associated domains (TADs), are cell-type invariant and conserved from mouse to human. The boundary elements insulating TADs are enriched for housekeeping genes, suggesting that other factors, in addition to CTCF, may serve as barriers in genomic organization. TAD boundaries are also known to play a role in gene expression, as deletion or disruption of boundary elements promotes aberrant enhancer-promoter interactions across adjacent TADs, leading to defective transcription (Nora et al. 2012; Lupianez et al. 2015). Indeed, disruption of TAD boundaries can produce severe physiological defects. Genomic deletions and inversions causing human limb malformation were shown to disrupt TAD boundary elements, an observation that was replicated in CRISPR-/Cas9-modified mice (Lupianez et al. 2015). Thus, TADs are evolutionarily conserved and regulate enhancer-promoter specificity.

3 Structural Organization of CREs

Within TADs, chromatin structure is variable depending on cell type and cell context. The functional diversity of cells is accomplished, in part, through regulatory elements residing in regions of open chromatin that determine isoform selection of genes during transcription and the timing and extent of their expression (Heinz et al. 2015). In 1981, Banerji and colleagues described a DNA element capable of enhancing transcription in HeLa cells (Banerji et al. 1981). This activity was mediated by a (virally derived) 72 bp repeat sequence that could act in either orientation and in multiple positions relative to the transcription start site (TSS). The same group subsequently described the first mammalian enhancer (Banerji et al. 1983). This element was shown to influence expression of the beta-globin gene and could function over a broad range of genomic distances (several hundred to several thousand bp, either up- or downstream of the promoter). The element also displayed cell-type specificity as, among the cell lines tested, it was only functional in lymphocyte derived cells. A subsequent boom in enhancer identification occurred, and their mechanism of action in numerous biological processes began to emerge (Shlyueva et al. 2014).

Although the human genome contains hundreds of thousands of potential regulatory elements, only a subset is active in a given cell type at any time (ENCODE Consortium 2012). The distribution of CREs throughout the genome, therefore, provides an additional layer of transcriptional regulation in eukaryotic cells, as loop structures within chromatin facilitate direct and specific interactions between otherwise distal DNA elements, often separated by many kilobases along the linear genome (Gaszner and Felsenfeld 2006; Sanyal et al. 2011, 2012; Benabdallah et al. 2016). The 3D genome differs between cell types, is dynamic, and can change both during development and in response to external stimuli. Regions of open chromatin are enriched in enhancers, which facilitate the specificity of gene expression through interactions with sequence-specific transcription factors (TFs) (Vernimmen and Bickmore 2015). Each gene may be regulated by multiple enhancers depending on the cellular context. Similarly, a single enhancer can also contribute to the expression of a number of different genes, in some cases even if those genes reside on different chromosomes (Spilianakis et al. 2005). Thus, the manner and extent to which the genome is utilized is pivotal in defining the specifics of cellular identity and function.

4 Implication of CREs in Disease

Numerous studies link dysregulation of the epigenome to disease (e.g., see Robertson 2005; Williamson et al. 2011; Ward and Kellis 2012; Spielmann and Klopocki 2013; Spielmann and Mundlos 2013; Shen et al. 2014; Albert and Kruglyak 2015; Mirabella et al. 2015; Chatterjee and Ahituv 2017) and genetic variation within the noncoding, regulatory regions of the genome have been associated with a growing list of diseases (Epstein 2009; Manolio et al. 2009; Parker et al. 2013; Scacheri and Scacheri 2015), including developmental disorders (Benko et al. 2009; Gordon et al. 2009), heart disease (Goring et al. 2007; Postma et al. 2016), cancer (Herz et al. 2014; Northcott et al. 2014; He et al. 2015; Khurana et al. 2016; Lin et al. 2016), diabetes (Scott et al. 2007), Crohn's (Libioulle et al. 2007), and asthma (Moffatt et al. 2007; Manolio et al. 2008). In addition, defects in higher-order chromatin structure have also been implicated in disease. For example, mutations within the cohesin complex (Deardorff et al. 2012; Gervasini et al. 2013), which functions to facilitate interactions between distal promoters and enhancers (Kagey et al. 2010), have been associated with Cornelia de Lange syndrome, which is characterized by physical and cognitive defects.

In the context of neuropsychiatric disease risk, both genetic and environmental factors are known to contribute (Gandal et al. 2016). Genome-wide association studies (GWAS) of complex neuropsychiatric diseases, such as schizophrenia (SCZ; Schizophrenia Working Group of the Psychiatric Genomics 2014), bipolar disorder (BD) (Stahl et al. 2018), Alzheimer's disease (AD) (Lambert et al. 2013), and major depressive disorder (Wray et al. 2018), have identified numerous risk loci, most of which are noncoding. Examining over 35,000 cases and 110,000 controls, the Schizophrenia Working Group of the Psychiatric Genomics Consortium identified 108 genome-wide significant loci associated with increased risk of SCZ (Schizophrenia Working Group of the Psychiatric Genomics 2014). This list has subsequently been expanded, to include over 200 risk loci (ongoing analysis by PGC3), and is expected to grow further with additional studies. The majority of the identified loci are noncoding and are enriched in functional elements including enhancer sequences detected in human brain tissue (Gusev et al. 2014; Roussos et al. 2014). SCZ risk loci also co-localize with expression quantitative trait loci (eQTLs) (in particular, brain-derived eQTLs) (Roussos et al. 2014; Fromer et al. 2016; Hauberg et al. 2017), thereby implicating specific genes. However, these efforts have limited spatiotemporal resolution as most studies performed, thus far, have been restricted to homogenate brain tissue or include only broadly defined neuronal and non-neuronal populations (Fullard et al. 2017). As it is known that CREs display tissue- and cell-type specificity, with variants therein often only affecting those cells and tissues relevant to a given disease (Maurano et al. 2012; Trynka et al. 2013), an imperative of future studies will be to employ cell-type-specific or single-cell assays in order to more comprehensively understand the impact of these variants. For instance, SCZ-associated genetic loci are enriched within promoter and enhancer regions of neuronal cells (Roussos et al. 2014; Fullard et al. 2017). Furthermore, variants located in H3K4me3 (a marker of active promoters) sites specific to neuronal cells were more abundant when compared to those of non-neurons (Tansey and Hill 2018). In the case of AD, single nucleotide polymorphisms (SNPs) associated with increased risk for disease (Lambert et al. 2013) have been shown to be enriched at sites of open chromatin in immune cells and microglia (Tansey et al. 2018). These open chromatin sites contain DNA-binding motifs for specific TFs, including SPI1 and MEF2. Taken together, these observations are consistent with the fact that regulatory elements display tissue- and cell-type specificity. As such, identifying the specific cell types affected by a given variant will be an important step toward a more comprehensive understanding of the genetic etiology of neuropsychiatric disease.

5 Large-Scale Efforts to Study the Noncoding Genome

A number of large-scale efforts have sought to examine the role played by the noncoding genome in the regulation of cellular function. The ENCyclopedia Of DNA Elements (ENCODE) project (ENCODE Consortium 2012), the NIH Roadmap Epigenome Mapping Consortium (REMC) (Bernstein et al. 2010; Roadmap Epigenomics et al. 2015), and FANTOM5 (FANTOM Consortium 2014) have made great strides toward systematically cataloguing CREs within the human genome. Unfortunately, much of this data is not readily applicable to the study of gene expression in neuropsychiatric disease. The ENCODE project did not include brain tissue but focused, instead, on a variety of actively dividing cell lines and tissues. Although both REMC and FANTOM5 did include brain specimens, their use was limited to homogenate tissue isolated from controls. Within brain tissue, however, the different subclasses of neurons coexist alongside other cell types, including microglia, oligodendrocytes, and astrocytes. Since CRE-mediated transcriptional regulation has been shown to be cell-type specific (Heintzman et al. 2009; Cheung et al. 2010; Maurano et al. 2012; Roadmap Epigenomics et al. 2015), data derived from such studies is of limited use when applied to the study of a tissue as complex as the human brain. SCZ-associated abnormalities have been demonstrated in specific populations of brain cells, including neocortical neurons (Benes and Berretta 2001), astrocytes (Schnieder and Dwork 2011; McCullumsmith et al. 2015), oligodendrocytes (Haroutunian et al. 2014; Roussos and Haroutunian 2014; Mighdoll et al. 2015), and microglia (Bernstein et al. 2015). As such, the study of homogenate tissue samples may fail to distinguish signals unique to specific cell types, potentially missing critical changes in less abundant cell populations. Large-scale collaborative projects such as PsychENCODE (Akbarian et al. 2015) will attempt to address this issue by applying state-of-the-art omics approaches to specific cell populations (neurons and non-neurons in the case of PsychENCODE), in an effort to further our understanding of the complex genetic mechanisms that contribute both to normal brain function and to disease (see below).

6 Approaches to Detect Enhancer Sequences

Although the full extent to which the 3D structure of chromatin influences cell function remains unclear (Cattoni et al. 2015), the advent of methodologies that incorporate chromatin immunoprecipitation and next-generation sequencing technologies (ChIP-seq) is likely to further our understanding of this critical regulatory process (Fullwood et al. 2009; Fullwood and Ruan 2009). ChIP-seq utilizes antibodies against epigenetic markers and can be used to assess enrichment for histone 3Lys4 tri-methylation (H3K4me3), histone 3Lys27 acetylation (H3K27ac), and histone H3 at lysine 4 mono-methylation (H3K4me1) to identify putative active promoters and enhancers and active or primed enhancers, respectively.

Cap analysis of gene expression (CAGE) facilitates fine mapping of TSSs and promoter regions by sequencing the 5′ end of mature RNA (Shiraki et al. 2003; Takahashi et al. 2012). In CAGE-seq, RNA is reverse-transcribed and the 3′ ends biotinylated. Nonhybridized, single-stranded RNAs are removed by digestion with RNase, leaving 5′ complete cDNAs that are captured using streptavidin and subjected to next-generation sequencing. The FANTOM5 project has applied CAGE-seq to a wide array of cells and tissues from human and mouse, including the brain. In so doing, they have identified and quantified the activity of at least one promoter for more than 95% of annotated protein-coding genes in the human reference genome (FANTOM Consortium 2014). In addition, using H3K27ac and H3K4me1 ChIP-seq data from the ENCODE project (ENCODE Consortium 2012), FANTOM5 also applied CAGEseq data to identify a range of enhancers across human cells, through the detection of enhancer RNAs (eRNA) (Andersson et al. 2014). eRNAs are transcribed in proportion to enhancer activity, and their levels correlate with those of mRNA from nearby genes and have been shown to be differentially expressed in SCZ (Hauberg et al. 2018).

The nucleosome is known to play a central role in mediating gene expression and exists in a dynamic equilibrium between open and closed states (Mellor 2005). Nucleosome rearrangement at promoters and enhancers leads to open chromatin states and results from the binding of specific regulatory factors (Henikoff 2008). Open or accessible regions of the genome are regarded as primary positions for regulatory elements and, as such, play a critical role in regulating transcription (John et al. 2011). Approaches such as DNase-seq (DNase I hypersensitivity regions) (Song et al. 2011) or FAIRE-seq (Formaldehyde-Assisted Isolation of Regulatory Elements) (Simon et al. 2012) have been utilized to map open chromatin (Maurano et al. 2012); however, these techniques have largely been superseded by methods that require lower amounts of input material, an important consideration when working with precious biological samples of limited availability. More recently, a tagmentation-based method called the Assay for Transposase-Accessible Chromatin followed by Sequencing (ATAC-seq) has been developed (Buenrostro et al. 2013). ATAC-seq employs a transposome complex to insert oligonucleotides into accessible regions of the genome. This facilitates the generation of sequencing libraries enriched for open chromatin with sufficient resolution to map TF occupancy and nucleosome positions in regulatory sites (Buenrostro et al. 2015a). Subsequent iterations of the method allow for chromatin structure to be profiled in as few as 500 cells (Corces et al. 2017) and those isolated from 50 μm sections and frozen archival material (including human brain (Egervari et al. 2017; Fullard et al. 2017)). In addition, the approach has been further optimized to allow chromatin profiling at the single-cell level (Buenrostro et al. 2015b; Lake et al. 2018).

7 Approaches to Detect the Long-Range Enhancers

Several methods based on chromosome conformation capture (3C) have been developed to assess the frequency at which any two loci in the genome are in close enough physical proximity to functionally interact (Dekker et al. 2002; Zhao et al. 2006; Dostie et al. 2006). 3C involves cross-linking of interacting DNA segments, followed by digestion with a frequently cutting restriction enzyme. Digested DNA is then religated, cross-links are reversed and the resulting 3C library subjected to PCR using primers that flank putative ligation junctions, thereby assessing the frequency at which otherwise distal genetic elements ligate to one another; a reflection of their physical proximity within chromatin. 3C can detect interactions between distal genetic elements, e.g., between a gene and an enhancer (Simonis et al. 2007; Naumova et al. 2012), and can range from target-specific (3C) to unbiased, genome-wide approaches (Hi-C) (Dekker et al. 2013), including those at the resolution of single cells (Flyamer et al. 2017; Ulianov et al. 2017). Importantly, a number of these approaches have been successfully applied to studies of the human brain (Mitchell et al. 2014; Roussos et al. 2014; Won et al. 2016).

Alterations in chromosomal loop structures have been implicated in neuropsychiatric disease (Bharadwaj et al. 2013, 2014; Roussos et al. 2014). Bharadwaj and colleagues mapped the 3D configuration of a 200 kb stretch of the human genome containing the GAD1 GABA synthesis enzyme gene locus (Bharadwaj et al. 2013), which has previously been implicated in neuropsychiatric disease (Addington et al. 2005; Straub et al. 2007). This led to the identification a 50 Kb loop structure between noncoding intergenic DNA elements and the TSS of the gene encoding GAD1. This loop was further identified in an induced pluripotent stem cell-derived neuronal cell line when compared to skin fibroblasts or undifferentiated pluripotent stem cells, indicating the cell-type specificity of the structure. The loop was decreased in the prefrontal cortex of subjects with SCZ with a concurrent decrease GAD1 expression when compared with controls.

Variations in DNA sequence can have a profound effect on the function of distal regulatory elements. Bharadwaj and co-workers have also described a conserved, methyltransferase dependent loop structure in the prefrontal cortex involved in regulating the expression of the NMDA glutamate receptor, GRIN2B (Bharadwaj et al. 2014). Disruption of the loop resulted in decreased GRIN2B expression and led to impaired cognitive function and working memory defects in a mouse model. In the above examples, the loop structure is conserved between human and mouse, indicating their functional importance.

Disruption of long-range enhancer function has also been implicated in SCZ. Promoter and enhancer sequences are enriched in SCZ variants associated with eQTL (Roussos et al. 2014). Putative physical interactions between noncontiguous proximal and distal regulatory elements have been identified, and subsequent validation experiments confirmed the existence of a functional loop structure between a distal regulatory element and the gene encoding the L-type calcium channel (CACNA1C). Moreover, this regulatory element overlaps a disease risk locus, confirming a functional link between SCZ-associated noncoding SNPs, the 3D genome, and the regulation of transcription in the brain. More recently, Won and colleagues generated comprehensive 3D maps of chromatin contacts during human corticogenesis, leading to the identification of hundreds of interactions between genes and distal regulatory elements (Won et al. 2016). Integrating such structural data with SCZ GWAS and eQTL data revealed a number of TF and signaling pathways associated with increased vulnerability to the disease.

Taken together, these studies demonstrate that distal regulatory elements appear to be highly dependent on 3D genomic structures to facilitate their interaction with target genes. Thus, there is a growing body of evidence to suggest that functional disruption of long-range intrachromosomal interactions might contribute to disease, thereby accounting for the observation that the majority of known neuropsychiatric disease risk loci reside within noncoding regions of the genome.

8 Refining the Search to Identify Active Enhancer Elements

All of the aforementioned approaches allow for the genome-wide identification of potential regulatory elements but fail to directly assess their functionality. Techniques such as STARR-seq (*s*elf-*t*ranscribing *a*ctive *r*egulatory *r*egion sequencing) (Arnold et al. 2013; Muerdter et al. 2015), and iterations thereof (Vanhille et al. 2015; Dao et al. 2017), and FIREWACh (*F*unctional *I*dentification of *R*egulatory *E*lements *W*ithin *A*ccessible *Ch*romatin) (Murtha et al. 2014) allow for the identification of active cell-type-specific regulatory elements, enabling the rapid screening of entire genomes, reviewed in Dailey (2015). STARR-seq and FIREWACh take advantage of the observation that enhancers can work independent of their relative locations and both methods use reporter assays to interrogate DNA populations for elements capable of driving transcription. By directly coupling candidate sequences to enhancer activity, these approaches enable the evaluation of millions of DNA fragments in a single experiment. A critical consideration with each approach, however, is the relevance of the cells used to carry out the assay.

9 Linking the Activity of Enhancers to Specific Genes

Having identified active *cis*-regulatory elements within a given cellular context, the next requirement would be to assign their activities to a specific gene or set of genes. 3C-based methodologies are a useful tool toward this purpose, as they allow for the identification of physical interactions between distal genetic elements.

The relevance of long-range chromatin interactions to human disease has been examined in a few recent studies. Jager and colleagues used an oligo capture-based approach to Hi-C (capture Hi-C) to identify chromatin interactions of 14 colorectal cancer risk (CRC) loci (Jager et al. 2015). They demonstrated that 3D contacts are enriched for enhancers and promoters and CRC-specific TF binding sites. The role of long-range chromatin interactions in neurodevelopmental disorders like SCZ has also been investigated (Bharadwaj et al. 2014; Roussos et al. 2014; Won et al. 2016; Zhang et al. 2018).

SNPs associated with SCZ are predominantly located within distal enhancer elements. Integrating Hi-C data with noncoding SCZ GWAS variants revealed that most SNPs interact with non-proximal genes that are specifically involved in pathways related to brain development and function (Won et al. 2016). Many of these long-range chromatin interactions are detected in neuronal cells, are enriched for specific neuronal functions such as axon guidance and synaptic transmission, but are not detected in non-neuronal cells. CRISPR-/Cas9-mediated genome editing in neural progenitors validated the functional relevance of an enhancer harboring a SCZ-associated SNP that interacts with *FOXG1,* a gene encoding a TF implicated in early brain development. Deletion of the region flanking the SNP led to decreased expression of *FOXG1*, but not of the nearby *PRKD1* locus.

The effect of large deletion copy number variants (CNVs) on both the local and global chromatin interactome was investigated in a recent study (Zhang et al. 2018). Zhang and co-workers examined specific changes in chromatin interactions, histone modifications, and gene expression caused by the 3 MB chromosome 22q11.2 deletion, containing more than 60 known genes, which was shown to be associated with several psychiatric disorders, in particular SCZ and autism spectrum disorders. Hi-C analysis on cell lines derived from patients harboring the 22q11.2 deletion CNV showed that intrachromosomal contacts between regions flanking the deletion were strengthened, accompanied by respective changes in H3K27ac and H3K27me3, and associated with increased and decreased transcription, respectively. In addition, there were pronounced changes in the A/B compartment structure and TAD structure spanning the deletion region. There were also changes in inter-chromosomal *trans*-contacts, some of which could be directly attributed to the deletion CNV itself.

10 Concluding Remarks

To better understand these relatively common neuropsychiatric disorders, therefore, requires the systematic study of the regulatory effects of noncoding mutations on gene expression. Toward this end, a number of large collaborative efforts including the CommonMind Consortium (CMC) (Fromer et al. 2016) and PsychENCODE (Akbarian et al. 2015) are underway that focus on the regulatory mechanisms driving gene expression in the human brain. The CommonMind Consortium (www.synapse.org/cmc) aims to generate and analyze large-scale transcriptome data from brain samples, including control specimens and cases with SCZ and BD. The PsychENCODE consortium was established for the purpose of studying the role played by the epigenome in neuropsychiatric diseases (Akbarian et al. 2015). The project primarily focuses on three neuropsychiatric diseases (autism spectrum disorder, SCZ, and BD). The goal of PsychENCODE is to chart the epigenomic landscape of the brain in a large cohort of cases and controls. This large, collaborative project will employ a variety of methodologies: ChIP-seq will be used to identify putative active promoters and enhancers. In turn, ATACseq will be performed to identify open chromatin regions (Buenrostro et al. 2013), and Hi-C will be used to identify direct interactions between CREs and their downstream target genes (Dekker et al. 2013). The functional relevance of these putative enhancers will be assessed using mouse transgenesis and the STARR-seq assay (Arnold et al. 2013). Importantly, these approaches will be applied to neuronal and non-neuronal nuclei isolated at a number of different developmental time points and from multiple neocortical brain regions. Findings will be validated using neurons derived from induced pluripotent stem cells (iPSCs) and cultured neuronal cells derived from olfactory neuroepithelium (CNON cells).

In summary, there are many exciting findings from recent studies that implicate a clear role for spatial organization of the genome in gene expression regulation, aberrations of which can lead to disease. Much work remains to be done to integrate chromatin organization with gene expression and epigenome data to elucidate the systems biology of neurodevelopmental disorders. However, recent technological advances, including the development of cell-type-specific and single-cell approaches, hold great promise toward identifying the biological pathways and mechanisms that underlie disease.

References

Addington AM et al (2005) GAD1 (2q31.1), which encodes glutamic acid decarboxylase (GAD67), is associated with childhood-onset schizophrenia and cortical gray matter volume loss. Mol Psychiatry 10:581–588

Akbarian S et al (2015) The PsychENCODE project. Nat Neurosci 18:1707–1712

Albert FW, Kruglyak L (2015) The role of regulatory variation in complex traits and disease. Nat Rev Genet 16:197–212

Andersson R et al (2014) An atlas of active enhancers across human cell types and tissues. Nature 507:455–461

Arnold CD et al (2013) Genome-wide quantitative enhancer activity maps identified by STARR-seq. Science 339:1074–1077

Banerji J et al (1981) Expression of a beta-globin gene is enhanced by remote SV40 DNA sequences. Cell 27:299–308

Banerji J et al (1983) A lymphocyte-specific cellular enhancer is located downstream of the joining region in immunoglobulin heavy chain genes. Cell 33:729–740

Benabdallah NS et al (2016) SBE6: a novel long-range enhancer involved in driving sonic hedgehog expression in neural progenitor cells. Open Biol 6:pii: 160197

Benes FM, Berretta S (2001) GABAergic interneurons: implications for understanding schizophrenia and bipolar disorder. Neuropsychopharmacology 25:1–27

Benko S et al (2009) Highly conserved non-coding elements on either side of SOX9 associated with Pierre Robin sequence. Nat Genet 41:359–364

Bernstein BE et al (2010) The NIH Roadmap Epigenomics Mapping Consortium. Nat Biotechnol 28:1045–1048

Bernstein HG et al (2015) Glial cells as key players in schizophrenia pathology: recent insights and concepts of therapy. Schizophr Res 161:4–18

Bharadwaj R et al (2013) Conserved chromosome 2q31 conformations are associated with transcriptional regulation of GAD1 GABA synthesis enzyme and altered in prefrontal cortex of subjects with schizophrenia. J Neurosci Off J Soc Neurosci 33:11839–11851

Bharadwaj R et al (2014) Conserved higher-order chromatin regulates NMDA receptor gene expression and cognition. Neuron 84:997–1008

Buenrostro JD et al (2013) Transposition of native chromatin for fast and sensitive epigenomic profiling of open chromatin, DNA-binding proteins and nucleosome position. Nat Methods 10:1213–1218

Buenrostro JD et al (2015a) ATAC-seq: a method for assaying chromatin accessibility genome-wide. Curr Protoc Mol Biol 109:21.29.1–9

Buenrostro JD et al (2015b) Single-cell chromatin accessibility reveals principles of regulatory variation. Nature 523:486–490

Bulger M, Groudine M (2010) Enhancers: the abundance and function of regulatory sequences beyond promoters. Dev Biol 339:250–257

Cattoni DI et al (2015) A matter of scale: how emerging technologies are redefining our view of chromosome architecture. Trends Genet 31:454–464

Chatterjee S, Ahituv N (2017) Gene regulatory elements, major drivers of human disease. Annu Rev Genomics Hum Genet 18:45–63

Cheung I et al (2010) Developmental regulation and individual differences of neuronal H3K4me3 epigenomes in the prefrontal cortex. Proc Natl Acad Sci U S A 107:8824–8829

ENCODE Consortium (2012) An integrated encyclopedia of DNA elements in the human genome. Nature 489:57–74

FANTOM Consortium (2014) A promoter-level mammalian expression atlas. Nature 507:462–470

Corces MR et al (2017) An improved ATAC-seq protocol reduces background and enables interrogation of frozen tissues. Nat Methods 14:959–962

Cremer T, Cremer C (2001) Chromosome territories, nuclear architecture and gene regulation in mammalian cells. Nat Rev Genet 2:292–301

Cremer T et al (1982) Analysis of chromosome positions in the interphase nucleus of Chinese hamster cells by laser-UV-microirradiation experiments. Hum Genet 62:201–209

Dailey L (2015) High throughput technologies for the functional discovery of mammalian enhancers: new approaches for understanding transcriptional regulatory network dynamics. Genomics 106:151–158

Dao LTM et al (2017) Genome-wide characterization of mammalian promoters with distal enhancer functions. Nat Genet 49:1073–1081

Deardorff MA et al (2012) HDAC8 mutations in Cornelia de Lange syndrome affect the cohesin acetylation cycle. Nature 489:313–317

Dekker J, Mirny L (2016) The 3D genome as moderator of chromosomal communication. Cell 164:1110–1121

Dekker J et al (2002) Capturing chromosome conformation. Science 295:1306–1311

Dekker J et al (2013) Exploring the three-dimensional organization of genomes: interpreting chromatin interaction data. Nat Rev Genet 14:390–403

Dixon JR et al (2012) Topological domains in mammalian genomes identified by analysis of chromatin interactions. Nature 485:376–380

Dixon JR et al (2016) Chromatin domains: the unit of chromosome organization. Mol Cell 62:668–680

Dostie J et al (2006) Chromosome conformation capture carbon copy (5C): a massively parallel solution for mapping interactions between genomic elements. Genome Res 16:1299–1309

Egervari G et al (2017) Striatal H3K27 acetylation linked to glutamatergic gene dysregulation in human heroin abusers holds promise as therapeutic target. Biol Psychiatry 81:585–594

Epstein DJ (2009) Cis-regulatory mutations in human disease. Brief Funct Genomic Proteomic 8:310–316

Flyamer IM et al (2017) Single-nucleus Hi-C reveals unique chromatin reorganization at oocyte-to-zygote transition. Nature 544:110–114
Fromer M et al (2016) Gene expression elucidates functional impact of polygenic risk for schizophrenia. Nat Neurosci. https://doi.org/10.1038/nn.4399
Fullard JF et al (2017) Open chromatin profiling of human postmortem brain infers functional roles for non-coding schizophrenia loci. Hum Mol Genet 26:1942–1951
Fullwood MJ, Ruan Y (2009) ChIP-based methods for the identification of long-range chromatin interactions. J Cell Biochem 107:30–39
Fullwood MJ et al (2009) An oestrogen-receptor-alpha-bound human chromatin interactome. Nature 462:58–64
Gandal MJ et al (2016) The road to precision psychiatry: translating genetics into disease mechanisms. Nat Neurosci 19:1397–1407
Gaszner M, Felsenfeld G (2006) Insulators: exploiting transcriptional and epigenetic mechanisms. Nat Rev Genet 7:703–713
Gervasini C et al (2013) Molecular characterization of a mosaic NIPBL deletion in a Cornelia de Lange patient with severe phenotype. Eur J Med Genet 56:138–143
Gordon CT et al (2009) Long-range regulation at the SOX9 locus in development and disease. J Med Genet 46:649–656
Goring HH et al (2007) Discovery of expression QTLs using large-scale transcriptional profiling in human lymphocytes. Nat Genet 39:1208–1216
Gusev A et al (2014) Partitioning heritability of regulatory and cell-type-specific variants across 11 common diseases. Am J Hum Genet 95:535–552
Habermann FA et al (2001) Arrangements of macro- and microchromosomes in chicken cells. Chromosome Res 9:569–584
Haroutunian V et al (2014) Myelination, oligodendrocytes, and serious mental illness. Glia 62:1856–1877
Hauberg ME et al (2017) Large-scale identification of common trait and disease variants affecting gene expression. Am J Hum Genet 100:885–894
Hauberg ME et al (2018) Differential activity of transcribed enhancers in the prefrontal cortex of 537 cases with schizophrenia and controls. Mol Psychiatry. https://doi.org/10.1038/s41380-018-0059-8
He H et al (2015) Multiple functional variants in long-range enhancer elements contribute to the risk of SNP rs965513 in thyroid cancer. Proc Natl Acad Sci U S A 112:6128–6133
Heintzman ND et al (2009) Histone modifications at human enhancers reflect global cell-type-specific gene expression. Nature 459:108–112
Heinz S et al (2015) The selection and function of cell type-specific enhancers. Nat Rev Mol Cell Biol 16:144–154
Henikoff S (2008) Nucleosome destabilization in the epigenetic regulation of gene expression. Nat Rev Genet 9:15–26
Herz HM et al (2014) Enhancer malfunction in cancer. Mol Cell 53:859–866
Jager R et al (2015) Capture Hi-C identifies the chromatin interactome of colorectal cancer risk loci. Nat Commun 6:6178
John S et al (2011) Chromatin accessibility pre-determines glucocorticoid receptor binding patterns. Nat Genet 43:264–268
Kagey MH et al (2010) Mediator and cohesin connect gene expression and chromatin architecture. Nature 467:430–435
Khurana E et al (2016) Role of non-coding sequence variants in cancer. Nat Rev Genet 17:93–108
Lake BB et al (2018) Integrative single-cell analysis of transcriptional and epigenetic states in the human adult brain. Nat Biotechnol 36:70–80
Lambert JC et al (2013) Meta-analysis of 74,046 individuals identifies 11 new susceptibility loci for Alzheimer's disease. Nat Genet 45:1452–1458
Libioulle C et al (2007) Novel Crohn disease locus identified by genome-wide association maps to a gene desert on 5p13.1 and modulates expression of PTGER4. PLoS Genet 3:e58

Lieberman-Aiden E et al (2009) Comprehensive mapping of long-range interactions reveals folding principles of the human genome. Science 326:289–293
Lin CY et al (2016) Active medulloblastoma enhancers reveal subgroup-specific cellular origins. Nature 530:57–62
Lupianez DG et al (2015) Disruptions of topological chromatin domains cause pathogenic rewiring of gene-enhancer interactions. Cell 161:1012–1025
Lupianez DG et al (2016) Breaking TADs: how alterations of chromatin domains result in disease. Trends Genet 32:225–237
Manolio TA et al (2008) A HapMap harvest of insights into the genetics of common disease. J Clin Invest 118:1590–1605
Manolio TA et al (2009) Finding the missing heritability of complex diseases. Nature 461:747–753
Maurano MT et al (2012) Systematic localization of common disease-associated variation in regulatory DNA. Science 337:1190–1195
McCullumsmith RE et al (2015) Cell-specific abnormalities of glutamate transporters in schizophrenia: sick astrocytes and compensating relay neurons? Mol Psychiatry. https://doi.org/10.1038/mp.2015.148
Meaburn KJ, Misteli T (2007) Cell biology: chromosome territories. Nature 445:379–781
Mellor J (2005) The dynamics of chromatin remodeling at promoters. Mol Cell 19:147–157
Mighdoll MI et al (2015) Myelin, myelin-related disorders, and psychosis. Schizophr Res 161:85–93
Mirabella AC et al (2015) Chromatin deregulation in disease. Chromosoma. https://doi.org/10.1007/s00412-015-0530-0
Mitchell AC et al (2014) The genome in three dimensions: a new frontier in human brain research. Biol Psychiatry 75:961–969
Moffatt MF et al (2007) Genetic variants regulating ORMDL3 expression contribute to the risk of childhood asthma. Nature 448:470–473
Muerdter F et al (2015) STARR-seq - principles and applications. Genomics. https://doi.org/10.1016/j.ygeno.2015.06.001
Murtha M et al (2014) FIREWACh: high-throughput functional detection of transcriptional regulatory modules in mammalian cells. Nat Methods 11:559–565
Naumova N et al (2012) Analysis of long-range chromatin interactions using chromosome conformation capture. Methods 58:192–203
Nora EP et al (2012) Spatial partitioning of the regulatory landscape of the X-inactivation centre. Nature 485:381–385
Northcott PA et al (2014) Enhancer hijacking activates GFI1 family oncogenes in medulloblastoma. Nature 511:428–434
Parada LA et al (2002) Conservation of relative chromosome positioning in normal and cancer cells. Curr Biol 12:1692–1697
Parada LA et al (2004) Tissue-specific spatial organization of genomes. Genome Biol 5:R44
Parker SC et al (2013) Chromatin stretch enhancer states drive cell-specific gene regulation and harbor human disease risk variants. Proc Natl Acad Sci U S A 110:17921–17926
Postma AV et al (2016) Genetics of congenital heart disease: the contribution of the noncoding regulatory genome. J Hum Genet 61:13–19
Rabl C (1885) Uber Zelltheilung. In: Gegenbaur C (ed) Morphologisches Jahrbuch, vol 10, pp 214–330
Rao SS et al (2014) A 3D map of the human genome at kilobase resolution reveals principles of chromatin looping. Cell 159:1665–1680
Roadmap Epigenomics C et al (2015) Integrative analysis of 111 reference human epigenomes. Nature 518:317–330
Robertson KD (2005) DNA methylation and human disease. Nat Rev Genet 6:597–610
Romanoski CE et al (2015) Exploiting genomics and natural genetic variation to decode macrophage enhancers. Trends Immunol 36:507–518

Roussos P, Haroutunian V (2014) Schizophrenia: susceptibility genes and oligodendroglial and myelin related abnormalities. Front Cell Neurosci 8:5

Roussos P et al (2014) A role for noncoding variation in schizophrenia. Cell Rep 9:1417–1429

Sanyal A et al (2011) Chromatin globules: a common motif of higher order chromosome structure? Curr Opin Cell Biol 23:325–331

Sanyal A et al (2012) The long-range interaction landscape of gene promoters. Nature 489:109–113

Scacheri CA, Scacheri PC (2015) Mutations in the noncoding genome. Curr Opin Pediatr 27:659–664

Schizophrenia Working Group of the Psychiatric Genomics C (2014) Biological insights from 108 schizophrenia-associated genetic loci. Nature 511:421–427

Schnieder TP, Dwork AJ (2011) Searching for neuropathology: gliosis in schizophrenia. Biol Psychiatry 69:134–139

Scott LJ et al (2007) A genome-wide association study of type 2 diabetes in Finns detects multiple susceptibility variants. Science 316:1341–1345

Shen E et al (2014) Regulation of histone H3K4 methylation in brain development and disease. Philos Trans R Soc Lond Ser B Biol Sci 369:20130514

Shiraki T et al (2003) Cap analysis gene expression for high-throughput analysis of transcriptional starting point and identification of promoter usage. Proc Natl Acad Sci U S A 100:15776–15781

Shlyueva D et al (2014) Transcriptional enhancers: from properties to genome-wide predictions. Nat Rev Genet 15:272–286

Simon JM et al (2012) Using formaldehyde-assisted isolation of regulatory elements (FAIRE) to isolate active regulatory DNA. Nat Protoc 7:256–267

Simonis M et al (2007) An evaluation of 3C-based methods to capture DNA interactions. Nat Methods 4:895–901

Song L et al (2011) Open chromatin defined by DNaseI and FAIRE identifies regulatory elements that shape cell-type identity. Genome Res 21:1757–1767

Spielmann M, Klopocki E (2013) CNVs of noncoding cis-regulatory elements in human disease. Curr Opin Genet Dev 23:249–256

Spielmann M, Mundlos S (2013) Structural variations, the regulatory landscape of the genome and their alteration in human disease. Bioessays 35:533–543

Spilianakis CG et al (2005) Interchromosomal associations between alternatively expressed loci. Nature 435:637–645

Stahl E et al (2018) Genomewide association study identifies 30 loci associated with bipolar disorder. *bioRxiv*. https://doi.org/10.1101/173062

Straub RE et al (2007) Allelic variation in GAD1 (GAD67) is associated with schizophrenia and influences cortical function and gene expression. Mol Psychiatry 12:854–869

Takahashi H et al (2012) 5′ end-centered expression profiling using cap-analysis gene expression and next-generation sequencing. Nat Protoc 7:542–561

Tanabe H et al (2002) Evolutionary conservation of chromosome territory arrangements in cell nuclei from higher primates. Proc Natl Acad Sci U S A 99:4424–4429

Tansey KE, Hill MJ (2018) Enrichment of schizophrenia heritability in both neuronal and glia cell regulatory elements. Transl Psychiatry 8:7

Tansey KE et al (2018) Genetic risk for Alzheimer's disease is concentrated in specific macrophage and microglial transcriptional networks. Genome Med 10:14

Trynka G et al (2013) Chromatin marks identify critical cell types for fine mapping complex trait variants. Nat Genet 45:124–130

Ulianov SV et al (2017) Single-cell Hi-C bridges microscopy and genome-wide sequencing approaches to study 3D chromatin organization. Bioessays 39:1700104

Vanhille L et al (2015) High-throughput and quantitative assessment of enhancer activity in mammals by CapStarr-seq. Nat Commun 6:6905

Vernimmen D, Bickmore WA (2015) The hierarchy of transcriptional activation: from enhancer to promoter. Trends Genet 31:696–708

Ward LD, Kellis M (2012) Interpreting noncoding genetic variation in complex traits and human disease. Nat Biotechnol 30:1095–1106
Williamson I et al (2011) Enhancers: from developmental genetics to the genetics of common human disease. Dev Cell 21:17–19
Won H et al (2016) Chromosome conformation elucidates regulatory relationships in developing human brain. Nature 538:523–527
Wray NR et al (2018) Genome-wide association analyses identify 44 risk variants and refine the genetic architecture of major depression. Nat Genet 50:668–681
Zhang X et al (2018) Local and global chromatin interactions are altered by large genomic deletions associated with human brain development. bioRxiv. https://doi.org/10.1101/182451
Zhao Z et al (2006) Circular chromosome conformation capture (4C) uncovers extensive networks of epigenetically regulated intra- and interchromosomal interactions. Nat Genet 38:1341–1347

DNA Methylation and Hydroxymethylation and Behavior

Sarah R. Rustad, Ligia A. Papale, and Reid S. Alisch

Contents

Abstract Environmentally sensitive molecular mechanisms in the brain, such as DNA methylation, have become a significant focus of neuroscience research because of mounting evidence indicating that they are critical in response to social situations, stress, threats, and behavior. The recent identification of 5-hydroxymethylcytosine (5hmC), which is enriched in the brain (tenfold over peripheral tissues), raises new questions as to the role of this base in mediating epigenetic effects in the brain. The development of genome-wide methods capable of distinguishing 5-methylcytosine (5mC) from 5hmC has revealed that a growing number of behaviors are linked to independent disruptions of 5mC and 5hmC levels, further emphasizing the unique importance of both of these modifications in the brain. Here, we review the recent

S. R. Rustad
Madison, WI, USA

L. A. Papale
Department of Neurological Surgery, University of Wisconsin, Madison, WI, USA

R. S. Alisch (✉)
Department of Neurological Surgery, University of Wisconsin, Madison, WI, USA

Department of Neurological Surgery, University of Wisconsin School of Medicine and Public Health, Madison, WI, USA
e-mail: alisch@wisc.edu

© Springer Nature Switzerland AG 2019

Curr Topics Behav Neurosci (2019) 42: 51–82
DOI 10.1007/7854_2019_104
Published Online: 8 August 2019

links that indicate DNA methylation (both 5mC and 5hmC) is highly dynamic and that perturbations in this modification may contribute to behaviors related to psychiatric disorders and hold clinical relevance.

Keywords 5-Hydroxymethylcytosine · 5-Methylcytosine · Behavior · DNA methylation · Epigenetics

1 Introduction

The most studied epigenetic modification in the mammalian genome is DNA methylation, which is the addition of a methyl group to the fifth carbon of cytosines (e.g., 5-methylcytosine (5mC)), and is predominantly found at cytosine-phosphate-guanine (CpG) dinucleotide sites (Fig. 1). This DNA modification is catalyzed by DNA methyltransferases (DNMTs), which utilize S-adenosyl-L-methionine as the methyl donor (Cheng 1995). Three active DNMTs have been identified in mammals, each displaying their own distinct functions. DNMT3a and DNMT3b initiate methylation, having an affinity toward unmethylated CpG sites, while DNMT1 preserves methylation, showing preference toward hemimethylated CpG sites (Okano et al. 1999). CpG-rich regions, known as CpG islands, and gene promoters have a significant reduction in 5mC levels, whereas the X chromosome has an overabundance of 5mC (Sharp et al. 2011; Ioshikhes and Zhang 2000). Generally speaking, this DNA modification functions in genomic imprinting, X-chromosome inactivation, chromatin structure, and gene silencing (Bird 2002; Sharma et al. 2010; Han et al. 2015; Suzuki and Bird 2008; Robertson 2005). Human studies support a role for 5mC in the development of behavior-related disorders, including bipolar disorder (BD), schizophrenia (SCZ), and major depressive disorder (MDD), often

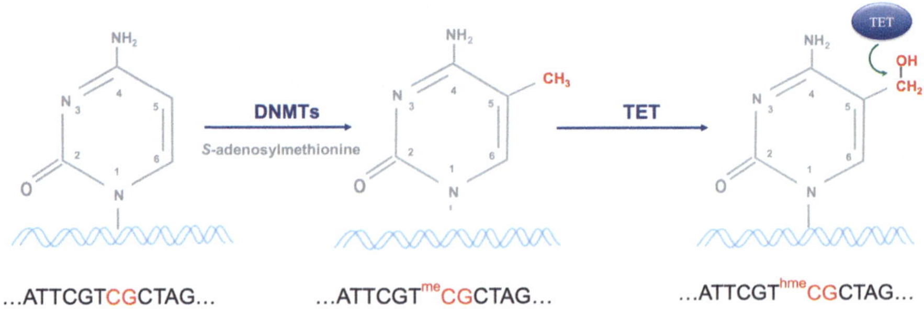

Fig. 1 The process of DNA methylation. Shown is the aromatic ring of cytosine. DNA methyltransferases (DNMTs) and S-adenosylmethionine facilitate the addition of a methyl group (CH_3) to the fifth position of cytosine in a CG (shown in red below; ^{me}CG) dinucleotide context. The ten-eleven translocation (TET) family of enzymes can oxidize the methyl group on cytosine, resulting in hydroxylation (OH; ^{hme}CG). Notably, the TET enzymes can further oxidize the methyl group, resulting in complete demethylation of the cytosine (not shown)

resulting in concomitant changes in gene expression (Abdolmaleky et al. 2006; Poulter et al. 2008; Kuratomi et al. 2008; Kappeler and Meaney 2010; Weaver et al. 2004). It was recently shown that 5mC can be oxidized to 5-hydroxymethylcytosine (5hmC) and that this modification is environmental sensitive (Wu and Zhang 2011), stable (Penn et al. 1972), and highly enriched in the brain (Kriaucionis and Heintz 2009; Sun et al. 2014; Wang et al. 2014; He et al. 2011; Ito et al. 2010; Tahiliani et al. 2009). This oxidation process is catalyzed by the ten-eleven translocation (TET) family of enzymes and can contribute to active demethylation, whereby two mechanisms can convert 5hmC back into cytosine; iterative oxidation by TET enzymes, which continuously oxidize 5hmC; and deamination by activation-induced cytidine deaminase/apolipoprotein B mRNA-editing enzyme complex AID/APOBEC (Ito et al. 2011; Guo et al. 2014).

While the full functional potential of 5hmC is yet to be determined, data supporting several putative molecular mechanisms have been found, including the regulation of transcription factor (TF) binding (Li et al. 2016), sex-specific development (Gross et al. 2015), and transcript diversity through interactions with the spliceosome (Feng et al. 2015) (Fig. 2). It is noteworthy that traditional DNA methylation detection methods utilizing sodium bisulfite treatment cannot distinguish between the methylated and hydroxymethylated forms of cytosine, meaning that past studies using such methods report a composite of 5mC and 5hmC but have attributed any findings solely to 5mC. This chapter will provide evidence to support the independent importance of 5mC and 5hmC in mental health and the development of mental illness and present several putative molecular functions of DNA methylation that may shed light on its promising clinical relevance. With recent findings implicating dynamic and unique roles for 5mC and 5hmC in behavior, DNA methylation brings a new frontier to the field of psychiatry.

2 Total DNA Methylation (5mC + 5hmC) and Behavior

A growing body of evidence suggests that the methylome varies dramatically throughout life, as the result of extrinsic influences such as environmental and physical factors (Tammen et al. 2013; King-Batoon et al. 2008; Park et al. 2012). One such factor with robust connections to DNA methylation is psychological adversity though most of these links were found using animal models. For example, rodent mothers that performed enhanced licking, grooming, and arched back nursing altered their offspring's DNA methylation levels on the glucocorticoid receptor (*NR3C1*) gene in the hippocampus, which is a gene best known for its role in the stress response. These mother-to-offspring interactions affected the development of their offspring's hypothalamic-pituitary-adrenal (HPA) axis, which is a dynamic metabolic system that regulates homeostatic mechanisms, including future reactions to new environments. Similarly, an abusive caregiver leads to lifelong increases in offspring DNA methylation on the brain-derived neurotrophic factor (*Bdnf*) gene in the prefrontal cortex of both male and female rodents, which can persist to the next

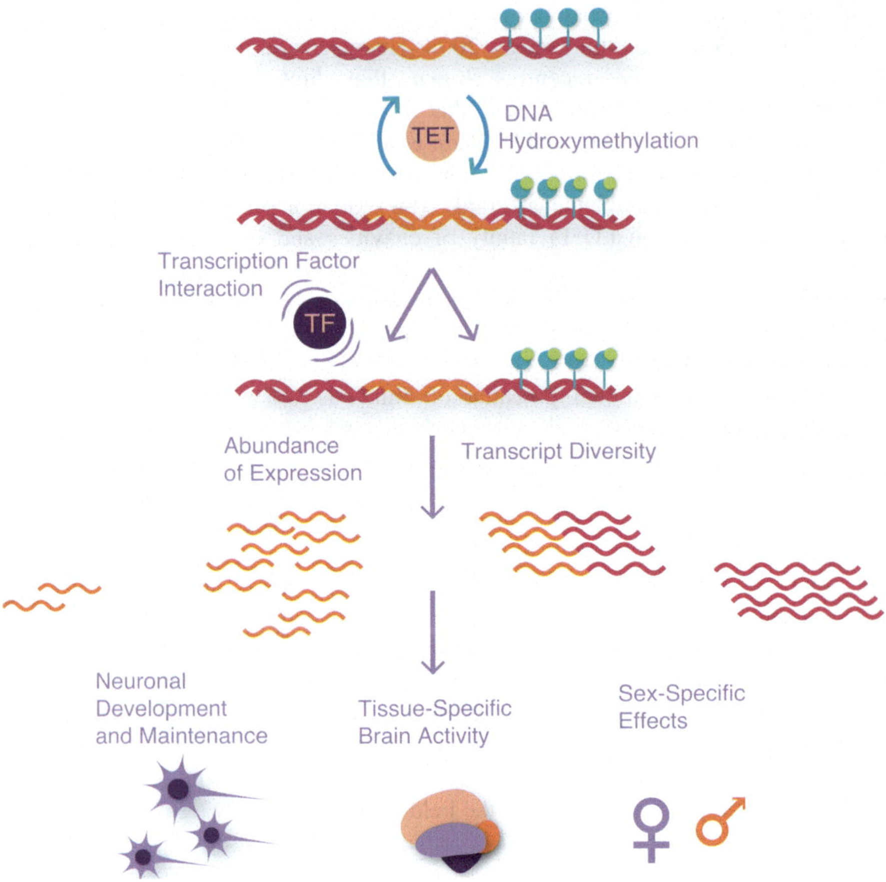

Fig. 2 Putative molecular mechanisms of 5hmC. Following oxidation by the TET enzymes, the 5hmC modification has been shown to interact with transcription factors (TF) and affect transcript abundance. In addition, 5hmC modifications have been implemented in the splicing process, resulting in transcript diversity. Ultimately, these changes in transcript expression and diversity can have lasting effects at the cellular and tissue level, in a sex-specific manner

generation (Roth et al. 2009). More recent work also found increased DNA methylation on *Bdnf* and concomitant decreased expression in the amygdala and hippocampus of rats exposed to prenatal stress (Boersma et al. 2014). Studies in young monkeys identified differentially methylated genes that are implicated as risk factors for developing anxiety and depressive disorders (Alisch et al. 2014). In addition, several other studies have shown evidence of altered DNA methylation and gene expression due to maternal separation or varied maternal care and prenatal stress (Beery et al. 2016; Murgatroyd et al. 2009; Mueller and Bale 2008; Novikova et al. 2008).

Human studies are obviously more difficult to conduct, due to limited access to brain tissue, making them slow to contribute to these findings. Nonetheless, more

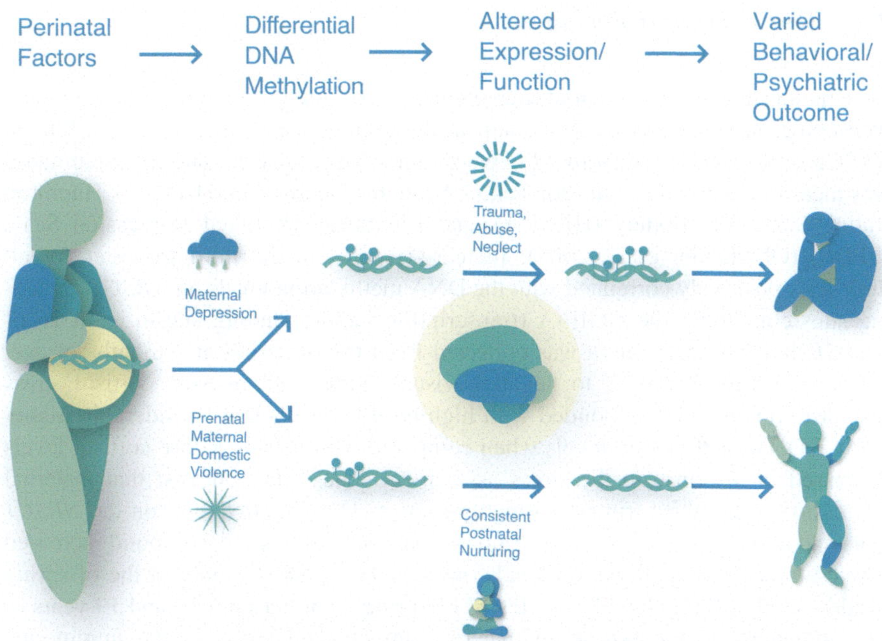

Fig. 3 Perinatal environmental adversities impact DNA methylation levels linked to human behavior. Perinatal exposure to maternal depression and negative experiences can alter DNA methylation in the perinatal brain, resulting in psychiatric disorders later in life. However, postnatal nurturing may reverse these effects and lead to healthier psychiatric outcomes

recent human studies also have shown that stress, anxiety, depression, neglect, socioeconomic status, violence, and maternal care all impact DNA methylation levels (Fig. 3) (Kim et al. 2016; Lam et al. 2012; King et al. 2015; Naumova et al. 2012), for example, suicide completers who were exposed to childhood abuse and neglect carried hypermethylation in the promoter regions of ribosomal RNAs and concomitant downregulated expression in the hippocampus (Mcgowan et al. 2008). However, aberrant DNA methylation patterns due to psychological influences were not restricted to promoter regions, as they also were linked to changes in gene bodies, intergenic regions, and 5′ and 3′ untranslated regions (UTRs) (Weder et al. 2014). Moreover, stress-dependent DNA methylation levels can be linked to genotype-specific gene transcription, resulting in a long-term dysregulation of the stress hormone system (Klengel et al. 2013). Together, these studies suggest that early-life adversities increase the risk of behavioral disturbances into adulthood and that these outcomes are governed by DNA methylation. Evidence of the environmental impact on DNA methylation patterns is often examined in the HPA axis. We start with *NR3C1*, the human glucocorticoid receptor gene, which is highly abundant in the HPA system.

2.1 The Perinatal Period

To examine the effects of offspring prenatal exposure to maternal psychological well-being on DNA methylation status at the promoter and exon 1F of the human *NR3C1* gene, cord blood from 82 newborn girls was collected, and maternal mood was measured using the Hamilton Rating Scale for Depression (HAM-D), Hamilton Rating Scale for Anxiety (HAM-A), and Edinburgh Postnatal Depression Scale (EPDS) (Oberlander et al. 2008). Increased scores in the third trimester for all three tests positively correlated with the DNA methylation levels on *NR3C1*, including sites containing the NGIF-A (transcription factor) binding site in exon 1F of *NR3C1*. Interestingly, saliva was collected from the offspring at 3 months of age, before and after exposure to negative visual stimuli, to measure cortisol stress response. Infants who responded with high cortisol levels had considerably higher *NR3C1* DNA methylation levels when compared with infants whose cortisol levels decreased in response to the stress assessment. These data suggest that maternal depression in the third trimester directly impacts DNA methylation on the *NR3C1* gene and alters the infant's stress response. Another study similarly found increased whole blood DNA methylation levels on exon 1F of *NR3C1* only in the offspring (Radtke et al. 2011). Notably, the tested offspring were between 10 and 19 years of age, indicating that exposure to maternal anxiety and stress in the intrauterine environment induced stable disruptions in DNA methylation. Similar to the licking and grooming studies conducted in rodents, maternal stroking in human infants leads to fluctuations of methylation of the *NR3C1* gene (Murgatroyd et al. 2015). In this study, DNA methylation levels were examined in saliva samples from 181 infants and showed that high maternal stroking of the infant face, back, abdomen, legs, and arms reduced *NR3C1* DNA methylation levels. This study highlighted the extreme vulnerability and sensitivity of the early postnatal period to extrinsic factors that may directly impact fetal development. While several groups have found links between maternal emotion, stress, childhood adversity, and the methylation status of exon 1F in the promoter of *NR3C1* (Mulligan et al. 2012; Hompes et al. 2013; Radtke et al. 2015; Braithwaite et al. 2015), sex-specific differences in *NR3C1* DNA methylation levels also have been shown, as male infants specifically exhibited elevated exon 1F methylation due to maternal depression during pregnancy (Braithwaite et al. 2015). This finding may shed mechanistic light on sex-specific behaviors, including psychological disorders.

Studies of extrinsic factors impacting development during the perinatal period also have focused on the neurotransmitter serotonin (*5-HT* or *SLC6A4*), which is linked with impulsive aggression as well as increased susceptibility for a lifetime risk of depression (Devlin et al. 2010; Feinn et al. 2005; Seo et al. 2008). A study of prenatal maternal depression in 82 pregnant women found that second trimester DNA methylation levels were lower on the *SLC6A4* promoter in mothers with higher depressed mood symptoms (Devlin et al. 2010). These decreased DNA methylation levels likely result in increased expression of *SLC6A4*, higher serotonin uptake, and decreased intrasynaptic serotonin, though such studies have not yet been conducted.

However, as serotonin modulates neuronal differentiation and growth, altered levels of the serotonin transporter during critical developmental periods likely affect brain development and may have long-term effects on subsequent child emotional development and susceptibility to affective disorders later in life (Ansorge et al. 2004, 2008; Gaspar et al. 2003).

Another mediator of neural function and plasticity is the brain-derived neurotrophic factor (*BDNF*), which is essential for neurogenesis and has been associated with a variety of affective disorders later in life (Mcewen 2007; Roth and Sweatt 2011a; Berton et al. 2006; Fuchikami et al. 2011; Keller et al. 2010; Post 2007). The genetic structure of the *BDNF* gene is complex, suggesting an intricate regulatory system. The gene consists of nine 5′ noncoding exons that are each linked to a distinct promoter that controls differential expression of exon-specific transcripts (Roth et al. 2009; Aid et al. 2007; Liu et al. 2006). Examination of *BDNF* DNA methylation levels related to prenatal maternal depression in 57 infants found reduced DNA methylation on the *BDNF* exon IV promoter. Interestingly, the reduced DNA methylation levels were located near the binding site of CREB, a transcription factor that regulates *BDNF* transcription via a DNA methylation-dependent mechanism, suggesting that prenatal maternal depression may affect this mechanism.

Oxytocin is a neuropeptide hormone that regulates reproductive physiology and contributes to maternal behaviors, pair bonding, and social interaction (Pedersen et al. 2006; Young and Wang 2004; Carter 2003; Winslow and Insel 2002; Carter et al. 1992; Pedersen and Boccia 2002; Popik and Van Ree 1991). An investigation into the potential interplay between oxytocin, DNA methylation, and the development of psychopathy examined the oxytocin receptor (*OXTR*) gene in blood from male children aged 4–16 years that had severe ratings of child conduct problems, as measured using the clinician rating scale called the Quality of Family Environment (Dadds et al. 2014; Rey et al. 1997). Interestingly, younger children (4–8 years) did not exhibit conduct-related DNA methylation levels on the *OXTR* gene; however, older children (9–16 years) had a positive correlation between conduct problems and DNA methylation levels on *OXTR*. This increased methylation in older children also had a direct impact on the level of circulating oxytocin, suggesting impairment of the entire oxytocin system, highlighted by the methylation-dependent downregulation of *OXTR* expression (Dadds et al. 2012, 2014). Indeed, childhood stress is linked to differential methylation of oxytocin (*OXT*) in the saliva of older girls (10–12 years) (Papale et al. 2018). Interestingly, a related study found that increased DNA methylation on the *OXTR* gene, and subsequent reduced expression, in the temporal cortex served as a valid indicator of autism spectrum disorder (ASD) (Gregory et al. 2009), differentiating between ASD and control subjects. Clearly the abundance of DNA methylation can be used to assess social behaviors in children.

While these aforementioned studies and many others were narrowly focused on a single promoter and gene, the use of genome-wide approaches has greatly improved our understanding of the impact of extrinsic factors on child susceptibility to affective disorders. One study collected saliva from 94 maltreated children and 96 controls (all 5–14 years of age) and found trauma-associated DNA methylation levels across the genome, including some on well-known stress-related genes, such as *NR3C1*, *BDNF*, and FK506 binding protein 51 (*FKBP5*), as well as on biologically relevant genes, including *ID3* (DNA-binding protein inhibitor ID-3), *TPPP* (tubulin polymerization promoting protein), and *GRIN1* (glutamate (NMDA) receptor NR1 subunit) (Weder et al. 2014). Moreover, since *ID3* expression is increased following stress (Konishi et al. 2010), the variation in salivary cortisol can predict *ID3* methylation-related diurnal cortisol secretion. Another study that examined saliva from 11 girls that experienced extremely high levels and 11 girls with normative levels of early childhood stress exposure (all 9–12 years of age) also found trauma-related DNA methylation levels across the genome at more than 100 genes, including on stress-related genes like *OXT* and a serotonin receptor: *HTR3A* (Papale et al. 2018). Together, these data suggest that DNA methylation is stable long after trauma exposure and may hold predictive information of mental well-being and prognosis.

2.2 The Adult Period

Variations in DNA methylation levels due to early-life adversity can persist into adulthood. Studies have linked altered DNA methylation levels and gene expression of hippocampal *NR3C1* and psychological illnesses, including mood disorders and SCZ, which are associated with suicide in adults; indeed, often suicide relates to a history of childhood abuse (Heim and Nemeroff 2001; Webster et al. 2002). In one such study, postmortem hippocampal samples were obtained from suicide completers that had traumatic childhoods, suicide completers that did not have a history of abuse, and controls that had sudden accidental deaths and did not experience abuse (Mcgowan et al. 2009). The suicide completers who had traumatic childhoods showed higher DNA methylation in exon 1F of *NR3C1*, compared to both controls (i.e., suicide completers without a history of abuse and victims of sudden accidental death). As predicted, the increased DNA methylation levels interrupted NGIF-A binding to *NR3C1* and gene transcription.

Suicide victims also show decreased *BDNF* expression levels in the hippocampus and prefrontal cortex, suggesting that *BDNF* plays a critical role in pathophysiological aspects of suicidal tendencies. Examination of *BDNF* DNA methylation levels in postmortem brain samples (Wernicke's area of the brain) from 44 suicide completers and 33 controls revealed hypermethylation in suicide completers, which was shown in vitro to result in lower *BDNF* mRNA levels (Keller et al. 2010). This study showed that a gene-specific increase of DNA methylation downregulates its expression, which can account for the pathophysiology observed the suicidal brain.

Another study examined the blood of 101 women aged 30 and 41 years that had borderline personality disorder (BPD) and had experienced extreme childhood sexual, physical, and emotional traumas; 99 of the women had MDD and 15 had MDD and post-traumatic stress disorder (PTSD) (Perroud et al. 2011). A positive correlation was found between sexual abuse severity and *NR3C1* DNA methylation levels. Similarly, physically and/or emotionally abused and neglected participants all had higher *NR3C1* DNA methylation than controls. In a related study, blood samples from 115 patients with BPD and 52 controls were collected for DNA methylation analysis, and the patients were found to exhibit higher *BDNF* DNA methylation compared to controls, which was positively correlated with childhood trauma such as sexual, physical, and emotional abuse and physical and emotional neglect (Perroud et al. 2013). Similarly, a positive association was observed between *BDNF* DNA methylation levels and hopelessness, depression severity, and impulsivity. Finding higher *BDNF* DNA methylation in BPD patients is consistent with the higher *BDNF* DNA methylation of suicide completers, and suicidal tendencies are common in BPD patients. Interestingly, these patients were subjected to intensive dialectical behavior therapy, which reduced *BDNF* DNA methylation and BPD symptoms in responding patients. While a similar increase in *BDNF* DNA methylation was observed in BD patients compared to controls, it was hypothesized that therapies such as antidepressants contributed to these elevated levels of *BDNF* DNA methylation. Together, these data suggested that *BDNF* DNA methylation levels might be a biomarker of BD therapy efficacy.

Childhood abuse also can stably alter DNA methylation levels that may contribute to the development of adult eating disorders, which are often comorbid with personality disorders. Examination of blood from 32 women with bulimia nervosa that had a history of childhood abuse, 32 women with bulimia nervosa that did not have a history of childhood abuse, and 32 controls revealed higher DNA methylation in exon 1C of *NR3C1* in women with bulimia nervosa and history of abuse compared to both controls (Steiger et al. 2013). While these findings do not definitively link DNA methylation to bulimia nervosa, they do suggest that the dysregulation of HPA activity, which contributes to the predisposition to psychological disorders observed in adults, originates early in life and is marked by stable changes in DNA methylation that are correlated to abuse severity. Other reports have corroborated these findings (Lawson et al. 2012; Maguire et al. 2013; Frieling et al. 2008; Toyokawa et al. 2012). In one such study, buccal cells from 15 anorexia nervosa patients and 36 controls were examined for DNA methylation levels on the *OXTR* gene (Kim et al. 2014). All participants were assessed using the Eating Disorder Examination Questionnaire (EDE-Q) (Fairburn and Beglin 1994), the Beck Depression Inventory (BDI), the Spielberger State and Trait Anxiety Inventory (STAI) (Speilberger et al. 1983), and the Autism Spectrum Quotient (ASQ) (Baron-Cohen et al. 2001), since eating disorders and autism have common traits including focus on detail, rigidity, and social cognition. *OXTR* DNA methylation levels were positively correlated with dietary restraint (EDE-Q), communication (ASQ), and depression and anxiety (BDI and STAI) and inversely correlated with body mass index (BMI). Notably, a linear regression of these data revealed that eating disorder psychopathology, BMI, and anxiety were the main determinants of the differential *OXTR* DNA methylation levels found in patients compared to controls. While it is not known whether DNA methylation levels in buccal cells truly reflect the levels in the brain, this study provides insight into a role for DNA methylation in eating disorder psychopathology.

One potential way to further link peripheral DNA methylation to brain function is by pairing DNA methylation data with functional magnetic resonance imaging (fMRI) data. One such study took this approach and showed that adverse childhood experiences can have long-lasting effects on DNA methylation-associated brain activity. This relationship was demonstrated in 25 mothers with interpersonal violence (IPV)-related PTSD, 9 with PTSD symptoms (non-IPV related), and 20 controls, revealing a positive correlation between *BDNF* DNA methylation and maternal brain activity in the ventromedial prefrontal cortex and anterior cingulate, which are regions associated with emotion regulation (Moser et al. 2015; Hu and Jiang 2014). Interestingly, there was a negative correlation in the right hippocampus and parahippocampus, left and right precuneus, left cerebellum, and right superior temporal gyrus. Together, these findings suggest that long-lasting changes in peripheral *BDNF* DNA methylation levels may reflect disrupted functions in many regions of the brain.

Using a similar approach, blood was collected from 42 participants that were subjected to social perception tasks during fMRI. These tasks recruit a network of brain structures that are vital to social perception and mentalizing abilities, such as the temporal parietal junction. A positive correlation was observed between *OXTR* DNA methylation levels and brain activity in the temporal parietal junction and dorsal anterior cingulate cortex (Jack et al. 2012). Moreover, the degree to which *OXTR* DNA methylation fluctuated was positively associated with BOLD activity (blood oxygenation level-dependent signal, which reflects neural activity) from the superior temporal gyrus into the supramarginal gyrus at the temporal parietal junction and the dorsal anterior cingulate cortex. The dorsal anterior cingulate cortex plays a central role in social and affective appraisals of motivationally salient stimuli. Functional impairment of this region is linked to emotional and social deficits in anxiety disorders and ASD (Etkin et al. 2011; Mcclure et al. 2007). Others also have shown that increased *OXTR* DNA methylation levels were concomitant with increased activity in the amygdala and high neural response, when observing negative facial expressions. Together, these studies highlighted that emotional and social perceptual processes involving the oxytocin system may be governed by DNA methylation.

Of course variations in DNA methylation also have been associated with psychological conditions in males, including anxiety, depression, hostility, happiness, and general life satisfaction. Examination of DNA methylation in blood of more than 500 men older than 70 years of age revealed psychological-related relationships to DNA methylation levels in several promoter regions of genes involved in immune/inflammatory processes related to atherosclerosis. For example, increased DNA methylation at the *ICAM-1* promoter was associated with increased anxiety, depression, and hostility, whereas happiness and life satisfaction showed an inverse correlation with DNA methylation. Notably, while higher promoter DNA methylation is expected to decrease *ICAM-1* expression, these psychological outcomes were not associated with lower serum *ICAM-1* levels. At the *TLR-2* promoter, DNA methylation levels were positively associated with hostility, while higher life satisfaction led to lower DNA methylation levels. At the coagulation factor III (*F3*) promoter, DNA methylation levels were positively correlated with depression, whereas life satisfaction and happiness were inversely associated with *F3* DNA methylation. Finally, at the iNOS promoter, DNA methylation levels were inversely correlated with depression and anxiety, but happiness and life satisfaction positively correlated with DNA methylation. Notably, significant correlations were not found at the *NR3C1* promoter, which is in contrast with several other studies that have found associations with negative psychological factors and DNA methylation on the *NR3C1* promoter (Braithwaite et al. 2015; Perroud et al. 2011). This was one of the very few studies that associated an optimistic psychological factor (happiness) with DNA methylation levels, demonstrating a role for DNA methylation across the psychological spectrum.

To examine DNA methylation levels on the serotonin transporter gene (*SLC6A4*) in adults related to childhood aggressive behaviors, blood was collected from 25 healthy adult males (mean age of 27 years). Here, a positive correlation was found between childhood aggression and DNA methylation in monocytes and T cells (Wang et al. 2012a). This study also used positron emission tomography (PET) measures of brain serotonin synthesis and found inverse associations between mean *SLC6A4* DNA methylation and *SLC6A4* expression in the lateral left and right orbitofrontal cortex. Notably, a specific *SLC6A4* genotype, which is known to affect gene expression, was not correlated to DNA methylation levels. On the other hand, in vitro analysis confirmed that the aggressiveness-related DNA methylation levels decreased *SLC6A4* transcriptional activity, suggesting that peripheral DNA methylation might be used to screen for serotonin-related psychological and psychiatric disorders and allow for preventive and corrective interventions. A subsequent study found that more severe physical childhood abuse along with reduced hippocampal volume was associated with higher DNA methylation on the *SLC6A4* promoter (Booij et al. 2015). This finding is consistent with the fact that the hippocampus is richly innervated with serotonin and is central in regulation of stress (Frodl and O'Keane 2013). DNA methylation on the *SLC6A4* promoter also has been studied in adult monozygotic twins that are discordant for MDD or PTSD (Zhao et al. 2013). Here genomic DNA was extracted from blood leukocytes of 84 monozygotic twin pairs (168 total participants), revealing that increased *SLC6A4* promoter DNA methylation was associated with depressive symptoms. Finding associations between childhood abuse, peripheral *SLC6A4* DNA methylation, brain *SLC6A4* expression, and hippocampal volumes further supports that peripheral DNA methylation levels may serve as a valuable biomarker for serotonin-associated stress-related psychopathology.

Social anxiety disorder (SAD) is characterized by fear, anxiety, and evasion of social situations for fear of being negatively evaluated or scrutinized and rejected (Labuschagne et al. 2010; Stein and Stein 2008). To characterize the role of DNA methylation in SAD patients, blood was collected from 110 patients and 110 age- and sex-matched controls. The severity of social anxiety in both patients and controls were assessed using the Social Interaction Anxiety Scale and Social Phobia Scale (SIAS and SPS) (Stangier et al. 1999). Analysis of the *OXTR* promoter revealed significantly lower DNA methylation compared to controls and a negative correlation between *OXTR* DNA methylation levels and SIAS and SPS scores (Ziegler et al. 2015). This finding was corroborated in independent saliva samples from 16 healthy females subjected to the Trier Social Stress Test that showed a negative correlation between *OXTR* DNA methylation and maximum salivary cortisol. An additional test on SAD patients using fMRI revealed increased amygdala activation and decreased *OXTR* DNA methylation during social phobia-related word processing. Together, this detailed study found that decreased *OXTR* DNA methylation was associated with SAD, stress-related cortisol levels, and heightened amygdala response, revealing the extensive role of DNA methylation in this chronic mental health condition.

Obsessive-compulsive disorder (OCD) is a severe disorder characterized by troubling thoughts and obsessions, such as the need for symmetry, exactness, and order and contamination fears and worries about harm to self and others (Barrett and Healy 2003; Lack 2012; Swedo et al. 1989). To investigate a role for DNA methylation in OCD, genomic DNA was obtained from the blood of 42 OCD patients that had Yale-Brown Obsessive-Compulsive Scale (Y-BOCS) scores of $\geq$16 for the combination of compulsions and obsessions or $\geq$10 for compulsions and obsessions alone and 31 controls (all 18–65 years old) (Cappi et al. 2016; Goodman et al. 1989). *OXTR* DNA methylation levels were higher in OCD patients compared to controls and negatively correlated with depression severity (Cappi et al. 2016); this relation to depression contrasts the findings of other studies (Kim et al. 2014; Ziegler et al. 2015). It is unknown whether the OCD-related hypermethylation of *OXTR* influences *OXTR* expression. A similar study of psychosis recruited 167 men and women with a psychotic disorder, including SCZ, BD, and schizoaffective disorder, and 75 controls (Rubin et al. 2016). Here, females displayed higher *OXTR* DNA methylation that was linked to poorer recognition of emotional expressions. Moreover, while a positive association between *OXTR* DNA methylation levels and oxytocin levels were found in females, a negative association was found in males. Together, these sex-specific differences in *OXTR* DNA methylation and oxytocin levels may hold key information into more personalized therapeutic treatment of SCZ and BD.

Studies of DNA methylation levels in novel genes related to psychological conditions are being reported on a continual basis. For example, DNA methylation levels in the promoters of *ZNF266*, *AGTR1*, *ASPH*, *PLAC1L*, and numerous other genes have been linked to chronic physical aggression in both males and females (Guillemin et al. 2014; Provencal et al. 2013). SCZ patients exhibit differential DNA methylation in the first intron of *RELN* (Aberg et al. 2014) and in several glutamate receptor genes (*GRIA2*, *GRIA3*, *GMR2*, *GMR5*, *GMR8*) (Aberg et al. 2014; Kordi-Tamandani et al. 2013) when compared to controls. In addition, BD patients have higher DNA methylation in *PRIMA1* compared to controls. It is of great interest to confirm the role of the aberrant DNA methylation levels of these genes and determine if they can be used for novel diagnostic, prognostic, and modifiable therapeutic targets.

2.3 *In the Pathogenesis of Neurodegeneration*

Dynamic changes in DNA methylation levels have been linked with aging and memory loss, leading to neurodegeneration (Johnson et al. 2012; Day and Sweatt 2010; Day et al. 2013; Sanchez-Mut et al. 2016; Alisch et al. 2012). Indeed, several groups have shown locus-specific and/or global disruptions of DNA methylation abundance in postmortem brain tissue of Alzheimer's disease (AD) patients, though findings differ between studies, suggesting a complex involvement of DNA methylation in AD pathogenesis (Bakulski et al. 2012; Lunnon et al. 2014; De Jager et al. 2014; Watson et al. 2016; Coppieters et al. 2014). One such study isolated genomic DNA from four postmortem brain regions of 122 AD patients and premortem whole blood when available ($n = 57$) and measured genome-wide DNA methylation levels associated with Braak staging, a standardized measure of neurofibrillary tangle burden determined at autopsy (Braak and Braak 1991). This approach revealed that the DNA methylation levels in the ankyrin 1 (*ANK1*) gene were associated with neuropathology in the entorhinal cortex, superior temporal gyrus, and prefrontal cortex, but not in the cerebellum or whole blood of the same AD patients. These data suggest that AD-associated variation in DNA methylation is consistent across pathologically relevant regions of the brain. In addition, recent epigenome-wide association studies (EWAS) detected both cross-tissue and tissue-specific DNA methylation profiles in brain tissue, identifying differential methylation in candidate genes of AD and in genes previously unassociated with AD (Bakulski et al. 2012; Lunnon et al. 2014; De Jager et al. 2014; Watson et al. 2016). Together, these studies underscore the utility of EWAS in identifying novel genes and pathways associated with AD pathogenesis that may otherwise be overlooked. Notably, however, these previous studies have relied heavily on the accessibility of brain tissue. Shifting studies of DNA methylation to peripheral whole blood provides a potential tool that may be utilized clinically to improve diagnosis and guide personalized treatment of AD. Support for this approach came from a recent study that extracted genomic DNA from the whole blood of 45 late-onset AD (LOAD) patients and 39 matched controls and found differentially methylated positions (DMPs) that distinguish people with and without LOAD (Madrid et al. 2018). Interestingly, subsequent independent comparison using six continuous clinical LOAD phenotypes as variables, comprising RAVLT scores, and CSF t-tau and p-tau$_{181}$ levels, or t-tau/Aβ_{42}, p-tau$_{181}$/Aβ_{42}, or Aβ_{42}/Aβ_{40} ratios, yielded a unique set of 17 DMPs that were all hypomethylated in two genes, *B3GALT4* (beta-1,3-galatcosyltransferase 4) and *ZADH2* (prostaglandin reductase 3). Taken together, these data reinforce the use of blood as an accessible tissue of value in the identification of DMPs associated with dementia onset and progression.

Parkinson's disease (PD) is the second most common chronic neurodegenerative disease in the elderly population. While the motor-related symptoms that characterize PD are bradykinesia, tremor, rigidity, and postural instability, the behavioral deficits include depression, anxiety, sleep disorders, and cognitive dysfunction. Collectively, these outcomes lead to severe impairment of the quality of life for

PD patients (Frucht 2004). The first genetic cause of PD was identified as a missense mutation in the alpha-synuclein (*SNCA*) gene (SNCAp.Ala53Thr), a locus that also was found to undergo duplications and triplications linked to PD (Polymeropoulos et al. 1997). *SNCA* gene dosage is critical for the development of PD, leading researcher to hypothesize that deregulation of *SNCA* may be a potential mechanism for PD. This hypothesis was confirmed when DNA methylation levels of *SNCA* were found to be reduced in the substantia nigra, putamen, and cortex of PD patients ($N = 12$) compared to controls ($N = 14$) and was linked to the increased expression of *SNCA* (Jowaed et al. 2010). In 2011, a comprehensive genomic study identified several PD risk loci in cerebellum and frontal cortex of PD brains, including *PARK16*, *GPNMB*, and *STX1B* genes, all of which were associated with differential DNA methylation at proximal CpG sites (International Parkinson's Disease Genomics C and Wellcome Trust Case Control C 2011). Examination of DNA methylation levels across the genome in the frontal cortex and blood leukocytes from the same PD patients ($N = 5$) and controls ($N = 6$) found a high concordance of genes differentially methylated in both tissues (Masliah et al. 2013). Taken together, these data reinforce the use of blood as an accessible tissue of value in the identification of differentially methylated sites associated with PD onset and progression and lend further support to the pathogenesis of PD, by providing novel diagnostic, prognostic, and modifiable therapeutic targets.

Huntington's disease (HD) is characterized by deficits in cognitive, psychiatric, and motor stability that is typically caused by a trinucleotide repeat (CAG) mutation in the *HTT* gene (Walker 2007). Expression of the disease protein, huntingtin, leads to extensive transcriptional dysregulation, and a growing body of evidence suggests that epigenetic modifications play a key role in HD pathogenesis (Lee et al. 2013; Reik et al. 1993). Although links between DNA methylation and HD are only beginning to emerge, several recent studies have reported HD-related changes in DNA methylation in rodents and humans. One such study targeted the *ADORA2A* gene, a G-protein-coupled receptor that decreases its expression in HD patients, and found increased levels of DNA methylation in the putamen brain tissue of HD patients ($N = 6$) compared to controls ($N = 5$) (Villar-Menendez et al. 2013). Interestingly, contradictory results were found when genome-wide approaches were employed to identify HD-related changes in DNA methylation throughout the genome. While one study found thousands of differentially methylated sites in a fibroblast cell line from an HD patient compared to control, another study failed to find differentially methylated sites in genomic DNA from the forebrain cortex brain tissue from seven HD patients and six controls (Jia et al. 2015; De Souza et al. 2016). Of course this latter study was more complex, profiling multiple cell types from both sexes, and the data was stringently normalized to account for this complexity, which may contribute for the lack of findings. Nonetheless, studies profiling rodent brain tissues have clearly shown that aberrant DNA methylation levels are associated with HD pathogenesis (Ng et al. 2013), indicating that larger human sample sizes are needed to definitely determine the role of DNA methylation in HD.

Finally, it was recently shown that patterns of DNA methylation correlate with chronologic age with high precision, thereby providing a "DNA methylation age"

(Horvath 2013); importantly, the gap between chronologic age and DNA methylation age widens with environmental exposures and clinical disease (Almen et al. 2014; Kananen et al. 2015; Beach et al. 2015). An accelerated DNA methylation age – that is, when estimated age is higher than expected (on the basis of chronological age) – predicts several age-related cognitive phenotypes. Notably, accelerated DNA methylation age has been associated with Alzheimer's, Parkinson's, and Huntington's disease (Horvath et al. 2016; Levine et al. 2015; Horvath and Ritz 2015; Chen et al. 2016). While it is currently unclear what these biomarkers can teach us about the biology of these disorders, their longitudinal investigation could help determine the impact of endogenous or exogenous stress factors on disease onset and progression. Perhaps the most exciting feature of identifying DNA methylation biomarkers is that epigenetic changes are reversible, raising the prospect that DNA methylation age estimates might be useful for identifying or validating interventions.

3 5hmC and Behavior

The elucidation of the link between 5hmC and behavior is still in its infancy, especially in humans. Aside from its relatively recent discovery in mammalian cells in 2009 (Kriaucionis and Heintz 2009; Tahiliani et al. 2009), the primary reason for this delay in humans, as compared to 5mC, is that 5hmC has a very low abundance in peripheral tissue, making access to brain tissue paramount for these studies. Nonetheless, in less than a decade, numerous studies in mice have led to the emergence of a putative role for 5hmC in behaviors related to psychiatric disorders. Accordingly, this section will feature more findings from rodent studies than humans, simply due to the limited number of human studies performed to date.

Studies have demonstrated an age-dependent accumulation of 5hmC in the brain throughout early-life development and into adulthood (Szulwach et al. 2011; Chen et al. 2012; Zampieri et al. 2015). Environmental stimuli can alter this age-associated accumulation. For example, calorically restricted mice exhibited an age-dependent reduction of 5hmC levels in hippocampal and cerebellar tissue (Chouliaras et al. 2012). In addition, mice exposed to an enriched environment had reduced 5hmC abundance in the hippocampus, primarily on genes involved in axon guidance (Irier et al. 2014). These alterations also were associated with increased learning and memory, suggesting that environmental enrichment might modulate the dynamics of 5hmC in the hippocampus and contribute to improved learning and memory. In contrast, mice subjected to a 30-min acute stress showed increased 5hmC levels on the glucocorticoid receptor gene (*Nr3c1*) in the hippocampus (Li et al. 2015). Genome-wide 5hmC analysis of these same mice revealed that short-term stress induced genome-wide disruptions of 5hmC and confirmed an overall increase in 5hmC following stress. Interestingly, altered 5hmC was found near several transcription factor (TF) binding sites of genes that were differentially expressed and have known roles in neurogenesis and neurological activities (Li et al. 2016), suggesting that in response to stress, the function of 5hmC may be to influence TF binding to provide appropriate levels of gene expression needed to cope with the stress. The fact that 5hmC changes were found within 1 h of a short stress highlights the potential for rapid changes of 5hmC within the brain. It will be interesting to examine the long-term effects of short stress (i.e., more than 1 h after exposure) or how chronic stress alters 5hmC levels. Many believe 5hmC to be important in long-term consequences of mental health, yet these studies indicate that alterations in 5hmC can occur rapidly and may impact the expression of key genes related to the origins of mental illness.

Environmental stimuli can affect brain regions other than the hippocampus. For example, mice exposed to repeated administrations of cocaine have increased 5hmC in the nucleus accumbens, primarily in coding regions and enhancer sequences of genes involved in drug addiction (Feng et al. 2015). Notably, 5hmC changes persisted for a minimum of 1 month after cocaine exposure in only a small subset of loci, suggesting that these epigenetic changes are largely reversible. The modulation of 5hmC also may mediate behavioral adaptations. For example, fear extinction, a form of reverse learning, results in dramatic 5hmC changes in the prefrontal cortex of mice (Li et al. 2014). These studies also support unique molecular roles for the *Tet* enzymes, as *Tet3*, but not *Tet1*, mediated the increased gene expression that was associated with rapid behavioral adaptation. Interestingly, another group found that mice lacking the expression of *Tet1* exhibited impaired memory extinction, coupled with long-term synaptic depression and downregulation of neuronal activity-related genes (Rudenko et al. 2013). Thus, while *Tet3* may solely facilitate the accumulation of 5hmC in the prefrontal cortex of mice during rapid behavior adaptation in response to fear, *Tet1* governs alterations in 5hmC on synaptic plasticity genes during behavioral adaptation in response to stressful environmental exposures.

Taken together, these studies open up the possibility that 5hmC may function in the development of environmentally sensitive neuronal dysfunction. It will be of great interest to investigate the role of each *Tet* enzyme coupled with the rapid and stable dynamics of 5hmC at different developmental time points to understand its role in synaptic plasticity, neuronal development, the maintenance of mental health, and the onset of mental illness.

3.1 In the Origins of Psychiatric Disorders

Several studies indicate that early-life experiences have a profound impact on brain development and subsequent adult behavior (Oberlander et al. 2008; Mcgowan et al. 2009; Roth and Sweatt 2011b; Labonte et al. 2012). One such example involves rhesus macaques that were deprived of early-life maternal interactions. As adults, these monkeys have altered 5hmC in the prefrontal cortex on the promoter regions of genes related to neurological functions and psychological disorders (e.g., D_3 dopamine receptor (*DRD3*), serotonergic transporter (*5-HTT*), and GABAergic receptor (*GABRA2*)) (Massart et al. 2014). Since these 5hmC disruptions were detected during adulthood, these findings suggest that early-life changes in 5hmC are stable throughout development and may represent the origins of developmental brain disorders such as SCZ, BD, and autism.

SCZ and BD are psychiatric disorders with shared and distinct clinical and genetic features; however, the majority of SCZ and BD cases *cannot* be explained by genetics alone. To investigate a role for 5hmC in SCZ and BD, genomic DNA and total RNA were obtained from the postmortem inferior parietal lobule (IPL) brain tissue of 10 SCZ patients, 9 BD patients, and 11 nonpsychiatric subjects, who had no history of psychiatric disorders. SCZ and BD patients exhibited increased 5hmC abundance and *TET1* expression, but not altered *TET2* or *TET3* expression (Dong et al. 2012). Remarkably, *TET1* was not altered in the cerebellum of these patients, suggesting that 5hmC may be involved in the development of psychosis through the inferior parietal lobule, but not the cerebellum, perhaps shedding light on the tissue-specific development of SCZ and BD. The increases of 5hmC in these patients were associated with reduced expression of biologically relevant genes including glutamic acid decarboxylase 67 (*GAD67*) and *APOBEC3A*, which plays a role in active DNA demethylation. Together, these findings suggest a common etiology in psychosis, one that includes genome-wide changes in 5hmC.

While associations between DNA methylation levels and MDD were described above, a recent study extracted genomic DNA from the inferior frontal gyrus of 19 clinically depressed suicide completers and 19 controls and measured genome-wide 5hmC levels. This approach revealed 550 differential hydroxymethylated sites in a plurality of genes, some of which also had differential expression, including myosin XVI (*MYO16*) and insulin-degrading enzyme (*IDE*), genes previously implicated in brain development and neurodegenerative disorders (Kurochkin and Goto 1994; Yui et al. 2015). These data shed light on an alternative molecular mechanism that may be involved in the development of MDD.

ASDs encompass a broad range of behaviorally related disorders with a high prevalence in children. Notably, only ~20% of ASD cases show a genetic etiology (Gaugler et al. 2014; Bulik-Sullivan et al. 2015). Prenatal factors shown to increase the risk of ASD in offspring include environmental influences such as multiple births, in vitro fertilization, and parental exposure to common drug treatments (e.g., antiepileptic drugs (e.g., valproate) or folic acid) (Gardener et al. 2009; Rogers 2008). Together, these findings effectively open the door for contributions from environmentally sensitive epigenetic modifications such as 5hmC to have an underlying role in the ASD etiology. Consistent with this hypothesis, genomic DNA extracted from one fetal and two human cerebellum brain tissues revealed that developmentally specific changes in 5hmC are highly enriched in known ASD genes (Wang et al. 2012b). In another study, the mRNA and 5hmC levels from the cerebellar cortex of ten ASD patients and ten controls were compared, and ASD patients had an enrichment of both *TET1* mRNA and 5hmC levels at the promoters of both *GAD67* and *RELN*, both candidate genes of ASD (Zhubi et al. 2014). More recently, genome-wide 5hmC levels were measured in cerebellum tissue from 17 ASD patients and 19 controls, revealing 797 age-dependent differentially hydroxymethylated regions (DhMRs) in the young group (age $\leq$ 18) and no significant DhMRs in groups over 18 years of age, suggesting 5hmC may be affected across the spectrum early development and could contribute to the pathogenesis of ASD (Cheng et al. 2018). However, since all of these findings were observed post-symptomatically and in postmortem human brain tissue, it is unclear if the altered 5hmC represents a cause or a consequence of having autistic-like behaviors. Thus, these findings warrant a deeper investigation at pre-symptomatic developmental time points.

Finally, it is notable that rodent models exposed to prenatal stress exhibit long-lasting neurological, endocrinological, and behavioral changes that are thought to mirror the development of mental illness (Koenig et al. 2005). For example, prenatal stress increased anxiety-like behaviors and altered gene expression in mice heterozygous for the serotonin transporter (*5-HTT*) gene (Van Den Hove et al. 2011). Interestingly, these findings were more pronounced in female offspring, suggesting a sex-specific development of these altered behaviors. This study demonstrates a gene by environment interaction, and while this study did not examine 5hmC, the 5hmC-related data discussed in this review supports a 5hmC contribution to these altered behaviors and gene expression. If so, do these changes in 5hmC mirror those found in the homozygous mutants with advanced behavioral endophenotypes? This finding could shed light on the molecular mechanisms exacerbating early-life, stress-related behavioral outcomes in offspring that may have a genetic predisposition, a likely scenario for many mental disorders.

3.2 In Aging and Neurodegeneration

The emerging link between brain development and the accumulation of 5hmC with age has led researchers to examine this epigenetic mark in neurodegenerative diseases (Al-Mahdawi et al. 2014). Indeed, age-associated genes that acquire 5hmC are associated with pathways related to neurodegenerative diseases (Song et al. 2011). Human studies found that 5hmC was depleted in the hippocampus, cerebellum, and entorhinal cortex of ten patients suffering from AD, compared to ten controls (Chouliaras et al. 2013; Condliffe et al. 2014). On the other hand, studies of the middle frontal gyrus from 13 AD patients and the middle temporal gyrus from 29 AD patients revealed that an enrichment of 5hmC was positively correlated with hallmarks of AD, including neurofibrillary tangles (NFT), amyloid beta, and ubiquitin load (Coppieters et al. 2014). Together, these findings suggest that 5hmC may be driving the primary molecular components of AD progression within distinct regions of the brain. Notably, AD-associated levels of 5hmC also can be detected at preclinical stages of AD, as observed by the comparison of postmortem hippocampus and cerebellum brain tissues from five controls and five preclinical AD subjects, defined as having sufficient AD pathologic alterations at autopsy to meet intermediate or high NIA-RI criteria and moderate or frequent neuritic plaque scores according to the Consortium to Establish a Registry for AD (CERAD) with Braak scores of III–IV and antemortem psychometric test scores in the normal range when corrected for age and education (Bradley-Whitman and Lovell 2013; Schmitt et al. 2000). These findings indicate that 5hmC may act as a viable biomarker of AD onset and progression. However, studies employing nuclear labeling or enzyme-linked immunosorbent assay (ELISA) were unable to find significant alterations in 5hmC associated with AD (Lashley et al. 2015), suggesting that higher-resolution methods are required to identify AD-associated changes in 5hmC. Connections between 5hmC and other neurodegenerative disorders including HD, ataxia-telangiectasia, and fragile X-associated tremor/ataxia syndrome have recently surfaced, and the link to 5hmC for each of these disorders is described below.

Recent epigenetic connections led to the investigation of the genome-wide distribution of 5hmC in a mouse model of HD, which found a deficiency of 5hmC in the striatum and cortex tissues (Wang et al. 2013). This altered distribution of 5hmC was generally associated with pathways involved in neuronal development/differentiation, axonal guidance, and neuronal function/survival. A specific example included a 5hmC reduction on the *ADORA2A* gene, a G-protein-coupled receptor that decreases its expression in HD patients ($N = 6$) compared to controls ($N = 5$), revealing that 5hmC may play a role in the pathological expression of *ADORA2A* found in HD patients (Villar-Menendez et al. 2013). Together, these results suggest that the loss of 5hmC explains the corresponding increase in total DNA methylation (5mC + 5hmC; see above) and provides a more complete understanding of a potential molecular mechanism underlying the cause of neuronal cell death associated with HD.

Ataxia-telangiectasia, a devastating neurodegenerative disorder that shows high cell specificity in its development, is caused by a mutation in the *ATM* gene (Mckinnon 2004). Despite this monogenic etiology, alterations in 5hmC were recently found in repeated sequences and regulatory elements in genomic DNA from *ATM*-deficient Purkinje cells of human patients ($N = 3$) compared to controls (Jiang et al. 2015). This 5hmC distribution is likely the result of TET1 responding to DNA damage, which is a hallmark of *ATM*-deficient neuronal cells. To test whether TET1 activity is related to Purkinje cell degeneration and behavioral deficits in mice, adenoviral particles encoding either human wild-type TET1 (TET1-WT) or kinase dead mutant TET1 (TET-KD) were infected into *ATM*-deficient mouse cerebellum slices and controls. Cerebellum samples lacking TET1 expression that were infected with TET1-WT showed no activation of *Caspase-3*, a cell death marker, while infection with TET1-KD activated *Caspase-3*. These findings suggest that TET1-mediated 5hmC production is essential for Purkinje cell viability and the prevention of ataxia-telangiectasia-like symptoms in mice, supporting the concept that vulnerability to neurodegeneration is linked to aberrant changes of 5hmC in neuronal cells.

Fragile X-associated tremor/ataxia syndrome (FXTAS), a late-onset neurodegenerative disorder, is one of the characterized fragile X disorders that are caused by a CGG expansion in the 5′UTR of the *FMR1* gene (Hagerman and Hagerman 2015; Santoro et al. 2012). Genome-wide analysis of 5hmC in the cerebellum of a FXTAS mouse model (rCGG mice) revealed an overall reduction in 5hmC at 16 weeks of age when compared to age-matched controls (Yao et al. 2014). Despite the overall reduction of 5hmC, these mice have an increase of 5hmC in repetitive sequences as well as in cerebellum-specific enhancers. Differential 5hmC between the rCGG and control mice were predominantly found in transcription factor (TF) binding sites that are located in genes essential for neuronal development. Finally, ribosomal profiling revealed that the differential 5hmC-associated genes often exhibited altered ribosomal processing in the rCGG mice, suggesting that 5hmC may somehow influence translational changes. In summary, this study links 5hmC to the etiology of FXTAS and implicates a role for 5hmC in transcription factor binding and in regulating ribosomal processing of mature RNA transcripts.

Despite that the above disease-associated disruptions in 5hmC represent changes across age, these examples clearly demonstrate a role for 5hmC in development and in the onset and progression of neurodegenerative diseases. These developmental and neurodegenerative disease-associated changes in 5hmC often arise within distinct cell types and brain regions, supporting cell- and tissue-specific development of these diseases. Since these studies have been largely descriptive, it is imperative that future studies determine the functional mechanism(s) played by 5hmC if we are to modify it toward healthy outcomes.

4 Clinical Utilities of DNA Methylation

Diagnosis and risk assessment of psychiatric disorders are typically performed using family history, subjective symptom reports, and behavioral observations, which are vulnerable to the variability in accurate patient reporting and consistent clinician interpretations. Thus, molecular biomarkers are being sought because they represent objective measures that can more accurately diagnose and stratify behavior disorders. For example, suicide biomarkers could accurately distinguish between a pathological mood disorder and a normal symptomatic response to an acute stress (e.g., bereavement). These biomarkers also may provide insight into the underlying molecular mechanisms contributing to specific mood-related behaviors. As another example, a recent study examined the peripheral blood from PTSD patients with matched adult trauma exposure and found distinct PTSD-associated DNA methylation profiles, which were related to their different childhood adverse events ($N = 32$ and 29) (Mehta et al. 2013). These findings suggested that DNA methylation levels across the genome might reflect differences in the pathophysiology of PTSD. Ultimately, this information can provide much needed insight into treatment response and the development of novel therapeutics aimed at alleviating mood disorders.

DNA methylation provides a highly sensitive and dynamic biomarker in real time, immediately marking exposures to various environmental stimuli such as stress, cognitive behavioral therapy (CBT), electroconvulsive therapy (ECT), antidepressant treatment, and exercise. Moreover, it is notable that environmentally sensitive DNA methylation levels can be stable across generations. Thus, the methylome is a promising candidate to be an objective biomarker of human behavior following environmental stimuli. Recent support for using DNA methylation as an objective biomarker of environmental stimuli came from a study that examined the genomic DNA from two cohorts of urban African Americans, from the Grady Trauma Project ($N = 421$) and the Johns Hopkins Center for Prevention Research Study ($N = 326$), and found an interaction between blood *SKA2* methylation and trauma scores (particularly childhood emotional abuse). This interaction was successful at predicting suicidal ideation and suicidal behavior with 71% and 73% accuracy, respectively (Clive et al. 2016). Notably, prediction of suicidal ideation and behavior from saliva samples also was promising, as *SKA2* methylation levels following childhood abuse had a 76% accuracy in predicting suicidal ideation. On the other hand, saliva *SKA2* methylation is linked to extreme anxiety and had 69% accuracy in predicting suicidal ideation, further supporting the sensitivity of DNA methylation to distinguish between experience-induced symptomatic responses. In the same study, the *SKA2* methylation prediction model was used to predict PTSD in the Grady Trauma Project cohorts. Interestingly, while *SKA2* methylation had 72% accuracy in predicting PTSD that was defined by the Child Trauma Questionnaire (CTQ) scores, *SKA2* methylation alone only had 55% accuracy in predicting PTSD. Together, these data suggested that combining DNA methylation information with rigorous behavioral assessments improves our power to diagnose mood disorders. Other researchers sought to further improve this suicide prediction model by determining if DNA methylation levels at other genes correlate with suicide. These studies revealed that the DNA methylation levels of three genes (*DDR1*, *ARHGEF10*, *SHP1*) correlated with suicide and *SKA2* methylation levels and together they could consistently predict suicide across all tested data sets, including youth at high risk for depression, pregnant women at risk for elevated postpartum depression, and middle-aged individuals with a high incidence trauma and PTSD. Together, these findings shed light on the possibility for standardized screening with a single objective test, perhaps without the need for a subjective behavioral assessment.

Another promising example of using DNA methylation to diagnose and stratify risk for behaviors includes recent work that investigated fear extinction in patients suffering from panic disorder. This study examined genomic DNA from blood of female patients that received CBT for 6 weeks ($N = 28$), panic disorder patients that did not receive CBT ($N = 20$), and controls ($N = 28$) and found that the DNA methylation on the monoamine oxidase A (*MAOA*) gene was inversely associated with panic disorder severity (as measured by either panic attacks or severity of agoraphobia symptoms). Perhaps most exciting, *MAOA* methylation levels responded to CBT, as the *MAOA* methylation in patients undergoing CBT matched the levels of controls. These findings indicate that *MAOA* methylation levels may be physiologically linked to the therapeutic effects of CBT, with the simplest interpretation being that increased *MAOA* methylation results in decreased *MAOA* gene expression and increased bioavailability of monoamines in the synaptic cleft. Clinically, this could allow for an objective screening tool for panic disorder symptoms and for monitoring the efficacy of panic disorder treatment.

The sensitivity of 5hmC levels following prenatal and/or acute stress underscores the potential for 5hmC as a novel biomarker in the diagnosis of mental health. In addition, the presence/absence of these epigenetic modifications is reversible; thus, they may become relevant in therapeutic interventions (Szyf 2015), especially if methods to selectively modulate 5hmC in vivo are developed at the nucleotide level. Finally, 5hmC has sex-specific profiles, which is of interest because the development of several psychiatric disorders is seemingly sex-specific. For example, females show an increased risk in developing anxiety and depression, while males show a disposition to development of attention deficit hyperactivity disorder (Nolen-Hoeksema 1987; Wooten et al. 2004). A recent study found sex-specific 5hmC on genes with ontological terms correlating with organ morphogenesis, system development, and development of anatomical structures (Gross et al. 2015), suggesting that 5hmC may differentially influence the development of organs (e.g., the brain) in the sexes. Together, these factors must be considered for clinical application and treatment endeavors.

5 Conclusions

The findings described in this chapter highlight the research that has been conducted toward improving our understanding of the molecular events contributing to the development of mental health and illness. Importantly, these studies promote scientific questions that are less limited by the bounds of environment versus genetic, or categories of biologic versus psychiatric, and may spawn productive directions toward improved standard of care in psychiatry. Perhaps most promising is the quantitative nature of 5mC and 5hmC detection, indicating that these marks show promise to serve as unifying links between tissue-level activity and complex behavior. Surprisingly, the methylome is highly sensitive to environmental stimuli, showing alterations in as little as 30 min. Moreover, these marks are both stable and reversible – as shown in suicide completers and in cocaine addiction model, respectively. Harnessing the methods to control this reversibility is of great interest, and some examples were presented, which included decreases in methylation levels in patients with BPD following dialectic behavioral therapy or rodents in response to arched back nursing or enhanced maternal care. Fortunately, brain imaging methodologies such as BOLD and fMRI have confirmed that methylation levels in the blood are linked to tissue activity in the brain. This finding has enabled other studies to test a variety of tissues from which samples can be examined – expanding from brain tissue to blood and buccal swabs – allowing for more flexibility including the ability to obtain epigenetic data from living human subjects. While we have yet to fully understand the functional roles of DNA methylation and hydroxymethylation in behavior, we are beginning to recognize that environmentally sensitive alterations in the methylome are linked to gene expression changes governing behavior. The need for such understanding is especially clear in conditions where pathogenesis is not explained by genetics alone: as in autism, which is thought to be only ~25% explained by genetics; SCZ, MDD, and PTSD, for which monozygous twins can be discordant; and suicide, where predisposition is known to have a significant interaction with environmental factors, such as life experiences and history of trauma. We hope that further understanding and study of the methylome will allow us to improve classification, detection, treatment, and prevention of psychiatric conditions.

Disease classification is an area of constant clinical and research interest. The *Diagnostic and Statistical Manual of Mental Disorders* (DSM-V) guides diagnosis of psychiatric conditions, almost exclusively using behavioral symptoms and signs. This manual, while carefully crafted and painstakingly edited longitudinally by experts in the field of psychiatry, remains vulnerable to subjectivity. It is well documented that many conditions can confound the clinical picture of others; for example, anxiety can share symptoms with ADHD, yet the treatment of one can exacerbate the other. In the future, we hope investigators seek objective biomarkers to aid in classification of mental health and believe that the methylome offers a potential means toward the goal of providing a more personalized diagnosis that will lead to more precise/effective treatments (Fig. 4). This scenario will eliminate the diagnostic odyssey that many mental illness patients endure and provide a rapid treatment strategy toward a healthy outcome.

We also anticipate that the early detection of mental illness will be an important clinical application of the methylome. An example where a molecular indicator, like the methylome, would be effective is in the "prodrome" phase of a disorder (e.g., SCZ), when a patient presents with a subtle abnormal behavior that does not indicate a clear diagnosis. In addition, early detection leads to early treatment, which can dramatically improve outcomes for most mental illness patients.

The current standard of care for the detection of a mental illness is limited, relying heavily on patient reporting and symptom recognition. Among future areas for research, we see an opportunity to identify novel quantitative indicators for risk, through identifying aberrations to the methylome which can be measured peripherally in a clinical setting, perhaps through leukocytes via blood draw or buccal swab. We already have observed a relationship between the methylation levels in the promoter region of the oxytocin receptor gene and autism and *BDNF* in suicide completers. However, much work remains to tease out correlations between genes in multiple disorders, such as why increased DNA methylation levels on the *OXTR* promoter region is associated with autism yet decreased DNA methylation levels on the *OXTR* promoter region is associated with social anxiety disorder – much like copy number variations? Can DNA methylation levels lead to different disorders based on too much or too little? We anticipate that improved understanding of the complete methylome may lead to clinically useful biomarkers for risk of a variety of psychiatric illnesses that span the lifetime. This knowledge of the methylome and its role in etiology of psychiatric conditions may allow us to identify unique physiologic targets for pharmacological intervention. We can likewise develop more targeted nonpharmacologic therapies and apply them with more specificity. In short, the methylome represents an exciting frontier that can lead to more effective collaboration between the individually developed fields of psychology, psychiatry, neuroscience, and genetics. This unification of understanding may potentially lead to more objective and precise disease classification, risk stratification, treatment, and ultimately improved quality of life for vulnerable individuals and populations.

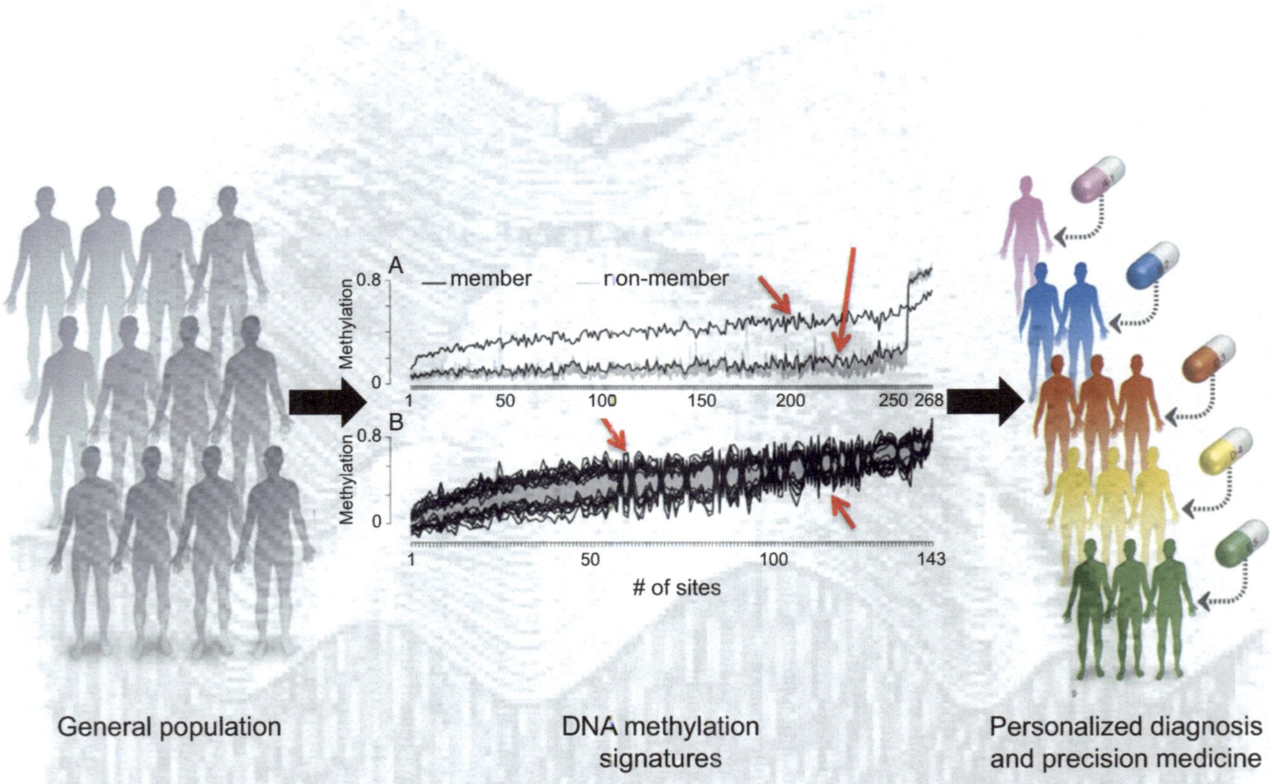

Fig. 4 Improving the standard of care in psychiatry. Genome-wide signatures of DNA methylation can be used to stratify the general population. This process can result in a personalized diagnosis that is more rapid and exact. Moreover, it will provide insight into the genes and pathways contributing to the individual diagnosis, which will guide more precise treatment for the individual

References

Abdolmaleky HM, Cheng KH, Faraone SV et al (2006) Hypomethylation of MB-COMT promoter is a major risk factor for schizophrenia and bipolar disorder. Hum Mol Genet 15(21):3132–3145
Aberg KA, Mcclay JL, Nerella S et al (2014) Methylome-wide association study of schizophrenia: identifying blood biomarker signatures of environmental insults. JAMA Psychiat 71(3):255–264
Aid T, Kazantseva A, Piirsoo M, Palm K, Timmusk T (2007) Mouse and rat BDNF gene structure and expression revisited. J Neurosci Res 85(3):525–535
Alisch RS, Barwick BG, Chopra P et al (2012) Age-associated DNA methylation in pediatric populations. Genome Res 22(4):623–632
Alisch RS, Chopra P, Fox AS et al (2014) Differentially methylated plasticity genes in the amygdala of young primates are linked to anxious temperament, an at risk phenotype for anxiety and depressive disorders. J Neurosci 34(47):15548–15556
Al-Mahdawi S, Virmouni SA, Pook MA (2014) The emerging role of 5-hydroxymethylcytosine in neurodegenerative diseases. Front Neurosci 8:397
Almen MS, Nilsson EK, Jacobsson JA et al (2014) Genome-wide analysis reveals DNA methylation markers that vary with both age and obesity. Gene 548(1):61–67
Ansorge MS, Zhou M, Lira A, Hen R, Gingrich JA (2004) Early-life blockade of the 5-HT transporter alters emotional behavior in adult mice. Science 306(5697):879–881
Ansorge MS, Morelli E, Gingrich JA (2008) Inhibition of serotonin but not norepinephrine transport during development produces delayed, persistent perturbations of emotional behaviors in mice. J Neurosci 28(1):199–207
Bakulski KM, Dolinoy DC, Sartor MA et al (2012) Genome-wide DNA methylation differences between late-onset Alzheimer's disease and cognitively normal controls in human frontal cortex. J Alzheimers Dis 29(3):571–588
Baron-Cohen S, Wheelwright S, Skinner R, Martin J, Clubley E (2001) The autism-spectrum quotient (AQ): evidence from Asperger syndrome/high-functioning autism, males and females, scientists and mathematicians. J Autism Dev Disord 31(1):5–17
Barrett PM, Healy LJ (2003) An examination of the cognitive processes involved in childhood obsessive-compulsive disorder. Behav Res Ther 41(3):285–299
Beach SR, Dogan MV, Lei MK et al (2015) Methylomic aging as a window onto the influence of lifestyle: tobacco and alcohol use alter the rate of biological aging. J Am Geriatr Soc 63:2519. https://doi.org/10.1111/jgs.13830
Beery AK, Mcewen LM, Macisaac JL, Francis DD, Kobor MS (2016) Natural variation in maternal care and cross-tissue patterns of oxytocin receptor gene methylation in rats. Horm Behav 77:42–52
Berton O, Mcclung CA, Dileone RJ et al (2006) Essential role of BDNF in the mesolimbic dopamine pathway in social defeat stress. Science 311(5762):864–868
Bird A (2002) DNA methylation patterns and epigenetic memory. Genes Dev 16(1):6–21
Boersma GJ, Lee RS, Cordner ZA et al (2014) Prenatal stress decreases Bdnf expression and increases methylation of Bdnf exon IV in rats. Epigenetics 9(3):437–447
Booij L, Szyf M, Carballedo A et al (2015) DNA methylation of the serotonin transporter gene in peripheral cells and stress-related changes in hippocampal volume: a study in depressed patients and healthy controls. PLoS One 10(3):e0119061
Braak H, Braak E (1991) Neuropathological stageing of Alzheimer-related changes. Acta Neuropathol 82(4):239–259
Bradley-Whitman MA, Lovell MA (2013) Epigenetic changes in the progression of Alzheimer's disease. Mech Ageing Dev 134(10):486–495
Braithwaite EC, Kundakovic M, Ramchandani PG, Murphy SE, Champagne FA (2015) Maternal prenatal depressive symptoms predict infant NR3C1 1F and BDNF IV DNA methylation. Epigenetics 10(5):408–417
Bulik-Sullivan B, Finucane HK, Anttila V et al (2015) An atlas of genetic correlations across human diseases and traits. Nat Genet 47(11):1236–1241

Cappi C, Brentani H, Lima L et al (2016) Whole-exome sequencing in obsessive-compulsive disorder identifies rare mutations in immunological and neurodevelopmental pathways. Transl Psychiatry 6:e764

Carter CS (2003) Developmental consequences of oxytocin. Physiol Behav 79(3):383–397

Carter CS, Williams JR, Witt DM, Insel TR (1992) Oxytocin and social bonding. Ann N Y Acad Sci 652:204–211

Chen H, Dzitoyeva S, Manev H (2012) Effect of aging on 5-hydroxymethylcytosine in the mouse hippocampus. Restor Neurol Neurosci 30(3):237–245

Chen BH, Marioni RE, Colicino E et al (2016) DNA methylation-based measures of biological age: meta-analysis predicting time to death. Aging 8(9):1844–1865

Cheng X (1995) Structure and function of DNA methyltransferases. Annu Rev Biophys Biomol Struct 24:293–318

Cheng Y, Li Z, Manupipatpong S et al (2018) 5-Hydroxymethylcytosine alterations in the human postmortem brains of autism spectrum disorder. Hum Mol Genet 27(17):2955–2964

Chouliaras L, Van Den Hove DL, Kenis G et al (2012) Age-related increase in levels of 5-hydroxymethylcytosine in mouse hippocampus is prevented by caloric restriction. Curr Alzheimer Res 9(5):536–544

Chouliaras L, Mastroeni D, Delvaux E et al (2013) Consistent decrease in global DNA methylation and hydroxymethylation in the hippocampus of Alzheimer's disease patients. Neurobiol Aging 34(9):2091–2099

Clive ML, Boks MP, Vinkers CH et al (2016) Discovery and replication of a peripheral tissue DNA methylation biosignature to augment a suicide prediction model. Clin Epigenetics 8:113

Condliffe D, Wong A, Troakes C et al (2014) Cross-region reduction in 5-hydroxymethylcytosine in Alzheimer's disease brain. Neurobiol Aging 35(8):1850–1854

Coppieters N, Dieriks BV, Lill C, Faull RL, Curtis MA, Dragunow M (2014) Global changes in DNA methylation and hydroxymethylation in Alzheimer's disease human brain. Neurobiol Aging 35(6):1334–1344

Dadds MR, Allen JL, Oliver BR et al (2012) Love, eye contact and the developmental origins of empathy v. psychopathy. Br J Psychiatry 200(3):191–196

Dadds MR, Moul C, Cauchi A et al (2014) Methylation of the oxytocin receptor gene and oxytocin blood levels in the development of psychopathy. Dev Psychopathol 26(1):33–40

Day JJ, Sweatt JD (2010) DNA methylation and memory formation. Nat Neurosci 13 (11):1319–1323

Day JJ, Childs D, Guzman-Karlsson MC et al (2013) DNA methylation regulates associative reward learning. Nat Neurosci 16(10):1445–1452

De Jager PL, Srivastava G, Lunnon K et al (2014) Alzheimer's disease: early alterations in brain DNA methylation at ANK1, BIN1, RHBDF2 and other loci. Nat Neurosci 17(9):1156–1163

De Souza RA, Islam SA, Mcewen LM et al (2016) DNA methylation profiling in human Huntington's disease brain. Hum Mol Genet 25(10):2013–2030

Devlin AM, Brain U, Austin J, Oberlander TF (2010) Prenatal exposure to maternal depressed mood and the MTHFR C677T variant affect SLC6A4 methylation in infants at birth. PLoS One 5(8):e12201

Dong E, Gavin DP, Chen Y, Davis J (2012) Upregulation of TET1 and downregulation of APOBEC3A and APOBEC3C in the parietal cortex of psychotic patients. Transl Psychiatry 2:e159

Etkin A, Egner T, Kalisch R (2011) Emotional processing in anterior cingulate and medial prefrontal cortex. Trends Cogn Sci 15(2):85–93

Fairburn CG, Beglin SJ (1994) Assessment of eating disorders: interview or self-report questionnaire? Int J Eat Disord 16(4):363–370

Feinn R, Nellissery M, Kranzler HR (2005) Meta-analysis of the association of a functional serotonin transporter promoter polymorphism with alcohol dependence. *Am J Med Genet B Neuropsychiatr Genet* 133B(1):79–84

Feng J, Shao N, Szulwach KE et al (2015) Role of Tet1 and 5-hydroxymethylcytosine in cocaine action. Nat Neurosci 18(4):536–544

Frieling H, Bleich S, Otten J et al (2008) Epigenetic downregulation of atrial natriuretic peptide but not vasopressin mRNA expression in females with eating disorders is related to impulsivity. Neuropsychopharmacology 33(11):2605–2609
Frodl T, O'Keane V (2013) How does the brain deal with cumulative stress? A review with focus on developmental stress, HPA axis function and hippocampal structure in humans. Neurobiol Dis 52:24–37
Frucht SJ (2004) Parkinson disease: an update. Neurologist 10(4):185–194
Fuchikami M, Morinobu S, Segawa M et al (2011) DNA methylation profiles of the brain-derived neurotrophic factor (BDNF) gene as a potent diagnostic biomarker in major depression. PLoS One 6(8):e23881
Gardener H, Spiegelman D, Buka SL (2009) Prenatal risk factors for autism: comprehensive meta-analysis. Br J Psychiatry 195(1):7–14
Gaspar P, Cases O, Maroteaux L (2003) The developmental role of serotonin: news from mouse molecular genetics. Nat Rev Neurosci 4(12):1002–1012
Gaugler T, Klei L, Sanders SJ et al (2014) Most genetic risk for autism resides with common variation. Nat Genet 46(8):881–885
Goodman WK, Price LH, Rasmussen SA et al (1989) The Yale-Brown Obsessive Compulsive Scale. II. Validity. Arch Gen Psychiatry 46(11):1012–1016
Gregory SG, Connelly JJ, Towers AJ et al (2009) Genomic and epigenetic evidence for oxytocin receptor deficiency in autism. BMC Med 7:62
Gross JA, Pacis A, Chen GG, Barreiro LB, Ernst C, Turecki G (2015) Characterizing 5-hydroxymethylcytosine in human prefrontal cortex at single base resolution. BMC Genomics 16:672
Guillemin C, Provencal N, Suderman M et al (2014) DNA methylation signature of childhood chronic physical aggression in T cells of both men and women. PLoS One 9(1):e86822
Guo F, Li X, Liang D et al (2014) Active and passive demethylation of male and female pronuclear DNA in the mammalian zygote. Cell Stem Cell 15(4):447–459
Hagerman PJ, Hagerman RJ (2015) Fragile X-associated tremor/ataxia syndrome. Ann N Y Acad Sci 1338:58–70
Han JA, An J, Ko M (2015) Functions of TET proteins in hematopoietic transformation. Mol Cells 38(11):925–935
He YF, Li BZ, Li Z et al (2011) Tet-mediated formation of 5-carboxylcytosine and its excision by TDG in mammalian DNA. Science 333(6047):1303–1307
Heim C, Nemeroff CB (2001) The role of childhood trauma in the neurobiology of mood and anxiety disorders: preclinical and clinical studies. Biol Psychiatry 49(12):1023–1039
Hompes T, Izzi B, Gellens E et al (2013) Investigating the influence of maternal cortisol and emotional state during pregnancy on the DNA methylation status of the glucocorticoid receptor gene (NR3C1) promoter region in cord blood. J Psychiatr Res 47(7):880–891
Horvath S (2013) DNA methylation age of human tissues and cell types. Genome Biol 14(10):R115
Horvath S, Ritz BR (2015) Increased epigenetic age and granulocyte counts in the blood of Parkinson's disease patients. Aging 7(12):1130–1142
Horvath S, Langfelder P, Kwak S et al (2016) Huntington's disease accelerates epigenetic aging of human brain and disrupts DNA methylation levels. Aging 8(7):1485–1512
Hu C, Jiang X (2014) An emotion regulation role of ventromedial prefrontal cortex in moral judgment. Front Hum Neurosci 8:873
International Parkinson's Disease Genomics C, Wellcome Trust Case Control C (2011) A two-stage meta-analysis identifies several new loci for Parkinson's disease. *PLoS Genet* 7(6):e1002142
Ioshikhes IP, Zhang MQ (2000) Large-scale human promoter mapping using CpG islands. Nat Genet 26(1):61–63
Irier H, Street RC, Dave R et al (2014) Environmental enrichment modulates 5-hydroxymethylcytosine dynamics in hippocampus. Genomics 104(5):376–382
Ito S, D'Alessio AC, Taranova OV, Hong K, Sowers LC, Zhang Y (2010) Role of Tet proteins in 5mC to 5hmC conversion, ES-cell self-renewal and inner cell mass specification. Nature 466(7310):1129–1133

Ito S, Shen L, Dai Q et al (2011) Tet proteins can convert 5-methylcytosine to 5-formylcytosine and 5-carboxylcytosine. Science 333(6047):1300–1303

Jack A, Connelly JJ, Morris JP (2012) DNA methylation of the oxytocin receptor gene predicts neural response to ambiguous social stimuli. Front Hum Neurosci 6:280

Jia H, Morris CD, Williams RM, Loring JF, Thomas EA (2015) HDAC inhibition imparts beneficial transgenerational effects in Huntington's disease mice via altered DNA and histone methylation. Proc Natl Acad Sci U S A 112(1):E56–E64

Jiang D, Zhang Y, Hart RP, Chen J, Herrup K, Li J (2015) Alteration in 5-hydroxymethylcytosine-mediated epigenetic regulation leads to Purkinje cell vulnerability in ATM deficiency. Brain 138(Pt 12):3520–3536

Johnson AA, Akman K, Calimport SR, Wuttke D, Stolzing A, De Magalhaes JP (2012) The role of DNA methylation in aging, rejuvenation, and age-related disease. Rejuvenation Res 15(5):483–494

Jowaed A, Schmitt I, Kaut O, Wullner U (2010) Methylation regulates alpha-synuclein expression and is decreased in Parkinson's disease patients' brains. J Neurosci 30(18):6355–6359

Kananen L, Nevalainen T, Jylhava J et al (2015) Cytomegalovirus infection accelerates epigenetic aging. Exp Gerontol 72:227–229

Kappeler L, Meaney MJ (2010) Epigenetics and parental effects. Bioessays 32(9):818–827

Keller S, Sarchiapone M, Zarrilli F et al (2010) Increased BDNF promoter methylation in the Wernicke area of suicide subjects. Arch Gen Psychiatry 67(3):258–267

Kim YR, Kim JH, Kim MJ, Treasure J (2014) Differential methylation of the oxytocin receptor gene in patients with anorexia nervosa: a pilot study. PLoS One 9(2):e88673

Kim D, Kubzansky LD, Baccarelli A et al (2016) Psychological factors and DNA methylation of genes related to immune/inflammatory system markers: the VA Normative Aging Study. BMJ Open 6(1):e009790

King K, Murphy S, Hoyo C (2015) Epigenetic regulation of Newborns' imprinted genes related to gestational growth: patterning by parental race/ethnicity and maternal socioeconomic status. J Epidemiol Community Health 69(7):639–647

King-Batoon A, Leszczynska JM, Klein CB (2008) Modulation of gene methylation by genistein or lycopene in breast cancer cells. Environ Mol Mutagen 49(1):36–45

Klengel T, Mehta D, Anacker C et al (2013) Allele-specific FKBP5 DNA demethylation mediates gene-childhood trauma interactions. Nat Neurosci 16(1):33–41

Koenig JI, Elmer GI, Shepard PD et al (2005) Prenatal exposure to a repeated variable stress paradigm elicits behavioral and neuroendocrinological changes in the adult offspring: potential relevance to schizophrenia. Behav Brain Res 156(2):251–261

Konishi H, Ogawa T, Nakagomi S, Inoue K, Tohyama M, Kiyama H (2010) Id1, Id2 and Id3 are induced in rat melanotrophs of the pituitary gland by dopamine suppression under continuous stress. Neuroscience 169(4):1527–1534

Kordi-Tamandani DM, Dahmardeh N, Torkamanzehi A (2013) Evaluation of hypermethylation and expression pattern of GMR2, GMR5, GMR8, and GRIA3 in patients with schizophrenia. Gene 515(1):163–166

Kriaucionis S, Heintz N (2009) The nuclear DNA base 5-hydroxymethylcytosine is present in Purkinje neurons and the brain. Science 324(5929):929–930

Kuratomi G, Iwamoto K, Bundo M et al (2008) Aberrant DNA methylation associated with bipolar disorder identified from discordant monozygotic twins. Mol Psychiatry 13(4):429–441

Kurochkin IV, Goto S (1994) Alzheimer's beta-amyloid peptide specifically interacts with and is degraded by insulin degrading enzyme. FEBS Lett 345(1):33–37

Labonte B, Suderman M, Maussion G et al (2012) Genome-wide epigenetic regulation by early-life TraumaGenome epigenetic regulation by early-life trauma. Arch Gen Psychiatry 69(7):722–731

Labuschagne I, Phan KL, Wood A et al (2010) Oxytocin attenuates amygdala reactivity to fear in generalized social anxiety disorder. Neuropsychopharmacology 35(12):2403–2413

Lack CW (2012) Obsessive-compulsive disorder: evidence-based treatments and future directions for research. World J Psychiatry 2(6):86–90

Lam LL, Emberly E, Fraser HB et al (2012) Factors underlying variable DNA methylation in a human community cohort. Proc Natl Acad Sci U S A 109(Suppl 2):17253–17260

Lashley T, Gami P, Valizadeh N, Li A, Revesz T, Balazs R (2015) Alterations in global DNA methylation and hydroxymethylation are not detected in Alzheimer's disease. Neuropathol Appl Neurobiol 41(4):497–506
Lawson EA, Holsen LM, Santin M et al (2012) Oxytocin secretion is associated with severity of disordered eating psychopathology and insular cortex hypoactivation in anorexia nervosa. J Clin Endocrinol Metab 97(10):E1898–E1908
Lee J, Hwang YJ, Kim KY, Kowall NW, Ryu H (2013) Epigenetic mechanisms of neurodegeneration in Huntington's disease. Neurotherapeutics 10(4):664–676
Levine ME, Lu AT, Bennett DA, Horvath S (2015) Epigenetic age of the pre-frontal cortex is associated with neuritic plaques, amyloid load, and Alzheimer's disease related cognitive functioning. Aging 7(12):1198–1211
Li X, Wei W, Zhao QY et al (2014) Neocortical Tet3-mediated accumulation of 5-hydroxymethylcytosine promotes rapid behavioral adaptation. Proc Natl Acad Sci U S A 111(19):7120–7125
Li S, Papale LA, Kintner DB et al (2015) Hippocampal increase of 5-hmC in the glucocorticoid receptor gene following acute stress. Behav Brain Res 286:236–240
Li S, Papale LA, Zhang Q et al (2016) Genome-wide alterations in hippocampal 5-hydroxymethylcytosine links plasticity genes to acute stress. Neurobiol Dis 86:99–108
Liu QR, Lu L, Zhu XG, Gong JP, Shaham Y, Uhl GR (2006) Rodent BDNF genes, novel promoters, novel splice variants, and regulation by cocaine. Brain Res 1067(1):1–12
Lunnon K, Smith R, Hannon E et al (2014) Methylomic profiling implicates cortical deregulation of ANK1 in Alzheimer's disease. Nat Neurosci 17(9):1164–1170
Madrid A, Hogan KJ, Papale LA et al (2018) DNA hypomethylation in blood links B3GALT4 and ZADH2 to Alzheimer's disease. J Alzheimers Dis 66:927
Maguire S, O'Dell A, Touyz L, Russell J (2013) Oxytocin and anorexia nervosa: a review of the emerging literature. Eur Eat Disord Rev 21(6):475–478
Masliah E, Dumaop W, Galasko D, Desplats P (2013) Distinctive patterns of DNA methylation associated with Parkinson disease: identification of concordant epigenetic changes in brain and peripheral blood leukocytes. Epigenetics 8(10):1030–1038
Massart R, Suderman M, Provencal N et al (2014) Hydroxymethylation and DNA methylation profiles in the prefrontal cortex of the non-human primate rhesus macaque and the impact of maternal deprivation on hydroxymethylation. Neuroscience 268:139–148
Mcclure EB, Monk CS, Nelson EE et al (2007) Abnormal attention modulation of fear circuit function in pediatric generalized anxiety disorder. Arch Gen Psychiatry 64(1):97–106
Mcewen BS (2007) Physiology and neurobiology of stress and adaptation: central role of the brain. Physiol Rev 87(3):873–904
Mcgowan PO, Sasaki A, Huang TC et al (2008) Promoter-wide hypermethylation of the ribosomal RNA gene promoter in the suicide brain. PLoS One 3(5):e2085
Mcgowan PO, Sasaki A, D'Alessio AC et al (2009) Epigenetic regulation of the glucocorticoid receptor in human brain associates with childhood abuse. Nat Neurosci 12(3):342–348
Mckinnon PJ (2004) ATM and ataxia telangiectasia. EMBO Rep 5(8):772–776
Mehta D, Klengel T, Conneely KN et al (2013) Childhood maltreatment is associated with distinct genomic and epigenetic profiles in posttraumatic stress disorder. Proc Natl Acad Sci U S A 110(20):8302–8307
Moser DA, Paoloni-Giacobino A, Stenz L et al (2015) BDNF methylation and maternal brain activity in a violence-related sample. PLoS One 10(12):e0143427
Mueller BR, Bale TL (2008) Sex-specific programming of offspring emotionality after stress early in pregnancy. J Neurosci Off J Soc Neurosci 28(36):9055–9065
Mulligan CJ, D'Errico NC, Stees J, Hughes DA (2012) Methylation changes at NR3C1 in newborns associate with maternal prenatal stress exposure and newborn birth weight. Epigenetics 7(8):853–857
Murgatroyd C, Patchev AV, Wu Y et al (2009) Dynamic DNA methylation programs persistent adverse effects of early-life stress. Nat Neurosci 12(12):1559–1566
Murgatroyd C, Quinn JP, Sharp HM, Pickles A, Hill J (2015) Effects of prenatal and postnatal depression, and maternal stroking, at the glucocorticoid receptor gene. Transl Psychiatry 5:e560

Naumova OY, Lee M, Koposov R, Szyf M, Dozier M, Grigorenko EL (2012) Differential patterns of whole-genome DNA methylation in institutionalized children and children raised by their biological parents. Dev Psychopathol 24(1):143–155

Ng CW, Yildirim F, Yap YS et al (2013) Extensive changes in DNA methylation are associated with expression of mutant huntingtin. Proc Natl Acad Sci U S A 110(6):2354–2359

Nolen-Hoeksema S (1987) Sex differences in unipolar depression: evidence and theory. Psychol Bull 101(2):259–282

Novikova SI, He F, Bai J, Cutrufello NJ, Lidow MS, Undieh AS (2008) Maternal cocaine administration in mice alters DNA methylation and gene expression in hippocampal neurons of neonatal and prepubertal offspring. PLoS One 3(4):e1919

Oberlander TF, Weinberg J, Papsdorf M, Grunau R, Misri S, Devlin AM (2008) Prenatal exposure to maternal depression, neonatal methylation of human glucocorticoid receptor gene (NR3C1) and infant cortisol stress responses. Epigenetics 3(2):97–106

Okano M, Bell DW, Haber DA, Li E (1999) DNA methyltransferases Dnmt3a and Dnmt3b are essential for de novo methylation and mammalian development. Cell 99(3):247–257

Papale LA, Seltzer LJ, Madrid A, Pollak SD, Alisch RS (2018) Differentially methylated genes in saliva are linked to childhood stress. Sci Rep 8(1):10785

Park LK, Friso S, Choi SW (2012) Nutritional influences on epigenetics and age-related disease. Proc Nutr Soc 71(1):75–83

Pedersen CA, Boccia ML (2002) Oxytocin links mothering received, mothering bestowed and adult stress responses. Stress 5(4):259–267

Pedersen CA, Vadlamudi SV, Boccia ML, Amico JA (2006) Maternal behavior deficits in nulliparous oxytocin knockout mice. Genes Brain Behav 5(3):274–281

Penn NW, Suwalski R, O'Riley C, Bojanowski K, Yura R (1972) The presence of 5-hydroxymethylcytosine in animal deoxyribonucleic acid. Biochem J 126(4):781–790

Perroud N, Paoloni-Giacobino A, Prada P et al (2011) Increased methylation of glucocorticoid receptor gene (NR3C1) in adults with a history of childhood maltreatment: a link with the severity and type of trauma. Transl Psychiatry 1:e59

Perroud N, Salzmann A, Prada P et al (2013) Response to psychotherapy in borderline personality disorder and methylation status of the BDNF gene. Transl Psychiatry 3:e207

Polymeropoulos MH, Lavedan C, Leroy E et al (1997) Mutation in the alpha-synuclein gene identified in families with Parkinson's disease. Science 276(5321):2045–2047

Popik P, Van Ree JM (1991) Oxytocin but not vasopressin facilitates social recognition following injection into the medial preoptic area of the rat brain. Eur Neuropsychopharmacol 1(4):555–560

Post RM (2007) Role of BDNF in bipolar and unipolar disorder: clinical and theoretical implications. J Psychiatr Res 41(12):979–990

Poulter MO, Du L, Weaver IC et al (2008) GABAA receptor promoter hypermethylation in suicide brain: implications for the involvement of epigenetic processes. Biol Psychiatry 64(8):645–652

Provencal N, Suderman MJ, Caramaschi D et al (2013) Differential DNA methylation regions in cytokine and transcription factor genomic loci associate with childhood physical aggression. PLoS One 8(8):e71691

Radtke KM, Ruf M, Gunter HM et al (2011) Transgenerational impact of intimate partner violence on methylation in the promoter of the glucocorticoid receptor. Transl Psychiatry 1:e21

Radtke KM, Schauer M, Gunter HM et al (2015) Epigenetic modifications of the glucocorticoid receptor gene are associated with the vulnerability to psychopathology in childhood maltreatment. Transl Psychiatry 5:e571

Reik W, Maher ER, Morrison PJ, Harding AE, Simpson SA (1993) Age at onset in Huntington's disease and methylation at D4S95. J Med Genet 30(3):185–188

Rey JM, Singh M, Hung SF et al (1997) A global scale to measure the quality of the family environment. Arch Gen Psychiatry 54(9):817–822

Robertson KD (2005) DNA methylation and human disease. Nat Rev Genet 6(8):597–610

Rogers EJ (2008) Has enhanced folate status during pregnancy altered natural selection and possibly Autism prevalence? A closer look at a possible link. Med Hypotheses 71(3):406–410

Roth TL, Sweatt JD (2011a) Annual research review: epigenetic mechanisms and environmental shaping of the brain during sensitive periods of development. J Child Psychol Psychiatry 52(4):398–408
Roth TL, Sweatt JD (2011b) Epigenetic marking of the BDNF gene by early-life adverse experiences. Horm Behav 59(3):315–320
Roth TL, Lubin FD, Funk AJ, Sweatt JD (2009) Lasting epigenetic influence of early-life adversity on the BDNF gene. Biol Psychiatry 65(9):760–769
Rubin LH, Connelly JJ, Reilly JL et al (2016) Sex and diagnosis specific associations between DNA methylation of the oxytocin receptor gene with emotion processing and temporal-limbic and prefrontal brain volumes in psychotic disorders. Biol Psychiatry 1(2):141–151
Rudenko A, Dawlaty MM, Seo J et al (2013) Tet1 is critical for neuronal activity-regulated gene expression and memory extinction. Neuron 79(6):1109–1122
Sanchez-Mut JV, Heyn H, Vidal E et al (2016) Human DNA methylomes of neurodegenerative diseases show common epigenomic patterns. Transl Psychiatry 6:e718
Santoro MR, Bray SM, Warren ST (2012) Molecular mechanisms of fragile X syndrome: a twenty-year perspective. Annu Rev Pathol 7:219–245
Schmitt FA, Davis DG, Wekstein DR, Smith CD, Ashford JW, Markesbery WR (2000) "Preclinical" AD revisited: neuropathology of cognitively normal older adults. Neurology 55(3):370–376
Seo D, Patrick CJ, Kennealy PJ (2008) Role of serotonin and dopamine system interactions in the neurobiology of impulsive aggression and its comorbidity with other clinical disorders. Aggress Violent Behav 13(5):383–395
Sharma RP, Gavin DP, Grayson DR (2010) CpG methylation in neurons: message, memory, or mask? Neuropsychopharmacology 35(10):2009–2020
Sharp AJ, Stathaki E, Migliavacca E et al (2011) DNA methylation profiles of human active and inactive X chromosomes. Genome Res 21(10):1592–1600
Song CX, Szulwach KE, Fu Y et al (2011) Selective chemical labeling reveals the genome-wide distribution of 5-hydroxymethylcytosine. Nat Biotechnol 29(1):68–72
Speilberger CDGR, Lushene R, Vagg PR, Jacobs GA (1983) Manual for the state-trait anxiety inventory. Consulting Pyschologists Press, Palo Alto
Stangier U, Heidenreich T, Berardi A, Golbs U, Hoyer J (1999) Assessment of social phobia by the Social Interaction Anxiety Scale and the Social Phobia Scale. Z Klin Psychol-Forsc 28(1):28–36
Steiger H, Labonte B, Groleau P, Turecki G, Israel M (2013) Methylation of the glucocorticoid receptor gene promoter in bulimic women: associations with borderline personality disorder, suicidality, and exposure to childhood abuse. Int J Eat Disord 46(3):246–255
Stein MB, Stein DJ (2008) Social anxiety disorder. Lancet 371(9618):1115–1125
Sun W, Zang L, Shu Q, Li X (2014) From development to diseases: the role of 5hmC in brain. Genomics 104(5):347–351
Suzuki MM, Bird A (2008) DNA methylation landscapes: provocative insights from epigenomics. Nat Rev Genet 9(6):465–476
Swedo SE, Rapoport JL, Leonard H, Lenane M, Cheslow D (1989) Obsessive-compulsive disorder in children and adolescents. Clinical phenomenology of 70 consecutive cases. Arch Gen Psychiatry 46(4):335–341
Szulwach KE, Li X, Li Y et al (2011) 5-hmC-mediated epigenetic dynamics during postnatal neurodevelopment and aging. Nat Neurosci 14(12):1607–1616
Szyf M (2015) Epigenetics, a key for unlocking complex CNS disorders? Therapeutic implications. Eur Neuropsychopharmacol 25(5):682–702
Tahiliani M, Koh KP, Shen Y et al (2009) Conversion of 5-methylcytosine to 5-hydroxymethylcytosine in mammalian DNA by MLL partner TET1. Science 324(5929):930–935
Tammen SA, Friso S, Choi SW (2013) Epigenetics: the link between nature and nurture. Mol Asp Med 34(4):753–764
Toyokawa S, Uddin M, Koenen KC, Galea S (2012) How does the social environment 'get into the mind'? Epigenetics at the intersection of social and psychiatric epidemiology. Soc Sci Med 74(1):67–74

Van Den Hove DL, Jakob SB, Schraut KG et al (2011) Differential effects of prenatal stress in 5-Htt deficient mice: towards molecular mechanisms of gene x environment interactions. PLoS One 6(8):e22715

Villar-Menendez I, Blanch M, Tyebji S et al (2013) Increased 5-methylcytosine and decreased 5-hydroxymethylcytosine levels are associated with reduced striatal A2AR levels in Huntington's disease. NeuroMolecular Med 15(2):295–309

Walker FO (2007) Huntington's disease. Lancet 369(9557):218–228

Wang D, Szyf M, Benkelfat C et al (2012a) Peripheral SLC6A4 DNA methylation is associated with in vivo measures of human brain serotonin synthesis and childhood physical aggression. PLoS One 7(6):e39501

Wang T, Pan Q, Lin L et al (2012b) Genome-wide DNA hydroxymethylation changes are associated with neurodevelopmental genes in the developing human cerebellum. Hum Mol Genet 21(26):5500–5510

Wang F, Yang Y, Lin X et al (2013) Genome-wide loss of 5-hmC is a novel epigenetic feature of Huntington's disease. Hum Mol Genet 22(18):3641–3653

Wang J, Tang J, Lai M, Zhang H (2014) 5-Hydroxymethylcytosine and disease. Mutat Res Rev Mutat Res 762:167–175

Watson CT, Roussos P, Garg P et al (2016) Genome-wide DNA methylation profiling in the superior temporal gyrus reveals epigenetic signatures associated with Alzheimer's disease. Genome Med 8(1):5

Weaver IC, Cervoni N, Champagne FA et al (2004) Epigenetic programming by maternal behavior. Nat Neurosci 7(8):847–854

Webster MJ, Knable MB, O'Grady J, Orthmann J, Weickert CS (2002) Regional specificity of brain glucocorticoid receptor mRNA alterations in subjects with schizophrenia and mood disorders. *Mol Psychiatry* 7(9):985–994, 924

Weder N, Zhang H, Jensen K et al (2014) Child abuse, depression, and methylation in genes involved with stress, neural plasticity, and brain circuitry. *J Am Acad Child Adolesc Psychiatry* 53(4):417–424 e415

Winslow JT, Insel TR (2002) The social deficits of the oxytocin knockout mouse. Neuropeptides 36(2–3):221–229

Wooten GF, Currie LJ, Bovbjerg VE, Lee JK, Patrie J (2004) Are men at greater risk for Parkinson's disease than women? J Neurol Neurosurg Psychiatry 75(4):637–639

Wu H, Zhang Y (2011) Mechanisms and functions of Tet protein-mediated 5-methylcytosine oxidation. Genes Dev 25(23):2436–2452

Yao B, Lin L, Street RC et al (2014) Genome-wide alteration of 5-hydroxymethylcytosine in a mouse model of fragile X-associated tremor/ataxia syndrome. Hum Mol Genet 23 (4):1095–1107

Young LJ, Wang Z (2004) The neurobiology of pair bonding. Nat Neurosci 7(10):1048–1054

Yui D, Nishida Y, Nishina T et al (2015) Enhanced phospholipase A2 group 3 expression by oxidative stress decreases the insulin-degrading enzyme. PLoS One 10(12):e0143518

Zampieri M, Ciccarone F, Calabrese R, Franceschi C, Burkle A, Caiafa P (2015) Reconfiguration of DNA methylation in aging. Mech Ageing Dev 151:60–70

Zhao J, Goldberg J, Bremner JD, Vaccarino V (2013) Association between promoter methylation of serotonin transporter gene and depressive symptoms: a monozygotic twin study. Psychosom Med 75(6):523–529

Zhubi A, Chen Y, Dong E, Cook EH, Guidotti A, Grayson DR (2014) Increased binding of MeCP2 to the GAD1 and RELN promoters may be mediated by an enrichment of 5-hmC in autism spectrum disorder (ASD) cerebellum. Transl Psychiatry 4:e349

Ziegler C, Dannlowski U, Brauer D et al (2015) Oxytocin receptor gene methylation: converging multilevel evidence for a role in social anxiety. Neuropsychopharmacology 40(6):1528–1538

Social Environment and Epigenetics

Sarah M. Merrill, Nicole Gladish, and Michael S. Kobor

Contents

Abstract Our social environment, from the microscopic to the macro-social, affects us for the entirety of our lives. One integral line of research to examine how interpersonal and societal environments can get "under the skin" is through the lens of epigenetics. Epigenetic mechanisms are adaptations made to our genome in response to our environment which include tags placed on and removed from the DNA itself to how our DNA is packaged, affecting how our genes are read, transcribed, and interact. These tags are affected by social environments and can persist over time; this may aid us in responding to experiences and exposures, both the enriched and the disadvantageous. From memory formation to immune function,

S. M. Merrill and N. Gladish
Centre for Molecular Medicine and Therapeutics, British Columbia Children's Hospital, Vancouver, BC, Canada

Department of Medical Genetics, University of British Columbia, Vancouver, BC, Canada

M. S. Kobor (✉)
Centre for Molecular Medicine and Therapeutics, British Columbia Children's Hospital, Vancouver, BC, Canada

Department of Medical Genetics, University of British Columbia, Vancouver, BC, Canada

Human Early Learning Partnership, University of British Columbia, Vancouver, BC, Canada
e-mail: msk@bcchr.ca

© Springer Nature Switzerland AG 2019

Curr Topics Behav Neurosci (2019) 42: 83–126
DOI 10.1007/7854_2019_114
Published Online: 5 September 2019

the experience-dependent plasticity of epigenetic modifications to micro- and macro-social environments may contribute to the process of learning from comfort, pain, and stress to better survive in whatever circumstances life has in store.

Keywords Adversity · Built environment · DNA methylation · Enriched environment · Epigenetics · Histone modifications · Interpersonal · Learning · Pain · Parenting · Social environment · Stress

1 Introduction

To be human is to have a fundamental need for love and belonging. Humans give birth to altricial young who depend on the effectiveness and attentiveness of social bonds from the moment of delivery, not only to thrive, but also to survive. Though this is a unique trait among mammals, it also means humans are exposed to their social environment much earlier in development than most offspring in the animal kingdom. Children's social relationships are arguably the most fundamental component of the early postnatal environment and facilitate both the beneficial and harmful effects of ecological factors. Even meeting necessary nutritional needs through breastfeeding is a social bonding experience with close proximity, ventro-ventral contact, body warmth, and soft touch. As the preeminent developmental psychologist Urie Bronfenbrenner stated: "No society can long sustain itself unless its members have learned the sensitivities, motivations, and skills involved in assisting and caring for other human beings" (Bronfenbrenner 2009).

All children develop within a dynamic social context of both interpersonal relationships and wider social structures, which can shape their cognitive, emotional, and biological processes for the remainder of their lives (Casper 2001). From the first moments with parents to city planning to feeling accepted by the community, the multifaceted nature of the social environment provides ample opportunity for both advantages and hindrances to be embedded "under the skin" (Boyce and Kobor 2015).

Though the debate between solitary contributions of nature and nurture is an academic artifact, there are a variety of overlapping conceptualizations of this embedding process. For example, the concept of vulnerability and resilience, or resistance, is commonly used in psychology and psychiatry (Ingram and Luxton 2005; Luthar 2003), as well as in engineering and ecology (Miller et al. 2010; Proag 2014). These terms have similar meanings even in these disparate fields with vulnerability referring to the underlying potential of an adverse reaction to a negative event, and resilience referring to the innate ability to withstand and recover from a negative event. This is the same principle as an individual genetic predisposition for a particular outcome, either beneficial or detrimental. Similarly, the psychological diathesis-stress model, or dual-risk model, presumes a predisposition that affects the likelihood of developing, or the trajectory of, a disorder (Ingram and Luxton 2005). This diathesis may be genetic, biological, environmental, or psychological and interacts with some form of biological or environmental stress to alleviate or exacerbate the effect. This also resembles the threshold model of neurobiological reactivity (Moore and Depue 2016). This model proposes that there is a biological constraint on the potency of external stimuli necessary to elicit an emotional response. This model presupposes that there is a predisposition to be stressed by an environment in the first place, in addition to the possibility that the subsequently elicited stress response may exacerbate yet another outcome. This cascade of interactions between innate differences and environmental exposures is addressed in the developmental origins of health and disease (DoHaD) hypothesis (Mandy and Nyirenda 2018; Suzuki 2018). This hypothesis refers to the potential biological programming from environmental exposures that may cause some of the intrinsic predispositions on which later environments may act (Wadhwa et al. 2009).

The most general and often used term for these relationships is gene-by-environment interactions. Ultimately, the distillation of these models equates to a basic understanding that the scaffolding of experience can be secured upon existing biological foundations. With poor foundations or inferior craftsmanship, the overall integrity of the structure may fail. One important aspect of how these pieces come together to build the human experience is through epigenetics (Meaney 2010).

2 Epigenetic Modifications

Epigenetics refers to the varied modifications to the underlying, permanent deoxyribose nucleic acid (DNA) sequence, which subsequently alter gene expression and, ultimately, phenotypes such as health and behaviors. The fundamental purpose of these epigenetic alterations is to achieve a diverse landscape of expression from a single DNA source (Boyce and Kobor 2015). These changes are likely involved in the biological embedding of environmental influence because of both their dynamic nature and sensitivity to experiential feedback. The field of epigenetics is a natural accretion of biological reductionism as it provides evidence that while we may be a product of our biology, our biology is partially a product of our environment.

The most common forms of epigenetic investigations in humans are correlational studies on DNA methylation (mC) and DNA hydroxymethylation (hmC), as discussed elsewhere in this volume. Due to its critical role in cell type differentiation, mC is highly affected by cell type differences and, thus, tissue types, as well as age, ethnicity, genotype, sex, and disease state (Edgar et al. 2017; Hannon et al. 2015; Husquin et al. 2018; Islam et al. 2019; Lienert et al. 2011; Turinsky et al. 2019; Wagner et al. 2015; Yousefi et al. 2015; Ziller et al. 2013). Three of these factors are especially important to highlight: genotype, tissue, and age. When considering genotype, de novo modifications may be affected by the underlying DNA sequence due to potential sites of interaction, a downstream consequence of other modifications, or the result of adaptation to environmental influences that can be passed on through cell mitosis to have long-lasting effects (Bock et al. 2006; Probst et al. 2009; Song et al. 2017). When considering tissue, the two most important aspects are differences in cell type and cell type proportions among the different tissues and the comparisons that can be made among and between different tissue types. Specifically, it is difficult to make inferences about epigenetic modifications in brain tissue when measuring more peripheral tissues such as blood or saliva; however, there are resources correlating some measures in these tissues that can assist in forming educated inferences (Braun et al. 2019; Edgar et al. 2017).

Finally, when considering age, both its potential as a confounder and variable of interest should be acknowledged. Patterns of epigenetic modifications change with age, including mC (Gopalan et al. 2017; Jones et al. 2015; Koch and Wanger 2011; McEwen et al. 2016, 2017; Sen et al. 2016). Therefore, it is crucial that age be accounted for, selected for, or counterbalanced across groups in the variable of interest. A subsequent benefit of these clear and replicable differences in ages, however, is the ability to predict age using DNA methylation "clocks" that can be both tissue specific and pan-tissue (Horvath et al. 2016; Horvath and Raj 2018; Liu et al. 2019; Marioni et al. 2015; Wagner 2017). By calculating age using DNA methylation, it is also possible to determine the acceleration or deceleration of a person's epigenetic age from their chronological age. Generally, when an adult is predicted to be older than their chronological age (i.e., epigenetic age acceleration), this suggests increased cellular aging and is associated with increases in morbidity and mortality (Fransquet et al. 2019; Horvath et al. 2015; McEwen et al. 2016).

While the field is acutely aware of the importance of accounting for these potential confounders, many early foundational studies in social epigenetics did not account for these differences and are primarily correlational, observational designs. Additionally, due to the difficulty, cost, and technological limitations in conducting epigenome-wide association studies (EWAS), much of the initial work focused on a candidate gene approach. The previous literature on mC in candidate genes should not be disregarded; however, new appreciation of the interconnectedness among mC indicates the benefit of EWAS due to accounting for differences among many correlated sites or regions simultaneously (Moore 2017). While there is crucial work to be done in understanding how social environments affect the underlying biology of developmental sequelae, a healthy dose of skepticism and a critical eye must be maintained both when evaluating past and current literature, as well as developing new experimental designs (Jones et al. 2018). There are also other possible DNA modifications that can affect gene expression that have been significantly less studied than mC. Although not much is known about their relationships with early social environments, additional cytosine modifications formylation and carboxylation, as well as the much rarer methylation at adenine sites on DNA, are ripe for future investigation (Wu et al. 2016b; Yao et al. 2017). Besides modifications directly onto DNA base pairs, there are other ways to affect the complex relationship between DNA structure and protein synthesis.

One such way is to affect the packing of DNA through modifications to chromatin, the condensed DNA-protein package that allows a structure as large as DNA to reside within the nucleus of a cell. Within the chromatin package are an octamer of proteins called histones. This package contains two each of four types of histones: H2A, H2B, H3, and H4, the tails of which have at least 14 possible modifications known to date (Bártová et al. 2008; Huang et al. 2014; Kouzarides 2007). By modifying chromatin structure, the accessibility of the DNA for gene transcription is significantly affected. This is a metric referred to as chromatin accessibility (Buenrostro et al. 2015, 2016; John et al. 2011). One modification that usually increases chromatin accessibility, thus increasing the ability of transcription factors to bind to the DNA, is histone acetylation (Görisch et al. 2005). This is when acetyl groups are deposited on lysines by histone acetyltransferaces (HAT). Alternatively, these groups that typically act to release compacted DNA to facilitate transcription initiation can be removed by histone deacetylases (HDAC), thus reducing chromatin accessibility (Chen and Townes 2000; Shahbazian and Grunstein 2007). There is significant evidence of histone acetylation and deacetylation having social behavioral effects in animal studies (Bukhari et al. 2017; Fitzsimons and Scott 2011; Hunter et al. 2012; Malik et al. 2014; Peixoto and Abel 2012; Saul et al. 2017; Shpigler et al. 2017).

Another possible histone modification is phosphorylation, which is often associated with acetylation, and functions primarily in the histone microenvironment by serving as a platform for communication between other histone modifications and downstream effects (Banerjee and Chakravarti 2011). Phosphate groups are customarily deposited by nuclear kinases and removed by protein phosphstases (Brami-Cherrier et al. 2009; Koshibu et al. 2009). Similar to DNA, histones can also be modified to be methylated, which, like DNA, can lead to both increased and decreased chromatin accessibility depending on the modified site (Kouzarides 2007). Methyl groups are deposited on histones by histone methyltransferases (HMT) and removed by histone demethylases (HDM) (Bártová et al. 2008; Zheng et al. 2015). In addition to their interactions with one another, these modifications are affected by the principle DNA structure, histone chaperones, age, and histone protein variants such as the H2A histone variant H2A.Z (Ausió and Abbott 2002; Bryois et al. 2018; Levine et al. 2012; Mcvicker et al. 2013; Stefanelli et al. 2018; Tessarz and Kouzarides 2014).

Once the DNA has been made available to transcription factors through variations of, and modifications to, DNA, chromatin, and histones, there is yet another epigenetic mechanism which takes place on ribonucleic acid (RNA). Messenger RNA (mRNA) acting as an intermediary between the underlying DNA structure and the protein synthesizing ribosomes, just like histones and DNA, can be modified by the addition or subtraction of hydroxymethyl, acetyl, phosphate, and the most frequently studied, methyl adenine (m^6A) groups (Boccaletto et al. 2017; Meyer and Jaffrey 2014; Saletore et al. 2012). These modifications, sometimes referred to as epitranscriptomics, can affect the structure, binding, transcription, and translational properties of RNA, which in turn has significant implications for downstream protein synthesis (Schwartz 2016). There are also two other families of RNAs that can have significant effects on gene transcription without affecting the underlying DNA sequence: microRNAs (miRNAs) and long non-coding RNAs (lncRNAs). Both miRNAs, which act as complimentary sequences to degrade mRNA and lncRNA, which regulate mRNA, generally reduce transcription and are affected by the underlying DNA sequence (Chu et al. 2011; Kim 2005; Kim and Nam 2006; Lee

2012; Mercer et al. 2009; Younger and Corey 2011). The direct relationship of the social environment on miRNAs and lncRNAs is currently unknown.

This diverse wealth of epigenetic mechanisms does not operate in a vacuum but functions together to have a cascade of comprehensive and varied influences. These modifications can be preserved for a lifetime through biological machinery that accurately maintains the epigenetic pattern (Probst et al. 2009), thus providing the opportunity for dynamic early environmental epigenetic adaption to be maintained throughout a lifetime. Therefore, there are many known epigenetic mechanisms that would allow for the possibility of early life social environments to stably modify phenotypes over the life course without altering the genotype. In this chapter, we discuss the scope of varied epigenetic investigations into the biological consequences of both interpersonal and structural social environments. To begin, we focus on the intersection of social behavioral neurogenomics: learning.

3 Epigenetics and Learning

The social environment is, fundamentally, an inescapable impetus for development and adaption through learning and memory, both positive and negative. Infants are innately attuned to social cues, even before birth. Infants have a strong preference for eyes, faces, and facial configurations (Dupierrix et al. 2014; Goren et al. 1975; Johnson et al. 1991, 2015; Turati et al. 2002). Even third trimester human fetuses preferentially orient toward a face shape when projected through maternal tissue (Reid et al. 2017). This strong preference can be interpreted as an indication of the importance of early exposure to faces and an early orientation toward social stimuli (Morton and Johnson 1991). In addition to the infant's predilection for faces, young infants develop in an especially face-dense environment, even in comparison to older infants (Jayaraman et al. 2017). As young infants have a low range of mobility, this demonstrates an implicit adult drive to place faces in front of developing infants as well. Infants also use touch to communicate with their caregivers and use a variety of movements depending on their needs and the caregiver's responsiveness (Moszkowski and Stack 2007). Additionally, evidence indicates maternal odors enhance neural signaling of facial categorization, but not general activation, in infant brains (Leleu et al. 2019).

Babies' relationship with the social environment is a multi-sensory process. In addition to sight, smell, and touch, infant's hearing is also highly attuned toward social environments and language development. Adults tend to use infant-directed speech (IDS) when with children, which incorporates a larger range of higher frequencies and has more rising contours than adult-directed speech (ADS). Infants are more attentive to, can discriminate sounds better in, and can be emotionally regulated by IDS as opposed to adult-directed speech (Cooper and Aslin 1990, 1994; Fernald 1992; Trehub et al. 1993). Even mothers of deaf children use exaggerated signs to their infants, who are more attentive to the exaggerated infant-directed signing (Masataka 1998). An infant's world is full of social stimuli to which they can experience, react, and learn.

This rich social environment is used by infants to both survive, via their caregivers, and begin to build their understanding of the world. Therefore, it logically follows that children's cognitive functional development, both normative and deviant, adapts to the social environment (Dishion 2016). Children show a novelty preference almost immediately and are able to respond to traditional behavioral conditioning paradigms (Hulsebus 1974; Thompson et al. 1991). Infants as young as 6 months can reliably show social learning through deferred imitation (Barr et al. 1996; Meltzoff 1988). Additionally, experiential learning generally is essential for normative development both pre- and postnatally (Bale et al. 2010; Mclaughlin et al. 2014; Perry 2002; Roth and Sweatt 2011a, b; Swain et al. 2007; Talge et al. 2007). Learning from and adapting to the early life environment as effectively as possible is crucial for both juvenile and adult survival. It is both demonstrable and logical that children would be learning from their environments, and it is clear that there is an emphasis on the social environment for those associations.

Some connections, such as those learned or imitated in an interpersonal social context, are more readily learned than others. This is called biological preparedness theory, and the belief is that there is an evolutionary advantage to more quickly making these environmentally and survival-relevant associations (Cummins and Cummins 2015; McNally 2016). It is also much more difficult to extinguish these connections once learned (Åhs et al. 2018). Most work has focused on preparedness role in phobias, anxiety disorders (including social anxiety), and taste aversion (de Silva 1988; Ohman and Mineka 2001). Due to the importance of social bonds to human survival, especially for young children, and the extreme innate infant preference for social stimuli, the assumption of an underlying biological preparedness to learn cues from the social environment is reasonable. Albert Bandura, a distinguished behaviorist, acknowledged the unique aspects of social environments. He adapted the behavioral conditioning theory to include the observational process of learning and noted that there are mediating processes between the experience of a stimuli and subsequent response (Bandura 1977). The cognitive ability of children to learn from, and disproportionate attunement toward, their early social environments indicate that these social factors are primed to make lasting neurological, epigenetic, and behavioral impressions.

In order to leave these lasting impressions, first the associations from the social environment need to be learned. The basis for most models of development, learning, and memory in neuroscience is synaptic plasticity, or the ability of synaptic connections to change their strength (Ehrlich and Malinow 2004). Hebbian theory, or cell assembly theory, posits that two mechanisms of synaptic plasticity are long-term potentiation (LTP) and long-term depression (LTD) (Ehrlich and Malinow 2004; Robert C Malenka and Nicoll 1999; Randic et al. 1993). Electrophysiological action potentials are primarily responsible for, or supportive of, cell-to-cell communication through the synapse in the nervous system through rapid cell depolarization that is passed along to adjacent cells.

LTP is one cellular mechanism through which the strength of a synapse can be amplified, while LTD is a mechanism for weakening a synapse. LTP primarily occurs through the action of presynaptic glutamate at two types of postsynaptic ionotropic glutamatergic receptors, N-methyl-D-aspartate (NMDA) receptors and α-amino-3-hydroxy-5-methyl-4-isoxazolepropionic acid (AMPA) receptors (Malenka and Nicoll 1999). Though glutamate binds to both NMDA and AMPA, NMDA receptors are gated by a magnesium ion block that prevents calcium ions, to which NMDA receptors are permeable, from entering the postsynaptic neuron. However, AMPA receptors do not have a magnesium block and are activated by presynaptic glutamate release, allowing sodium ions to flow through. When enough AMPA receptors have been activated and sodium ions have entered the postsynaptic neuron, the charge, or potential, of the neuron changes, releasing the magnesium ion block from the NMDA receptors. The NMDA receptors then allow calcium ions into the cell, which target calcium/calmodulin-regulated protein kinase II (CaMKII) (Malenka et al. 1989). CaMKII phosphorylates AMPA receptors, increasing their current, a mechanism of early LTP, and initiates protein synthesis through the MAPK (mitogen-activated protein kinase) pathway and CREB/CRE (cAMP-responsive element binding protein/cAMP response element) mechanisms to begin structural changes in and around the synapse (Giese et al. 1998).

Early LTP results in the addition of more AMPA receptors postsynaptically and a retrograde signal of nitric oxide to the presynaptic neuron, while late LTP adds an entirely new synapse between the two neurons (Sandkühler and Gruber-Schoffnegger 2012). BDNF (Brain-derived neurotrophic factor) is crucial in this process, with reduction leading to insufficient drive for synthesis of synaptic proteins, thus contributing to cognitive dysfunction (Wu et al. 2016a). Both adding more AMPA receptors, thus depolarizing and triggering NMDA receptor activation in the postsynaptic neuron, or adding a new synaptic cleft, increases the strength of the connection between the two neurons. This strengthened connection is sensitized to be activated again and the presynaptic cell is poised to interact with the postsynaptic neuron more quickly and strongly than before.

LTD, on the other hand, takes place when there is consistent low frequency activation. With constant low activation, only some, but not all, AMPA receptors are activated, which is not enough to remove the magnesium ion block from the majority of NMDA receptors. In this case, calcineurin (protein phosphatase 2B) is the target, which results in the removal of AMPA receptors by endocytosis (Malenka and Bear 2004). Though LTD is the weakening of synaptic strength, it plays an integral role in memory formation, most likely through preparing potential pathways for new connections. Additionally, spike-timing dependent potentiation (STDP) adds even more specificity to this relationship by altering the strength of LTP and LTD effects depending on the timing of electrical signals (Fiete et al. 2010). The strengthened relationships among neurons happen within milliseconds and can last from 30 min to years. LTP in multiple synapses can create engrams, or biophysical manifestations of memories, stored as cognitive units of interconnected cells (Poo et al. 2016). The maintenance of these engram connections composes long-term memory. How are these changes generated in the first place? How are they maintained over multiple cell turnovers and conceivably for an entire lifetime? One possibility is through epigenetic mechanisms that alter gene transcription to allow for stable, structural modifications to the synapse.

Contributing to the neuroarchitectural changes associated with learning, brief and distinct changes to neural gene expression are observed, termed genomic action potentials (gAP) (Clayton 2000; Clayton et al. 2019). The pathways activated by calcium interacting with CaMKII and calcineurin lead to increased gene transcription of elements necessary for cytoskeletal changes, termed the immediate early gene (IEG) response (Clayton et al. 2019). This dynamic change results in epigenetic modifications at the transcriptional and chromatin packaging levels (see Clayton et al. 2019 for review). One example is *ARC*, an IEG, which is involved in the endocytosis of AMPA receptors seen following calcineurin activation in LTD (Chowdhury et al. 2006; Rial Verde et al. 2006). Another example of IEG action is *FOS*, which interacts with histone methylation to prime histone acetylation based on neuronal activity. These common IEGs are poised to have quick transcriptional changes in multiple domains such as histone lysine modifications and mRNA interactions (D'Urso and Brickner 2014; D'Urso et al. 2016; Rye et al. 2014). It is also reasonable to hypothesize that, as there are differences in synaptic plasticity depending on the types and locations of neurons, as well as temporal differences such as early vs. late LTP or STDP, this is also true for the IEG associated with these changes (Tyssowski et al. 2018). Additionally, like the pattern of synaptic connections, the gAP differs when encoding different learning experiences (Mukherjee et al. 2018).

Histone and mC modifications, in particular, have been associated with synaptic plasticity and learning outcomes (Blaze and Roth 2013; Chwang et al. 2006; D'Urso and Brickner 2014; Dias et al. 2015; Feng et al. 2010; Kim and Kaang 2017; Levenson and Sweatt 2011; Peixoto and Abel 2012) as well as affording significant contribution to differences in the neural architecture of neonatal socioemotional learning (Ong et al. 2019). Consistently, pharmacological inhibition of mC modifications significantly impairs memory, and many EWAS identify mC modifications in genes involved in neurotransmitter pathways and learning (Day and Sweatt 2011; Mill et al. 2008).

Additionally, increases in histone acetylation and alterations in histone methylation are associated with memory formation, synaptic plasticity enhancement, and increased gene expression (Guan et al. 2002; Pang et al. 2019; Schaefer et al. 2009). For example, oral administration of the HDAC inhibitor suberoylanilide hydroxamic acid (SAHA) restored spatial memory and reduced inflammation in an aging animal model (Benito et al. 2015). There is also evidence that histone lysine methyltransferase complex G9a/GLP facilitates LTD maintenance in the hippocampus and inhibiting HATs significantly impairs learning potential (Oliveira et al. 2007; Sharma and Sajikumar 2018). A specific example of the role of histone methylation is the reduction of H3 histone dimethyl occupancy in the promoter region of a gene associated with long-term memory formation, *HOMER1A*, and an increase in its transcription in the amygdala during auditory fear conditioning (Mahan et al. 2012). Conversely, an increase in HDAC activity results in reduced synaptic plasticity and memory impairment, while pharmacological inhibition of HDAC improves memory formation (Guan et al. 2009; Levenson et al. 2004). Also, histone acetylation in the hippocampus is a consequence of contextual fear conditioning in rodent models (Levenson et al. 2004).

Another aspect of biological memory beyond encoding and maintenance of explicit and implicit learning is homeostatic conditioning (Clayton et al. 2019). Homeostasis is a summary mechanism maintaining a steady biological equilibrium to best preserve an optimal internal environment (Martin 2008). Depending on the changes to internal and external states, it is sometimes necessary to rebalance these homeostatic systems to reach a new equilibrium. In this case, a synapse changes because of homeostatic mechanisms. A common and illustrative neuronal example of homeostatic compensation is gaining tolerance to opiates (Koob 1996). By activating mu-opiate receptors in the brain at a constant, elevated level, these receptors become internalized and, eventually, broken down, while receptors responsible for opposite reactions, like norepinephrine receptors, are upregulated through increased transcription (Finn and Whistler 2001). This new receptor balance is a reaction to the ingestion of exogenous opiates and is responsible for both increasing tolerance to and withdrawal symptoms from opiates. Similar to Hebbian learning, these fluctuations could not occur without dynamic and maintained changes in gene transcription. Therefore, Clayton et al. (2019) posit an additional function of the gAP which could be regulating homeostatic compensation in a similar fashion. Homeostatic mechanisms working through the gAP at the network level may be more likely in response to large variations that activate homeostatic receptors such as changes in developmental stage, drug ingestion, or significant environmental experiences (Hrvatin et al. 2018; Miyashita et al. 2009; Miyatake et al. 2005; Tyssowski et al. 2018).

One example in humans of genetic deficits in epigenetic modifications resulting in profound cognitive and social dysfunction is Rett Syndrome (Jiang et al. 2004). Rett Syndrome is an intellectual disability disorder that results from a mutation in the *MeCP2* gene, which checks and binds areas of DNA with methylation and ensures DNA is packaged appropriately (Amir et al. 1999). *MeCP2* also affects synaptic formation and hippocampal memory (Jiang et al. 2004; Li et al. 2012; Skene et al. 2010). Extensive histone acetylation increases are a consequence of this *MeCP2* mutation found in mice that are associated with increased stress, social withdrawal, and profound cognitive impairment (Shahbazian et al. 2002). Rett Syndrome is an example of how disrupted mC machinery affects the regulation of histone modifications, and thus gene expression and cognitive and social phenotypes. Another example of the manifest impact of histone modification variations in humans is the intellectual disability disorder Rubinstein-Taybi Syndrome (RTS) (Alarcón et al. 2004). RTS individuals have a mutation in their *CBP* gene, which transcribes for a protein that promotes histone acetylation and, thus, gene expression. Because of this mutation, RTS individuals and animal models of the disease present with profound cognitive impairments and substantial decreases in histone acetylation throughout the genome (Josselyn 2005; Kalkhoven et al. 2003; Korzus et al. 2004). In mouse models of RTS, the epigenetic action of *CBP* is disrupted specifically after separation of the pup from their mother (Wang et al. 2010). Taken together, these two mutations provide evidence for the weighty impact of epigenetic mechanisms in learning and the ensuing social consequences.

There is mounting evidence suggesting the importance of epigenetic modifications specifically affecting *BDNF* gene transcription, necessary for activity-dependent neuronal plasticity. Histone deacetylase 2 (HDAC2), methyl-cytosinephosphate-guanine–binding protein 2 (MeCP2), and DNA methyltransferase 1 (DNMT1) can substantially induce chromatin remodeling in the promoter regions of the *BDNF* gene, which can subsequently modulate hippocampal synaptogenesis and cognitive function (Chen et al. 1967; Kavalali et al. 2011). Transcriptionally, with neuronal activity there is also an increase in mRNA levels and consistent decrease in mC changes at *BDNF* in the CA1 hippocampal region after fear conditioning. Additionally, there is a reduction in the long noncoding antisense RNA that downregulates *BDNF* expression (Lipovich et al. 2012; Lubin et al. 2008). *BDNF* promotes local synaptic protein mRNA translation and plays a necessary role in neuronal development, synaptic plasticity, and learning (Minichiello 2009; Park and Poo 2013; Takei et al. 2004). Epigenetic differences affecting *BDNF* have been associated with early-life adverse experiences, early-life stress, neglectful parenting, depression, posttraumatic stress disorder, and may be a valid biomarker for some psychiatric diseases due to its unique role in their pathology (de Lima et al. 2011; Kang et al. 2013; Roth and Sweatt 2011a, b; Roth et al. 2011; Seo et al. 2016; Zheleznyakova et al. 2016).

Epigenetics acts as the synaptic connection between genes and the environment, even at the synapse itself (Boyce and Kobor 2015). While there is evidence that the prenatal environment affects epigenetic differences, the clear role of epigenetics in the encoding of long-term memories indicates the importance of differences in these mechanisms as a response to postnatal experiences as well. Thus, as expected, there is an abundance of examples in human and animal models of environmental exposures correlating with differences in epigenetic mechanisms, much like environmental exposures correlate with differences in neural architecture and behavior (Araki et al. 2016; Bedrosian et al. 2018; Boyce and Kobor 2015; Essex et al. 2013; Gassen et al. 2017; Klengel et al. 2013; Mcgowan et al. 2009, 2011; Turecki and Meaney 2016; Weaver et al. 2004). Even epigenetic clocks are sensitive to environmental differences throughout the lifespan (Jylhävä et al. 2019). Also, similar to neural development, early life social experiences can sometimes more strongly relate to epigenetic differences than adult experiences, possibly due to the critical period of development and learning occurring at that time (Boyce and Kobor 2015; Bush et al. 2018; Gassen et al. 2017; Klengel et al. 2013; Lam et al. 2012; Wagner et al. 2015). Using the theoretical framework that changes in neural architecture and genomic transcription are reactions to the environment, it naturally follows that these long-term adaptations would happen early in childhood development in order to increase the chance of survival in whatever positive or negative circumstances are present. Learning from the social environment is one mechanism of behavioral neurogenomics where changes to neuroanatomy, epigenetic modifications, and behavior work in concert to adapt to an external stimulus. There are multiple types of stimuli that can act as an impetus for this learning, such as social reward, pain, and stress. We will discuss each of these possible provocations in the relevant social environments.

4 Interpersonal Environments

4.1 Parenting and Enriched Environments

The first social environment a human infant is born into is the uniquely vital and interpersonal relationship with a caregiver. Human altricial young need constant care and affection, which they hopefully receive from their primary caregiver, often a parent. Having a positive, loving relationship in infancy with this person, who responds immediately and effectively to the child's needs, has lifelong effects (Zayas and Hazan 2015). Attachment theory, one of the most well-established social psychological theories, provides a foundation to understand both the impact and quality of early close relationships amid development.

In development, establishing successful relationships with adults and peers provides a foundation of capacities that children will use for a lifetime (Belsky and Cassidy 1994; Pietromonaco and Barrett 2000; Thompson 2000; Zayas and Hazan 2015). Thus, Bowlby, father of attachment theory, referred to these attachment patterns as being "from cradle to grave" (Bowlby 1979, p. 129), possibly similar to the epigenetic modifications to be established in infancy and lasting through mortality. This attachment theoretical foundation provides a comprehensive account of the ontogeny and developmental sequelae of infant caregiver bonds, as well as a framework for investigating how perturbations of this system may result in individual differences. Additionally, animal models of pair bonding and advances in fMRI technology have contributed to a rich literature delineating the neural systems that underlie attachment behaviors (see Beckes et al. 2015 for review). These manifestations of social bonds in terms of physiology, emotion, and behavior are assumed to reflect the functioning of mental representations.

A defining principle of attachment theory is that past relationships and interactions with the social environment are stored in memory (e.g., Bowlby 1979, 1982; Bretherton and Munholland 2008; Collins et al. 2004; Pietromonaco et al. 2000; Zayas et al. 2011). Mental representations consist of detailed memories of interactions with, and conscious and nonconscious affective evaluations of, attachment figures (Zayas and Shoda 2005, 2015), as well as strategies to regulate negative affect in stressful and threatening situations (Collins et al. 2004; Pietromonaco et al. 2006; Zayas et al. 2009). From a psychological perspective, mental representations are impactful because they implicitly affect perceptions and expectations about likely events and patterns (Baldwin et al. 1993; Günaydin et al. 2012; Zayas et al. 2009). From a neuroscience perspective, these mental representations are akin to the engrams that are formed from a collection of neuronal connections built by LTP and recalled to regulate affect in times of need. Finally, from an epigenetic perspective, these mental representations may reflect the adaptations to early caregiver relationships and social environments that ultimately result in differing phenotypes. In regard to the later, there is a wealth of literature exhibiting reported epigenetic differences correlated with early caregiver relationships.

In animal research, the majority of studies have used rat models from the first 10 days of life, which represents a sensitive period in rat development known to facilitate early learning and infant–caregiver attachment (Roth and Sweatt 2011a, b). To continue the understanding of social epigenetics as adaptation in memories and the capacity for learning, one study found that poor early maternal caregiving in mice led to difficulties in spatial cognitive tasks (Bredy et al. 2004). More recent work in mice also found that expression of glutamate receptors necessary for LTP and LTD such as NMDA and AMPA in the ventromedial prefrontal cortex, the cognitive component of the socioemotional circuit, decreased in socially isolated adolescent, but not adult, mice (Lander et al. 2017). It has also been found that young mice which are not isolated, but receive less sensitivity and care from their mothers have abnormal hippocampal and amygdala *BDNF* gene expression necessary for learning and synaptic plasticity (Macrì et al. 2010). Similarly, amounts of BDNF decreased in the prefrontal cortex and hippocampus of juvenile socially isolated rodents (e.g., Branchi et al. 2004; Branchi 2009; Chatterjee et al. 2007; Choy et al. 2008; Fumagalli et al. 2007; Lippmann et al. 2007; Nair et al. 2007; Zimmerberg et al. 2009). These findings speak simultaneously to both the effect of social relationships on learning and the sensitive developmental period in which the plasticity of this adaptation takes place. In general, these findings are illustrative of the wider, rich literature on the epigenetic effects on plasticity in the prefrontolimbic and hippocampal regions of offspring exposed to reduced, or absent, maternal care (Branchi et al. 2006; Matas et al. 2016).

Stress effects also relate to a poor versus positive early caregiver relationship. An often-cited paper on the topic is by Weaver et al. (2004), on the relationship between maternal licking and grooming with mC differences in the promoter region of the glucocorticoid receptor gene (*NR3C1*). While they found that more licking and grooming from maternal rats lead to hypomethylation in the promoter region of the *NR3C1* and patterns of methylation in a broader surrounding area (Mcgowan et al. 2011), this study has yet to be replicated and is met with some skepticism in the field (Jones et al. 2018). However, more recent work in mice has also found different biological reactions to maternal caregiving in the hippocampus, a region associated with both learning and reactions to stress, and the dorsal raphe nucleus (DRN), the brain center for serotonin production and distribution (Araki et al. 2016; Bedrosian et al. 2018). In the DRN specifically, Araki and colleagues found hypomethylation affecting the GABA(B) receptor, which is a common pharmacological target for depression and anxiety relief (Araki et al. 2016; Felice et al. 2016).

Similar positive relationships between parental nurturance and memory development have also been found in humans (Farah et al. 2008). Generally, increased positive, engaged social environments increase memory formation in both humans and nonhuman primate models (Farah et al. 2008; Kozorovitskiy et al. 2005). Epigenetically, human studies have primarily focused on the mC of a few genes of specific and special prominence in the research, namely *BDNF* (see Zheleznyakova et al. 2016 for review), *NR3C1* (see Turecki and Meaney 2016 for review), *SLC6A4* (see Moore and Kobor 2018 for review), and *OXTR* (see Maud et al. 2018 for review). All of the proteins these genes encode have many functions across learning and homeostatic change, especially in relation to neurodevelopment and stress. For example, the *BDNF* gene codes for brain-derived neurotrophic factor, the canonical neuronal growth factor in the brain widely involved in the formation of any neuroarchitectural changes. Additionally, *NR3C1* is the most widely researched gene in regards to fMRI and stress-causing environments, such as poor maternal care, in both the animal and human literature (Jones et al. 2018; Turecki and Meaney 2016). One recent study found that increased maternal responsiveness and touch were correlated with hypomethylation in *NR3C1* exon 1F in female children (Ostlund et al. 2016). This sex-dependent response has been replicated in other studies as well, such as Garg et al. (2018) finding the greatest DNA methylation differences among attachment styles in females. Additionally, they found that across the sexes, attachment behavior patterns were correlated to over 10% of global mC differences in children, suggesting large biological responses to the sensitivity and consistency of the parental care environment (Garg et al. 2018).

Another nuanced aspect of the caregiver relationship is soft touch, which is incredibly important in healthy, normative infant development (Barnett 2005). There were significant mC differences between children who received high amounts of soft touch and those who did not as infants. Additionally, infants who were more distressed, yet received lower amounts of touch, were epigenetically younger, possibly indicating a biological developmental delay (Moore et al. 2017). This is most likely due to the social buffering effects of both mental representations and human touch. For example, holding the hand of a stranger can reduce both the subjective experience of pain and its neural signature, but holding the hand of a close relationship partner reduces pain in these areas to an even greater extent (Coan et al. 2006).

Ultimately, the research indicates that having a healthy, supportive early social environment leads to positive epigenetic, neurological, and psychological outcomes. This is most clear when examining the literature on the benefits to enriched environments both in buffering stress and in rescuing memory formation. In humans, for example, having a supportive family environment during development protected against harmful cellular and epigenetic aging due to racial prejudice experiences. However, individuals without a supportive family environment did experience biological weathering (Brody et al. 2016). In animals, enriched environments including both positive caregivers as well as social play and peer interactions associated with global brain differences such as larger total cortical weight, especially the dorsal cortex, and larger ratio of cortex weight to overall brain weight (Rosenzweig and Renner 1986). Although less than some of the more negative aspects of development, there is a respectable amount of literature on the effects of this socially and cognitively enhanced environment in model animals. One such study found that, even with poor early caregiving, having an enriched environment rescued the gene expression of the NR2A and NR2B subunits of the NMDA receptor and AMPA receptor, both suggesting that poor early caregiving is associated in a reduction of learning potential through a reduction of NMDA receptor activity, and that an enriched environment with peer sociality and cognitive stimulation can combat these learning effects (Bredy et al. 2004).

This effect is most likely related to the demonstrated modulation of synaptogenesis by exposure to environmental stimuli (Eckert and Abraham 2010; Fischer et al. 2007; Ramírez-Rodríguez et al. 2014). Being housed in a communal nest as a juvenile, a form of enriched environment, is linked to increased sociocognitive functioning and increased expression of BDNF in rodents (Branchi 2009). In fact, in enriched environments with dynamic social play, overall memory function is improved in rats, and even exposure to chemically induced short-term memory impairment in neonatal rodents can at least be partially rescued through enriched environment alone with no other pharmacological interventions (Shen et al. 2013; Shih et al. 2012). The cognitive disruption of this chemical impairment has been shown to derive through epigenetic protein interactions including such proteins as histone deacetylase 2, methyl-cytosine-phosphate-guanine–binding protein 2, and DNA methyltransferase 1 in the *BDNF* promoter region inhibiting BDNF expression necessary for synaptogenesis during development, which are specifically attenuated by an enriched environmental intervention alone (Wu et al. 2016a). Even in aging mice, which had altered H3 histone acetylation and histone methylation in hippocampal tissue, intervention with an enriched environment improved long-term memory deficits by reversing histone methylation around the *BDNF* gene in rodent hippocampal tissue (Morse et al. 2015). This study also indicated that histone lysine methylation may be a necessary transcriptional mechanism by which environmental enrichment rescues memory formation replicating previous literature in young rodent populations (Kuzumaki et al. 2011). Another pathway through which enriched environment improves memory is by preventing epigenetic changes, especially mC and histone deacetylation, driving oxidative stress (Griñan-Ferré et al. 2016). These environments do not need to be lifelong to have strong effects either. Even a relatively brief exposure to an enriched environment including dynamic social stimulation in juvenile mice enhances LTP through a cAMP/p38 MAP kinase-dependent signaling cascade (Arai et al. 2009).

However, not all early relationships are positive experiences. For example, both maternal and paternal life stress during early life was correlated with adolescent differences in mC in humans (Essex et al. 2013). In rats, newborns exposed to a stress-abusive mother showed increased methylation in the promoter region, and decreased expression, of the *BDNF* gene (Huang and Reichardt 2001). This difference in BDNF concentrations for abused versus non-abused rats appears to persist through adulthood (Roth et al. 2009). In this same study, a different group of newborn rats was also exposed to positive caregiving mothers. Both the maltreatment and beneficial caregiving mothers initially caused an increase in *BDNF* mRNA levels in the hippocampus (Roth et al. 2009). Both experiences, regardless of valance, equitably guided the growth of new neuronal connections in the memory center of the brain. Adaptationally, negative relationships and social environments are just as powerful as positive social learning experiences.

4.2 Pain, Stress, and Trauma

In early care environments, the perception of safety is the most critical component (Porges 2011). Breaching this trust results in negative social experiences that can be both stressful and painful. There are many ways to experience pain, such as acutely, chronically, physically, and emotionally; in abusive early social environments, children are exposed to all four kinds of pain. Pain is such a powerful motivator for learning that is often used in animal models for fear conditioning, which is quick to establish and difficult to extinguish (Hermans et al. 2006). This particular form of learning has been found to require epigenetic modifications to take place and results in epigenetic differences (see Dias et al. 2015 for review). Thus, the epigenetic consequences of such early-life adversity are, undoubtedly, affected by learning and the physiological consequences of experiencing pain.

It is first important to establish that the literature indicates neural reactions to physical and socioemotional pain are exceptionally similar, specifically in regards to the affective processing of pain (Eisenberger and Lieberman 2004; Papini et al. 2015). However, though pain is processed in the same regions, socioemotional pain, such as social rejection or isolation, is more potent on a chronic timescale because it is much more easily relived and remembered than physical pain once the original source of pain has subsided (Meyer et al. 2015). Animal studies have also repeatedly found that any unpredicted reward devaluation, such as through a sudden or bewildering social rejection, triggers the brain circuits involved in pain and stress (Papini et al. 2015). Individuals who experience abuse, especially from a caregiver, may experience the physical pain of abuse, but most certainly experience the social pain of rejection and betrayal in that moment and for years later.

Pain is the hedonically aversive conscious experience of the nociception response to damage (Moseley and Butler 2015). When physical damage to peripheral tissues occurs, this initiates an immune cascade of inflammatory mediators (Benzon et al. 2011). For example, bradykinin produces inflammatory pain and hyperalgesia through activation of G-protein-coupled receptors. Additionally, cytokines such as tumor necrosis factor alpha and interleukin 6 are released to moderate the inflammatory process and promote pain signaling by sensitizing nociceptors (Benzon et al. 2011). While these inflammatory mediators directly cause pain as a signal of tissue damage, they also modify sensory neurons, amplifying pain signal during transmission to the spinal cord, additionally motivating the desire to minimize injury and remove the aversive stimulus. Through subsequent changes in reaction to this acute event, such as gene regulation, receptor expression, glial activation, and sensitization, this pain may be maintained to become chronic pain (Denk and McMahon 2012). This immune response alone may result in epigenetic modifications, especially as immune responses lead to different cell type compositions in the blood as well as differentiation in epigenetic markers through the adaptive immune response (Janeway 2001).

This immune response is an important aspect of the sensory discriminant pain pathway. The input from this pathway passes through the nerves, spine, and brain stem to the thalamus and insular cortex where the homeostatic relevance and intensity of the signal is discerned (Craig 2003). However, there is also a learning aspect of pain through the formation of a threat memory. This is referred to as the affective motivational pain pathway and requires calcitonin gene-related peptide (CGRP) activity in the amygdala (Han et al. 2015). In fact, CGRP activity in the parabrachial nucleus, a junction between the cerebellum and brain stem, is both necessary and sufficient for pain responses and fear conditioning due to its role in transmitting pain information to the central amygdala (Han et al. 2015). The central amygdala, in turn, sends input to the anterior cingulate cortex, which governs the level of unpleasantness derived from the signal and the quality of the motivational response (i.e., the more unpleasant, the greater the aversive motivation) (Craig 2003). This affective motivational pain pathway is necessary for learning to avoid noxious and harmful stimuli in the future, whether those stimuli are physical or social, while the discriminant response is necessary for acute treatment of the damage in the present.

Both the discriminate and affective pain pathways show clear associations with epigenetic modifications, though most investigations have focused on evaluating potential affective pain mitigation (Odell 2018). For example, one investigation of chronic pain detected 1,147 genes with differing RNA expression enriched for pathways involved in neuronal development and cell differentiation (Alvarado et al. 2015). HDAC levels were shown to be increased in the spinal cord and of critical importance to the induction and maintenance of inflammatory hyperalgesia (Bai et al. 2010). Studies have also been performed demonstrating the efficacy of HDAC inhibitors in improving stress-induced visceral hypersensitivity (Cao et al. 2016; Maloney et al. 2015). Interestingly, although administration of HDAC inhibitors reduces mechanical and thermal hypersensitivity by half, this is only true when HDAC administration occurs preemptively (Denk et al. 2013). Histone methylation also changes with pain, such as the increase in expression of proinflammatory cytokine monocyte chemotactic protein 3 in response to pain correlating with a decrease of histone lysine methylation in that protein's gene promoter region (Imai et al. 2013). Additionally, a reduction of miRNA in the dorsal root ganglia significantly decreased pain-related gene transcription and inflammation, though the affective motivational response to pain was unaffected (Zhao et al. 2010).

DNA methylation differences have also been associated with pain exposure. The promoter region for cystathionine-beta-synthetase, an enzyme in the nociceptive signaling pathway, was demethylated and the protein upregulated when experiencing pain (Qi et al. 2013). A rodent model for neuropathic pain showed that chronic painful neuropathy led to global changes in the degree of mC in the brain. About 6 months following peripheral nerve injury, decreases in global mC were found in the prefrontal cortex and amygdala (Tajerian et al. 2013). In another rodent neuropathic pain model, increased methylation of the mu-opioid receptor gene proximal promoter in the dorsal root ganglion, a key region in pain processing, was demonstrated (Zhou et al. 2014). There is also evidence for miRNA regulation of opioid tolerance in this pathway (He et al. 2010). Mu-opioid receptors, especially, are integral to the pain pathways due to their ability to reduce the affective motivational component of pain (Simons et al. 2014), even in the absence of any visceral tissue damage (Papini 2009).

Ultimately, pain biology is both a unique, intense, and specific immune response and learning experience. These epigenetic changes related to the experience and sensitization of the pain response, both through immune cascades and neural receptor modifications, should be considered as a possible consequence of socially and physically painful social environments. However, the benefit of pain's relationship to these systems is discovering potential protective environments as well. For example, in a rodent model, increased maternal licking and grooming associated with mC in the interleukin-10 gene correlate with an increase in expression in the nucleus accumbens reward center. This, in turn, increased interleukin-10 protein in nucleus accumbens glial cells and reduced mu-opioid receptor glial activation to exogenous opioids, which resulted in less drug abuse (Schwarz et al. 2011). The inextricable link between pain and learning allows other positive environments such as sensitive maternal care or an enriched social play environment to reduce pain and the necessity of the pain-mitigating pathway response.

The possibility of experiencing pain, because it is such a noxious experience, is highly motivating to mitigate or prevent injury whenever and wherever possible. This causes a combination of constant uncertainty and the need for hypervigilance – a recipe for chronic stress. Though intimately intertwined with the pain pathway, the stress pathway is rooted in the effects of uncertainty. Whereas fear and pain generally have specific and proximal stimuli that trigger these reactions, stress requires no such unique or immediately relevant stimulus. The stress response is the collection of immune, neural, and homeostatic mechanisms that shift into a long-term state of hyperawareness with the anticipation of threats that could appear at any time. While this state can be lifesaving when triggered appropriately, being in a constant state of fearful uncertainty is not a healthy ideal and has lasting deleterious biological effects.

The stress system is extraordinarily far-reaching and complex, but importantly to the current discussion, involves an interaction of epigenetic, neural, and behavioral responses through both the immune system and learning and memory. When there is uncertainty about a potentially aversive or harmful outcome, the downstream stress response is activated by the hypothalamus. The hypothalamus secretes corticotropin-releasing hormone (CRH), which stimulates release of adrenocorticotropic hormone (ACTH) from the pituitary, which in turn signals two different stress pathways, one fast and one slow (Gunnar and Quevedo 2007). The two major stress pathways are the sympathetic-adrenomedullary (SAM) response and the hypothalamic-pituitary-adrenocortical (HPA) response (Gunnar and Quevedo 2007). In the fast, immediate SAM response, the ACTH triggers the adrenal medulla to release norepinephrine and epinephrine, neurotransmitters required to cause a rapid and intense nervous system response and hypervigilant attention (Benarroch 2007). In the slower, long-lasting HPA response, ACTH triggers the adrenal cortex to signal the release of the stress hormones glucocorticoids, the most important of which is cortisol, thus dysregulating metabolism, suppressing the immune system, and disrupting homeostasis through glucocorticoid receptor binding systemically (Gunnar and Quevedo 2007). During the stress response, homeostatic mechanisms attempt to maintain equilibrium over a wide range adaptive circumstances in order to respond to any possible challenge. Stress is, in essence, a "ready" state from which a large, quick biological response is primed at a moment's notice and equipped with constant vigilance. Therefore, this cascade of biological effects both elicits a physiological and behavioral response, and poises the requisite systems for future environmental reactivity. Unfortunately, when chronically stressed, there is a burden placed on these biological systems such as the immune system and metabolism that becomes harmful, referred to as allosteric load (Gunnar and Quevedo 2007; Gunnar 2017). An excellent theoretical understanding of this facet of the stress response system is to return to the gAP (Clayton 2000; Clayton et al. 2019). The brain has a response to an environmental stressor that leads to transcriptional and epigenetic changes. These changes then trigger a "neuroendocrine action potential" as this neural response triggers both immediate and long-lasting changes throughout the limbic system and multiple organs (Clayton et al. 2019).

The clear, widespread effect of stress throughout all physiological systems makes it an understandable and unique candidate for understanding the epigenetics of social environment. Many adverse environments, both in the interpersonal, such as traumatic and abusive relationships with caregivers and peers, and the societal, such as minority stress and socioeconomic status stress, can trigger these same underlying processes, as all are sources of aversive, potentially harmful uncertainty. It is then understandable why the most commonly researched gene in social epigenetics is *NR3C1*, the glucocorticoid receptor gene (Turecki and Meaney 2016). In Turecki and Meaney's study (2016), they found consistent mC at the *NR3C1* in exon 1F/17 regarding parental stress, but inconsistent in other types and later life stressors. Additionally, recent work found that stress affected mRNA methylation in a region-specific manner, ultimately altering fear learning and synaptic plasticity (Engel et al. 2018).

The literature does appear to indicate that chronic stress in early life has a greater impact on mC patterns than those that occur in later life, but more research in this area is needed to make a definitive statement (Austin et al. 2018; Esposito et al. 2016). Work on cumulative stress, as opposed to early life or later life considered separately, also indicates an association with accelerated epigenetic aging (Zannas et al. 2015a). If this stronger association with stress and mC at younger ages is robust, it may be due to differences in the immune system's environmental sensitivity during early development (Miller et al. 2011). Additionally, a key component of the attachment relationship is to learn and support affect regulation (Hazan and Shaver 1987); having a responsive, consistent caregiver helps children express and deal with their negative emotions without triggering the endocrine stress response as if an uncertain threat had been detected; however, in abusive relationships, the parent is the source of uncertainty and danger (Repetti et al. 2014). While it is true that a recent, large study in humans did not replicate findings from smaller studies that reported correlations between mC patterns and chronic social stress (Marzi et al. 2018), this may be due to differences in sample size, population, and consistency among ecologically valid measures of the type, context, and experience of stress. More replications with standardized measures and large sample sizes are needed to make any definitive statements about detectable mC differences among those who have experienced chronic stress.

In addition to *NR3C1*, there is a robust literature associating early-life adversity such as social, physical, or parental stressors, with epigenetic changes in *BDNF* gene expression (Roth and Sweatt 2011a, b). These differences may correlate with a reduction in socioemotional learning and plasticity, and have shown an increased capacity for fear learning (Dias et al. 2015). Impacted learning has also been implicated in more specific associations than general stress, such as the several reported associations between trauma, abuse, and differences in epigenetic modifications (e.g., Dickson et al. 2018; Klengel et al. 2014; Lutz and Turecki 2014; Mehta et al. 2013; Roberts et al. 2018; Suderman et al. 2014; Weder et al. 2014) as well as associations specifically with posttraumatic stress disorder (see Zannas et al. 2015b for review). Though there is a range in the findings of epigenetics in regards to early life adversity from both candidate gene approaches and EWAS approaches, overlapping genes associated with stress, pain, learning, and the immune system were common. For example, one study found immune cell differences and accelerated epigenetic age associated with lifetime PTSD severity (Rutering et al. 2016).

It is possible that trauma and pain in early life leads to learning and adaptation to a harsher world that requires more vigilance instead of a conservative homeostasis, a molecular push toward fear conditioning instead of socioemotional development, and a greater sensitivity to pain in order to more quickly identify threats. Those who develop in positive, enriched environments, on the other hand, thrive with reduced allosteric load and have resilience that seems especially prominent in stress coping and synaptic plasticity. Instead of being able to learn more and put their energy toward other endeavors, these adaptations in a world of agonizing uncertainty could be primed or activated by epigenetic modifications for the sole purpose of survival. A plausible model of chronic life stressors proposes a similar line of reasoning and theorizes with significant evidence that epigenetic modifications set into motion by the cascade of stress hormones both affect and prime a traumatized individual for accelerated aging and biological weathering (Gassen et al. 2017). Supporting this hypothesis, a high number of sites used in mC epigenetic clocks are located within glucocorticoid response elements (Zannas et al. 2015a).

However, though it may not undo the harmful developmental environment, social support, especially touch, has been associated with stress buffering in many studies (Coan et al. 2006; Cohen and Wills 1985; Matthews and Gallo 2011; Ozbay et al. 2007, 2008). This is most likely through the dual mechanisms of affect regulation through a social buffer so as to not trigger the stress hormone cascade, and through the mu-opiates that are released in social reward reducing affective motivational pain in the nociceptive pathways and the intensity of perceived threats (LaGraize et al. 2006; Troisi et al. 2011). This may be why we see rescue effects of social, enriched environments for memory deficits as discussed above. This may also contribute to the difficulty in reproducing many social epigenetic findings, as adverse effects are often accounted for, but buffering and resiliency effects are not.

5 Societal Environments

5.1 Minority Status and Socioeconomic Status

Chronic stress has clear physiological, psychological, neurological, and epigenetic effects. However, not all chronic stress stems from abuse or trauma. In fact, not all chronic stress stems from interpersonal relationships at all. As a social species, we have developed a society with biases, prejudices, and hierarchies. Our modern social environments remain embedded by historical power relationships. Though there is much work being done to correct these injustices, racial, ethnic, gender, sexuality, or religious minorities, as well as those with a low socioeconomic status (SES) and social position, face considerably more stress from societal pressures and inequities than their counterparts. This stress may then exacerbate these inequalities through cellular means, as well as societal.

A minority group that is illustrative of how social structures can affect epigenetics and, ultimately, associate with extreme behavioral phenotypes are people living with schizophrenia. Schizophrenia is a mental illness often characterized by flat affect, hallucinations, and psychosis for which the polygenetic burden is significantly associated with epigenetic variation, suggesting that regulatory variation of the disorder stems from both the genome and environment (Cromby et al. 2019). Schizophrenia is the most thoroughly studied psychiatric disorder from an epigenetic perspective, mostly likely due to the clear environmental effects in its presence, onset, and trajectory. Specifically, there is evidence associating schizophrenia presence and onset with low SES, ethnic differences and racial discrimination, immigration, urban living, childhood adversity and trauma, and parental absence (Cromby et al. 2019). Though there is an underlying genetic component to the disorder clear from family clustering, this risk is compounded by exposure to these social environmental factors, such as minority status (Hutchinson et al. 1997). There is significant evidence that minority status, or seeming apart from society in any way, increases the likelihood of developing schizophrenia most likely due to the chronic societally based stress of social position (Bourque et al. 2011; Cromby et al. 2019; Selten 2005; Van Os et al. 2010). For example, Black Caribbean immigrants to the UK who grew up in the Caribbean do not express any increased risk of psychosis; however, their children, born and raised as Black Caribbean immigrants in the UK, had a seven times higher risk of psychosis (Hutchinson et al. 1997). This was familial risk within an ethnically homogenous sample, pointing toward a significant association with the chronic stressors of immigrant and racial minority status being significant drivers of schizophrenia risk. Even more physical variables, such as urban neighborhood residence, that have been found to increase psychosis risk, appear to be more driven by social fragmentation than physical environment (Zammit et al. 2010).

Often compounding the chronic stress associated with minority status is the stress associated with class and SES. Most likely due to many social, physical, and biological factors encompassed in SES, there is a plethora of associations between SES and epigenetic changes. As SES is societally constructed, it is difficult to create an ecologically valid animal model with which to investigate epigenetic modifications; therefore, the majority of epigenetic research is focused on mC associations in humans. Low SES, especially during youth, has a significant and robust association with age acceleration and sites connected to immune function, development, and age-related diseases (Austin et al. 2018; Bush et al. 2018; Chen et al. 2011; Fiorito et al. 2017; Lam et al. 2012; McCrory et al. 2019; Mcdade et al. 2019; Simons et al. 2016; Tehranifar et al. 2013). Even with low-SES youth, greater self-control associates with improved socioemotional functioning and general success, but also epigenetic age acceleration, supporting the idea that increased allosteric load may contribute to worse health outcomes among the disadvantaged (Miller et al. 2015).

Another aspect of the relationship between SES and epigenetic modifications is that lower SES correlates with both smoking and drinking behavior (Sweeting and Hunt 2015; Van Oers et al. 1999). Not surprisingly, social stress also triggers the urge to smoke and drink (Fouquereau et al. 2003; Niaura et al. 2002). There is evidence that drinking increases emotional experiences and smoking temporarily reduces arousal as evidenced by reduced neurological event-related potentials, supporting a self-medication hypothesis of legal drug use (Choi et al. 2015; Sayette 2017). This relationship of SES with smoking and drinking, possibly as a way to deal with stress, may exacerbate epigenetic disparities due to the strong, reproducible effects of smoking and drinking on mC, especially on sites related to age, immune, and cardiovascular function (Brückmann et al. 2017; Goldowitz et al. 2014; Hillemacher et al. 2008; Mahnke et al. 2017; McCartney et al. 2018; Ponomarev et al. 2017; Tulisiak et al. 2017).

In addition to smoking and drinking, it is important in every mC investigation to account for cell type proportions, as these are the primary drivers of variation, but this is especially true in explorations of SES due to the significant immune system effects of the chronic stress system. For example, one study found that leukocyte composition of peripheral blood covaried with patterns of mC at many sites and mC was strongly associated with the monocyte inflammatory response (Lam et al. 2012). Monocytes also epigenetically aged faster in those exposed to low SES in early life (Austin et al. 2018). These immune responses, most likely from stress, may contribute, at least in part, to the association of SES, especially early life SES, with epigenetic age acceleration and aging-related disease risk, even controlling for related factors such as smoking and drinking (Austin et al. 2018; Simons et al. 2016). When Simons et al. (2016) investigated the main environmental driver of epigenetic age acceleration in a low SES sample, they found that it was the stress of financial insecurity that drove the SES and accelerated aging association, providing further evidence for the link between early life stress, immune response, epigenetic change, and health outcomes (reviewed in Miller et al. 2011). However, similar to other early exposures to stress, the effects of low SES on immune response can be buffered through social support in the form of warm, positive caregiving (Chen et al. 2011).

5.2 Social Effects on Physical Environment

Though it may not be interpersonal or even seem entirely social, there is a social environment in the construction of our societies in the types of nutrients we can access, the configuration of neighborhoods in which we live, the services to which we have access, and the physical environments to which we are exposed. Both physical and social environments can affect our epigenetic modifications (Mcdade et al. 2017). The social administration of our physical environments is yet another form of social environmental influence to which we are able to learn and adapt psychologically and biologically, whether it is conscious or unconscious.

One major aspect in the construction of socially administered environments is the spatial sorting of people based on their SES, race, or ethnicity. As discussed above, there is evidence chronic stress that accompanies being in a reduced societal position, whether through racism or classism, as well as the stress of deprivation, associates with epigenetic change. However, in addition to this divide, health differences among neighborhoods persist even after adjusting for SES and demographic factors, most likely due to the impact of broad environmental factors such as access to nutrition or exposure to pollution (Mair et al. 2008; Paczkowski and Galea 2010; Pickett and Pearl 2001; Roux and Mair 2010). Factors linked to differences in physical environment most likely contribute to and reinforce the detrimental effects of chronic societal stress on low SES and minority communities (Bleich et al. 2012; LaVeist et al. 2011). Unsurprisingly then, physical environment and location are also tied to risk of schizophrenia (March et al. 2008).

One example of how physical environments may perpetuate the biological differences among classes and races are food deserts. These are areas, either particularly urban or rural, where fresh produce and other healthy foods are either not available or too expensive to be purchased as an everyday source of caloric intake. Low SES neighborhoods are especially likely to be located in a food desert (Ghosh-Dastidar et al. 2014). Food availability and food advertising, which is different for lower SES neighborhoods, influence energy intake and the nutritional value of foods consumed (Grier and Kumanyika 2008; Harris et al. 2009). The wealth of literature on the epigenetics of nutrition, especially prenatal nutrition, pales only in comparison to epigenetic work in cancer (Anderson et al. 2012). The importance of a balanced, healthy diet from conception and throughout life on epigenetic modifications is an incredibly robust finding, as are similar results for morbidity and mortality (e.g., Anderson et al. 2012; Gabbianelli and Damiani 2018; Lillycrop and Burdge 2012; Mathers 2006; Milagro et al. 2013; Navarro et al. 2017; Zhang 2015). Along similar lines, the structure of a socially administrated physical environment can also be linked to differences in children's physical activity (Bringolf-Isler et al. 2010; Davison and Lawson 2006; Galvez et al. 2010; Sallis and Glanz 2006). Physical activity is often linked to morbidity and mortality, as well as epigenetic modifications, learning, and aging (e.g., Kaliman et al. 2011; Kashimoto et al. 2016; Kirchner et al. 2013; Ling and Rönn 2014; Mikkelsen et al. 2017; Moylan et al. 2013; Rodrigues et al. 2015; Zimmer et al. 2016). Another example is physical proximity to hazardous sites and pollution, which tend to be more prevalent in low-income or minority neighborhoods (Brulle and Pellow 2006; Mohai et al. 2009; Morello-Frosch et al. 2011). The effects of exposure to air pollution are well evidenced in both morbidity and epigenetics research (e.g., Barouki et al. 2018; Brook et al. 2010; Chen et al. 2016; Clifford et al. 2017; Gref et al. 2017; Laumbach and Kipen 2012; Luyten et al. 2018; Mustafíc et al. 2012; Rider et al. 2016; Somineni et al. 2016; Tzivian 2011). This is most likely due to the immune responses to breathing in toxic exogenous factors (Tzivian 2011).

Additionally, neighborhood conditions can create stress, such as feeling unsafe, as well as acting as social buffers against adverse effects of stress such as social cohesion or integration into the neighborhood or environments such as work or school (Cutrona et al. 2006; De Silva et al. 2005; Do et al. 2011; Mair et al. 2008; Merkin et al. 2009). One possibility of why some immigrant groups have better morbidity and mortality than other groups in the same city is the social support and cohesion within the community (Matthews et al. 2010). As discussed above, the stress response affects many systems and may lead to widespread epigenetic modifications, especially during early development. Once again, even the midst of possible adaptation to a harmful environment, a positive, enriched social environment shows ecological rescue effects for health. Our epigenetic mechanisms modify, our neurological mechanisms encode, and our psychological mechanisms learn from our social environments.

6 Conclusion

From the moment we are born, our social relationships are a key component of how the world affects us. They are one of the first postnatal inputs afforded to the rapidly developing neonatal biology and are essential for survival. While data on epigenetics and the social environment have been spread out across disciplines, one can imagine potential examples for stringing these findings together cohesively. One example of this cell-to-society effect could be a child born in a family with insensitive caregivers who consistently do not respond to the child's needs or give contact comfort. This, in turn, could lead to the hypermethylation of the promoter region for the *BDNF* gene, less *BDNF* transcription, and less BDNF present in the hippocampus and socioemotional circuit of the brain. This would then decrease the ability of the child to learn from their social environment and affect their ability to have successful interpersonal relationships. This lack of social efficacy could then make it more likely the individual would experience social pain and more difficult to receive social support to buffer stressful experiences throughout their lives. Without social buffering, the stress response could be triggered more often, resulting in reduced immune responses and homeostatic compensation through epigenetic modifications.

Over time, these cumulative biological and behavioral responses could increase allosteric load and possibly lead to accelerated aging and health decline, which may also affect access to services, financial earning ability, and physical environment to further exacerbate biological and sociological disparities seen at a population level.

These social environments are an opportunity for adaptation through experience seen pandiscipline through concepts such as mental representations, LTP and memory engrams, and epigenetic change such as histone lysine trimethylation affecting the promoters of synaptic plasticity-related genes. Fundamentally, all psychological, neurobiological, and epigenetic reactions to these social environments are opportunities to learn from and adapt to them in order to best thrive in the world as it is, whether that environmental situation be ideal, violent, or deprived. The experience-dependent plasticity gained from the interaction between neuroscience and epigenetics is integral to this adaptation (Clayton et al. 2019). These systems may be working as a new state of vigilance. Although the context of the social environment is paramount in the specific reaction and modification, the ultimate goal and underlying mechanistic interplay remain largely the same – to learn from our social world to better survive in the environment we find ourselves in.

References

Åhs F, Rosén J, Kastrati G, Fredrikson M, Agren T, Lundström JN (2018) Biological preparedness and resistance to extinction of skin conductance responses conditioned to fear relevant animal pictures: a systematic review. Neurosci Biobehav Rev. https://doi.org/10.1016/j.neubiorev.2018.10.017

Alarcón JM, Malleret G, Touzani K, Vronskaya S, Ishii S, Kandel ER, Barco A (2004) Chromatin acetylation, memory, and LTP are impaired in CBP+/− mice: a model for the cognitive deficit in Rubinstein-Taybi syndrome and its amelioration. Neuron. https://doi.org/10.1016/j.neuron.2004.05.021

Alvarado S, Tajerian M, Suderman M, Machnes Z, Pierfelice S, Millecamps M et al (2015) An epigenetic hypothesis for the genomic memory of pain. Front Cell Neurosci 9:1–10. https://doi.org/10.3389/fncel.2015.00088

Amir RE, Van Den Veyver IB, Wan M, Tran CQ, Francke U, Zoghbi HY (1999) Rett syndrome is caused by mutations in X-linked MECP2, encoding methyl-CpG-binding protein 2. Nat Genet 23(2):185–188. http://genetics.nature.com

Anderson OS, Sant KE, Dolinoy DC (2012) Nutrition and epigenetics: an interplay of dietary methyl donors, one-carbon metabolism and DNA methylation. J Nutr Biochem. https://doi.org/10.1016/j.jnutbio.2012.03.003

Arai JA, Li S, Hartley DM, Feig LA (2009) Transgenerational rescue of a genetic defect in long-term potentiation and memory formation by juvenile enrichment. J Neurosci 29(5):1496–1502. https://doi.org/10.1523/JNEUROSCI.5057-08.2009

Araki R, Hiraki Y, Nishida S, Kuramoto N, Matsumoto K, Yabe T (2016) Epigenetic regulation of dorsal raphe GABAB1a associated with isolation-induced abnormal responses to social stimulation in mice. Neuropharmacology. https://doi.org/10.1016/j.neuropharm.2015.09.013

Ausió J, Abbott DW (2002) The many tales of a tail: carboxyl-terminal tail heterogeneity specializes histone H2A variants for defined chromatin function. Biochemistry 41(19):5946–5949. https://doi.org/10.1021/bi020059d

Austin MK, Chen E, Ross KM, Mcewen LM, Maclsaac JL, Kobor MS, Miller GE (2018) Early-life socioeconomic disadvantage, not current, predicts accelerated epigenetic aging of monocytes. Psychoneuroendocrinology 97:131–134. https://doi.org/10.1016/j.psyneuen.2018.07.007
Bai G, Wei D, Zou S, Ren K, Dubner R (2010) Inhibition of class II histone deacetylases in the spinal cord attenuates inflammatory hyperalgesia. Mol Pain 6(1):51–63. http://www.ucsc
Baldwin MW, Fehr B, Keedian E, Seidel M, Thomson DW (1993) An exploration of the relational schemata underlying attachment styles: self-report and lexical decision approaches. Personal Soc Psychol Bull 19:746
Bale TL, Baram TZ, Brown AS, Goldstein JM, Insel TR, Mccarthy MM et al (2010) Early life programming and neurodevelopmental disorders. Biol Psychiatry 68(4):314–319. https://doi.org/10.1016/j.biopsych.2010.05.028
Bandura A (1977) Social learning theory. Prentice Hall, Englewood Cliffs
Banerjee T, Chakravarti D (2011) A peek into the complex realm of histone phosphorylation. Mol Cell Biol 31(24):4858–4873. https://doi.org/10.1128/MCB.05631-11
Barnett L (2005) Keep in touch: the importance of touch in infant development. Infant Observation 8(2):115–123. https://doi.org/10.1080/13698030500171530
Barouki R, Melén E, Herceg Z, Beckers J, Chen J, Karagas M et al (2018) Epigenetics as a mechanism linking developmental exposures to long-term toxicity. Environ Int. https://doi.org/10.1016/j.envint.2018.02.014
Barr R, Dowden A, Hayne H (1996) Developmental changes in deferred imitation by 6-to 24-month-old infants. Infant Behav Dev 19. https://s3.amazonaws.com/academia.edu.documents/32906896/1barrdowdenhayne.pdf?response-content-disposition=inline%253Bfilename%253DDevelopmental_changes_in_deferred_imitat.pdf&X-Amz-Algorithm=AWS4-HMAC-SHA256&X-Amz-Credential=AKIAIWOWYYGZ2Y53UL3A%252F201
Bártová E, Krejčí J, Harničarová A, Galiová G, Kozubek S (2008) Histone modifications and nuclear architecture: a review. J Histochem Cytochem 56(8):711–721. https://doi.org/10.1369/jhc.2008.951251
Beckes L, IJzerman H, Tops M (2015) Toward a radically embodied neuroscience of attachment and relationships. Front Hum Neurosci 9:1–18. https://doi.org/10.3389/fnhum.2015.00266
Bedrosian TA, Quayle C, Novaresi N, Gage FH (2018) Early life experience drives structural variation of neural genomes in mice. Science 359(6382):1395–1399. https://doi.org/10.1126/science.aah3378
Belsky J, Cassidy J (1994) Attachment: theory and evidence. In: Rutter M, Hay D (eds) Development through life. Blackwell Scientific, Oxford, pp 373–402
Benarroch EE (2007) Enteric nervous system: functional organization and neurologic implications. Neurology 69(20):1953–1957
Benito E, Dean C, Fischer A (2015) HDAC inhibitor-dependent transcriptome and memory reinstatement in cognitive decline models. J Clin Invest 125(9):3572. https://doi.org/10.1172/JCI79942
Benzon H, Raja SN, Fishman SE, Liu SS, Cohen SP (2011) Essentials of pain medicine. Elsevier Health Sciences, Philadelphia. E-book
Blaze J, Roth TL (2013) Epigenetic mechanisms in learning and memory. Wiley Interdiscip Rev Cogn Sci. https://doi.org/10.1002/wcs.1205
Bleich SN, Jarlenski MP, Bell CN, LaVeist TA (2012) Health inequalities: trends, progress, and policy. Annu Rev Public Health 33:7–40
Boccaletto P, Machnicka MA, Purta E, Piatkowski P, Baginski B, Wirecki TK et al (2017) MODOMICS: a database of RNA modification pathways. 2017 update. Nucleic Acids Res 46(2):303–307. https://doi.org/10.1093/nar/gkx1030
Bock C, Paulsen M, Tierling S, Mikeska T, Lengauer T, Walter J (2006) CpG island methylation in human lymphocytes is highly correlated with DNA sequence, repeats, and predicted DNA structure. PLoS Genet 2(3):243–252. https://doi.org/10.1371/journal.pgen.0020026.eor

Bourque F, Van Der Ven E, Malla A (2011) A meta-analysis of the risk for psychotic disorders among first- and second-generation immigrants. Psychol Med 41(5):897–910. https://doi.org/10.1017/S0033291710001406

Bowlby J (1979) The making and breaking of affectional bonds. Tavistock, London

Bowlby J (1982) Attachment and loss: retrospect and prospect. Am J Orthopsychiatry 52:664

Boyce WT, Kobor MS (2015) Development and the epigenome: the 'synapse' of gene-environment interplay. Dev Sci 18(1):1–23. https://doi.org/10.1111/desc.12282

Brami-Cherrier K, Roze E, Girault J-A, Betuing S, Caboche J (2009) Role of the ERK/MSK1 signalling pathway in chromatin remodeling and brain responses to drugs of abuse. J Neurochem 108:1323–1335. https://doi.org/10.1111/j.1471-4159.2009.05879.x

Branchi I (2009) The mouse communal nest: investigating the epigenetic influences of the early social environment on brain and behavior development. Neurosci Biobehav Rev 33:551–559. https://doi.org/10.1016/j.neubiorev.2008.03.011

Branchi I, Francia N, Alleva E (2004) Epigenetic control of neurobehavioural plasticity: the role of neurotrophins. Behav Pharmacol 15(5–6):353–362. https://doi.org/10.1097/00008877-200409000-00006

Branchi I, D'Andrea I, Fiore M, Di Fausto V, Aloe L, Alleva E (2006) Early social enrichment shapes social behavior and nerve growth factor and brain-derived neurotrophic factor levels in the adult mouse brain. Biol Psychiatry. https://doi.org/10.1016/j.biopsych.2006.01.005

Braun PR, Han S, Hing B, Nagahama Y, Gaul LN, Heinzman JT et al (2019) Genome-wide DNA methylation comparison between live human brain and peripheral tissues within individuals. Transl Psychiatry 9(47). https://doi.org/10.1038/s41398-019-0376-y

Bredy TW, Tie YZ, Grant RJ, Diorio J, Meaney MJ (2004) Peripubertal environmental enrichment reverses the effects of maternal care on hippocampal development and glutamate receptor subunit expression. Eur J Neurosci 20(5):1355–1362. https://doi.org/10.1111/j.1460-9568.2004.03599.x

Bretherton I, Munholland KA (2008) Internal working models in attachment relationships: elaborating a central construct in attachment theory. In: Cassidy J, Shaver PR (eds) Handbook of attachment: theory, research, and clinical applications, 2nd edn. Guilford, New York, pp 102–127

Bringolf-Isler B, Grize L, Mäder U, Ruch N, Sennhauser FH, Braun-Fahrländer C (2010) Built environment, parents' perception, and children's vigorous outdoor play. Prev Med 50(5–6):251–256

Brody GH, Miller GE, Yu T, Beach SRH, Chen E (2016) Supportive family environments ameliorate the link between racial discrimination and epigenetic aging: a replication across two longitudinal cohorts. Psychol Sci 27(4):230–541. https://doi.org/10.1177/0956797615626703

Bronfenbrenner U (2009) The ecology of human development. Harvard University Press, Boston

Brook RD, Rajagopalan S, Pope CA, Brook JR, Bhatnagar A, Roux ADV et al (2010) Particulate matter air pollution and cardiovascular disease. Circulation 121(21):2331–2378

Brückmann C, Islam SA, Macisaac JL, Morin AM, Karle KN, Di Santo A et al (2017) DNA methylation signatures of chronic alcohol dependence in purified CD3 + T-cells of patients undergoing alcohol treatment. Sci Rep 7(1):6605. https://doi.org/10.1038/s41598-017-06847-z

Brulle RJ, Pellow DN (2006) Environmental justice: human health and environmental inequalities. Annu Rev Public Health 27:103–124

Bryois J, Garrett ME, Song L, Safi A, Giusti-Rodriguez P, Johnson GD et al (2018) Evaluation of chromatin accessibility in prefrontal cortex of individuals with schizophrenia. Nat Commun 9(1):3121. https://doi.org/10.1038/s41467-018-05379-y

Buenrostro JD, Wu B, Litzenburger UM, Ruff D, Gonzales ML, Snyder MP et al (2015) Single-cell chromatin accessibility reveals principles of regulatory variation. Nature 523(7561):486–490. https://doi.org/10.1038/nature14590

Buenrostro J, Wu B, Chang H, Greenleaf W (2016) ATAC-seq: a method for assaying chromatin accessibility genome-wide. Curr Protoc Mol Biol 109:21.29.1–21.29.9. https://doi.org/10.1002/0471142727.mb2129s109

Bukhari SA, Saul MC, Seward CH, Zhang H, Bensky M, James N et al (2017) Temporal dynamics of neurogenomic plasticity in response to social interactions in male threespined sticklebacks. PLoS Genet 13(7):1–21. https://doi.org/10.1371/journal.pgen.1006840

Bush NR, Edgar RD, Park M, Macisaac JL, Mcewen LM, Adler NE et al (2018) The biological embedding of early-life socioeconomic status and family adversity in children's genome-wide DNA methylation. Epigenomics 10(11):1445–1461. https://doi.org/10.2217/epi-2018-0042

Cao L, Mcdonnell A, Nitzsche A, Alexandrou A, Saintot P-P, Loucif AJC et al (2016) Pharmacological reversal of a pain phenotype in iPSC-derived sensory neurons and patients with inherited erythromelalgia. Sci Transl Med 8(335):335ra56. www.ScienceTranslationalMedicine.org

Casper M (2001) A definition of "social environment". Am J Public Health 9(3):465

Chatterjee D, Chatterjee-Chakraborty M, Rees S, Cauchi J, de Medeiros CB, Fleming AS (2007) Maternal isolation alters the expression of neural proteins during development: "stroking" stimulation reverses these effects. Brain Res. https://doi.org/10.1016/j.brainres.2007.04.069

Chen WY, Townes TM (2000) Molecular mechanism for silencing virally transduced genes involves histone deacetylation and chromatin condensation. Proc Natl Acad Sci 97 (1):377–382. https://doi.org/10.1073/PNAS.97.1.377

Chen WG, Chang Q, Lin Y, Meissner A, West AE, Griffith EC et al (1967) Derepression of BDNF transcription involves calcium-dependent phosphorylation of MeCP2. Annu Rev Neurosci 35:648. https://doi.org/10.1126/science.1087846

Chen E, Miller GE, Kobor MS, Cole SW (2011) Maternal warmth buffers the effects of low early-life socioeconomic status on pro-inflammatory signaling in adulthood. Mol Psychiatry 16 (7):729–737. https://doi.org/10.1038/mp.2010.53

Chen R, Meng X, Zhao A, Wang C, Yang C, Li H et al (2016) DNA hypomethylation and its mediation in the effects of fine particulate air pollution on cardiovascular biomarkers: a randomized crossover trial. Environ Int. https://doi.org/10.1016/j.envint.2016.06.026

Choi D, Ota S, Watanuki S (2015) Does cigarette smoking relieve stress? Evidence from the event-related potential (ERP). Int J Psychophysiol. https://doi.org/10.1016/j.ijpsycho.2015.10.005

Chowdhury S, Shepherd JD, Okuno H, Lyford G, Petralia RS, Plath N et al (2006) Arc/Arg3.1 interacts with the endocytic machinery to regulate AMPA receptor trafficking. Neuron. https://doi.org/10.1016/j.neuron.2006.08.033

Choy KHC, De Visser Y, Nichols NR, Van Den Buuse M (2008) Combined neonatal stress and young-adult glucocorticoid stimulation in rats reduce BDNF expression in hippocampus: effects on learning and memory. Hippocampus 18:655–667. https://doi.org/10.1002/hipo.20425

Chu C, Qu K, Zhong FL, Artandi SE, Chang HY (2011) Genomic maps of long noncoding RNA occupancy reveal principles of RNA-chromatin interactions. Mol Cell. https://doi.org/10.1016/j.molcel.2011.08.027

Chwang WB, O'riordan KJ, Levenson JM, Sweatt JD (2006) ERK/MAPK regulates hippocampal histone phosphorylation following contextual fear conditioning. Learn Mem 13:322–328. https://doi.org/10.1101/lm.152906

Clayton DF (2000) The genomic action potential. Neurobiol Learn Mem. https://doi.org/10.1006/nlme.2000.3967

Clayton DF, Anreiter I, Aristizabal M, Frankland PW, Binder EB, Citri A (2019) The role of the genome in experience-dependent plasticity: extending the analogy of the genomic action potential. Proc Natl Acad Sci:201820837. https://doi.org/10.1073/pnas.1820837116

Clifford RL, Jones MJ, MacIsaac JL, McEwen LM, Goodman SJ, Mostafavi S et al (2017) Inhalation of diesel exhaust and allergen alters human bronchial epithelium DNA methylation. J Allergy Clin Immunol. https://doi.org/10.1016/j.jaci.2016.03.046

Coan JA, Schaefer HS, Davidson RJ (2006) Lending a hand: social regulation of the neural response to threat. Psychol Sci 17(12):1032–1039. https://doi.org/10.1111/j.1467-9280.2006.01832.x

Cohen S, Wills TA (1985) Stress, social support, and the buffering hypothesis: a theoretical analysis. Psychol Bull 98(2):310–357. https://s3.amazonaws.com/academia.edu.documents/40586205/PsychBull1985.pdf?AWSAccessKeyId=AKIAIWOWYYGZ2Y53UL3A&Expires=1516704768&Signature=PkHVhko7UgSkjnTelcj58YnnHR8%3D&response-content-disposition=inline%3Bfilename%3DStress_Social_Support_and_the_Buff

Collins NL, Guichard AC, Ford MB, Feeney BC (2004) Working models of attachment: new developments and emerging themes. In: Rholes S, Simpson JA (eds) Adult attachment: theory, research, and clinical implications. Guilford, New York, pp 196–239

Cooper RP, Aslin RN (1990) Preference for infant-directed speech in the first month after birth. Child Dev 61(5):1584–1595. https://www.jstor.org/stable/pdf/1130766.pdf?casa_token=VZ2upZuEDGMAAAAA:qTh21VOjCOOUEZAFbyw9AnykclAg6BbG_urpmOg7HL3nL6tcNeLf4cWPW0MQSh-EW72hLYcwxVBWS0_9b1xk9bXC4AK8unwHfBMncOJHWmOeun8rdw

Cooper RP, Aslin RN (1994) Developmental differences in infant attention to the spectral properties of infant. Child Dev 65(6):1663–1667. https://www.jstor.org/stable/pdf/1131286.pdf?casa_token=Y98xWKNBA4sAAAAA:4axq7Qw8WbU6_TeZp2HbVSXlgdRVL5ZTAc3nWis_xCXSCKSdp3v0enhcB-9-rpHfmjW0Z05zOgKCZX5KdLaLgIdBSur3bty4odWm0TVvpxfK6vaIVw

Craig AD (2003) A new view of pain as a homeostatic emotion. Trends Neurosci. https://doi.org/10.1016/S0166-2236(03)00123-1

Cromby J, Chung E, Papadopoulos D, Talbot C (2019) Reviewing the epigenetics of schizophrenia. J Ment Health 28(1):71–79. https://doi.org/10.1080/09638237.2016.1207229

Cummins DD, Cummins R (2015) Biological preparedness and evolutionary explanation. In: The world in the head. https://doi.org/10.1093/acprof:osobl/9780199548033.003.0013

Cutrona CE, Wallace G, Wesner KA (2006) Neighborhood characteristics and depression: an examination of stress processes. Curr Dir Psychol Sci 15(4):188–192

D'Urso A, Brickner JH (2014) Mechanisms of epigenetic memory. Trends Genet. https://doi.org/10.1016/j.tig.2014.04.004

D'Urso A, Takahashi Y, Xiong B, Marone J, Coukos R, Randise-Hinchliff C et al (2016) Set1/COMPASS and mediator are repurposed to promote epigenetic transcriptional memory. elife 5:1–29. https://doi.org/10.7554/elife.16691

Davison K, Lawson C (2006) Do attributes in the physical environment influence children's physical activity? A review of the literature. Int J Behav Nutr Phys Act 3(1):19

Day JJ, Sweatt JD (2011) Epigenetic treatments for cognitive impairments. Neuropsychopharmacology 37:247–260. https://doi.org/10.1038/npp.2011.85

de Lima MNM, Presti-Torres J, Vedana G, Alcalde LA, Stertz L, Fries GR et al (2011) Early life stress decreases hippocampal BDNF content and exacerbates recognition memory deficits induced by repeated d-amphetamine exposure. Behav Brain Res. https://doi.org/10.1016/j.bbr.2011.05.022

de Silva P (1988) Phobias and preparedness: replication and extension. Behav Res Ther. https://doi.org/10.1016/0005-7967(88)90036-8

De Silva MJ, McKenzie K, Harpham T, Huttly SRA (2005) Social capital and mental illness: a systematic review. J Epidemiol Community Health 59(8):619–627

Denk F, McMahon SB (2012) Chronic pain: emerging evidence for the involvement of epigenetics. Neuron. https://doi.org/10.1016/j.neuron.2012.01.012

Denk F, Huang W, Sidders B, Bithell A, Crow M, Grist J et al (2013) HDAC inhibitors attenuate the development of hypersensitivity in models of neuropathic pain. Pain. https://doi.org/10.1016/j.pain.2013.05.021

Dias BG, Maddox SA, Klengel T, Ressler KJ (2015) Epigenetic mechanisms underlying learning and the inheritance of learned behaviors. Trends Neurosci. https://doi.org/10.1016/j.tins.2014.12.003

Dickson DA, Paulus JK, Mensah V, Lem J, Saavedra-Rodriguez L, Gentry A et al (2018) Reduced levels of miRNAs 449 and 34 in sperm of mice and men exposed to early life stress. Transl Psychiatry 8:101. https://doi.org/10.1038/s41398-018-0146-2

Dishion TJ (2016) Social influences on executive functions development in children and adolescents: steps toward a social neuroscience of predictive adaptive responses. J Abnorm Child Psychol 44:57–61. https://doi.org/10.1007/s10802-015-0117-5

Do DP, Roux AVD, Hajat A, Auchincloss AH, Merkin SS, Ranjit N et al (2011) Circadian rhythm of cortisol and neighborhood characteristics in a population-based sample: the multi-ethnic study of atherosclerosis. Health Place 17(2):625–632

Dupierrix E, de Boisferon AH, Méary D, Lee K, Quinn PC, Di Giorgio E et al (2014) Preference for human eyes in human infants. J Exp Child Psychol. https://doi.org/10.1016/j.jecp.2013.12.010

Eckert MJ, Abraham WC (2010) Physiological effects of enriched environment exposure and LTP induction in the hippocampus in vivo do not transfer faithfully to in vitro slices. Learn Mem 17:480–484. https://doi.org/10.1101/lm.1822610

Edgar RD, Jones MJ, Meaney MJ, Turecki G, Kobor MS (2017) BECon: a tool for interpreting DNA methylation findings from blood in the context of brain. Transl Psychiatry 7. https://doi.org/10.1038/tp.2017.171

Ehrlich I, Malinow R (2004) Cellular/molecular postsynaptic density 95 controls AMPA receptor incorporation during long-term potentiation and experience-driven synaptic plasticity. J Neurosci 24(4):916–927. https://doi.org/10.1523/JNEUROSCI.4733-03.2004

Eisenberger NI, Lieberman MD (2004) Why rejection hurts: a common neural alarm system for physical and social pain. Trends Cogn Sci. https://doi.org/10.1016/j.tics.2004.05.010

Engel M, Eggert C, Kaplick PM, Eder M, Röh S, Tietze L et al (2018) The role of m6A/m-RNA methylation in stress response regulation. Neuron. https://doi.org/10.1016/j.neuron.2018.07.009

Esposito EA, Jones MJ, Doom JR, Macisaac JL, Gunnar MR, Kobor MS (2016) Differential DNA methylation in peripheral blood mononuclear cells in adolescents exposed to significant early but not later childhood adversity. Dev Psychopathol 28:1385–1399. https://doi.org/10.1017/S0954579416000055

Essex MJ, Boyce WT, Hertzman C, Lam LL, Armstrong JM, Neumann SMA, Kobor MS (2013) Epigenetic vestiges of early developmental adversity: childhood stress exposure and DNA methylation in adolescence. Child Dev 84(1):58–75. https://doi.org/10.1111/j.1467-8624.2011.01641.x

Farah MJ, Betancourt L, Shera DM, Savage JH, Giannetta JM, Brodsky NL et al (2008) Environmental stimulation, parental nurturance and cognitive development in humans. Dev Sci 11 (5):793–801. https://doi.org/10.1111/j.1467-7687.2008.00688.x

Felice D, O'Leary OF, Cryan JF (2016) Targeting the GABAB receptor for the treatment of depression and anxiety disorders. In: GABAB receptor. Humana Press, Cham, pp 219–250

Feng J, Zhou Y, Campbell SL, Le T, Li E, Sweatt JD et al (2010) Dnmt1 and Dnmt3a are required for the maintenance of DNA methylation and synaptic function in adult forebrain neurons. Nat Neurosci 13(4):423–430. https://doi.org/10.1038/nn.2514

Fernald A (1992) Human maternal vocalizations to infants as biologically relevant signals: an evolutionary perspective. In: Barkow JH, Cosmides L, Tooby J (eds) The adapted mind: evolutionary psychology and the generation of culture. Oxford University Press, New York, pp 391–428

Fiete IR, Senn W, Wang CZH, Hahnloser RHR (2010) Spike-time-dependent plasticity and heterosynaptic competition organize networks to produce long scale-free sequences of neural activity. Neuron 65(4):563–576. https://doi.org/10.1016/j.neuron.2010.02.003

Finn AK, Whistler JL (2001) Endocytosis of the mu opioid receptor reduces tolerance and a cellular hallmark of opiate withdrawal. Neuron. https://doi.org/10.1016/S0896-6273(01)00517-7

Fiorito G, Polidoro S, Dugué P-A, Kivimaki M, Ponzi E, Matullo G et al (2017) Social adversity and epigenetic aging: a multi-cohort study on socioeconomic differences in peripheral blood DNA methylation. Sci Rep 7:16266. https://doi.org/10.1038/s41598-017-16391-5

Fischer A, Sananbenesi F, Wang X, Dobbin M, Tsai LH (2007) Recovery of learning and memory is associated with chromatin remodelling. Nature 447(7141):178–182. https://doi.org/10.1038/nature05772
Fitzsimons HL, Scott MJ (2011) Genetic modulation of Rpd3 expression impairs long-term courtship memory in drosophila. PLoS One 6(12):29171. https://doi.org/10.1371/journal.pone.0029171
Fouquereau E, Fernandez A, Mullet E, Sorum PC (2003) Stress and the urge to drink. Addict Behav. https://doi.org/10.1016/S0306-4603(01)00276-3
Fransquet PD, Wrigglesworth J, Woods RL, Ernst ME, Ryan J (2019) The epigenetic clock as a predictor of disease and mortality risk: a systematic review and meta-analysis. Clin Epigenetics 11(62). https://doi.org/10.1186/s13148-019-0656-7
Fumagalli F, Di Pasquale L, Caffino L, Racagni G, Riva MA (2007) Repeated exposure to cocaine differently modulates BDNF mRNA and protein levels in rat striatum and prefrontal cortex. Eur J Neurosci 26(10):2756–2763. https://doi.org/10.1111/j.1460-9568.2007.05918.x
Gabbianelli R, Damiani E (2018) Epigenetics and neurodegeneration: role of early-life nutrition. J Nutr Biochem. https://doi.org/10.1016/j.jnutbio.2018.01.014
Galvez MP, Pearl M, Yen IH (2010) Childhood obesity and the built environment. Curr Opin Pediatr 22(2):202–207
Garg E, Chen LI, Nguyen TTT, Pokhvisneva I, Chen LM, Unternaehrer E et al (2018) The early care environment and DNA methylome variation in childhood. Dev Psychopathol 30:891–903. https://doi.org/10.1017/S0954579418000627
Gassen NC, Chrousos GP, Binder EB, Zannas AS (2017) Life stress, glucocorticoid signaling, and the aging epigenome: implications for aging-related diseases. Neurosci Biobehav Rev. https://doi.org/10.1016/j.neubiorev.2016.06.003
Ghosh-Dastidar B, Cohen D, Hunter G, Zenk SN, Huang C, Beckman R, Dubowitz T (2014) Distance to store, food prices, and obesity in urban food deserts. Am J Prev Med. https://doi.org/10.1016/j.amepre.2014.07.005
Giese KP, Fedorov NB, Filipkowski RK, Silva AJ (1998) Autophosphorylation at Thr 286 of the calcium-calmodulin kinase II in LTP and learning. Science 279(5352):870–873. http://science.sciencemag.org/
Goldowitz D, Lussier AA, Boyle JK, Wong K, Lattimer SL, Dubose C et al (2014) Molecular pathways underpinning ethanol-induced neurodegeneration. Front Genet 5:1–13. https://doi.org/10.3389/fgene.2014.00203
Gopalan S, Carja O, Fagny M, Patin E, Myrick JW, McEwen LM et al (2017) Trends in DNA methylation with age replicate across diverse human populations. Genetics 206:1659–1674. https://doi.org/10.1534/genetics.116.195594
Goren CC, Sarty M, Wu PVK (1975) Visual following and pattern discrimination of face-like stimuli by newborn infants. Pediatrics 56(4):544–549. www.aappublications.org/news
Görisch SM, Wachsmuth M, Tóth KF, Lichter P, Rippe K (2005) Histone acetylation increases chromatin accessibility. J Cell Sci 118(24):5825–5834. https://doi.org/10.1242/jcs.02689
Gref A, Merid SK, Gruzieva O, Ballereau S, Becker A, Bellander T et al (2017) Genome-wide interaction analysis of air pollution exposure and childhood asthma with functional follow-up. Am J Respir Crit Care Med 195:1373–1383. https://doi.org/10.1164/rccm.201605-1026OC
Grier SA, Kumanyika SK (2008) The context for choice: health implications of targeted food and beverage marketing to African Americans. Am J Public Health 98(9):1616–1629
Griñan-Ferré C, Puigoriol-Illamola D, Palomera-ávalos V, Pérez-Cáceres D, Companys-Alemany J, Camins A et al (2016) Environmental enrichment modified epigenetic mechanisms in SAMP8 mouse hippocampus by reducing oxidative stress and inflammaging and achieving neuroprotection. Front Aging Neurosci 8:1–12. https://doi.org/10.3389/fnagi.2016.00241
Guan Z, Giustetto M, Lomvardas S, Kim JH, Miniaci MC, Schwartz JH et al (2002) Integration of long-term-memory-related synaptic plasticity involves bidirectional regulation of gene expression and chromatin structure. Cell. https://doi.org/10.1016/S0092-8674(02)01074-7

Guan J-S, Haggarty SJ, Giacometti E, Dannenberg J-H, Joseph N, Gao J et al (2009) HDAC2 negatively regulates memory formation and synaptic plasticity. Nature 459(7243):55–60. https://doi.org/10.1038/nature07925
Günaydin G, Zayas V, Selcuk E, Hazan C (2012) I like you but I don't know why: objective facial resemblance to significant others influences snap judgments. J Exp Soc Psychol 48:250–353. https://doi.org/10.1037/a0028125
Gunnar MR (2017) Social buffering of stress in development: a career perspective. Perspect Psychol Sci 12(3):355–373. https://doi.org/10.1177/1745691616680612
Gunnar M, Quevedo K (2007) The neurobiology of stress and development. Annu Rev Psychol 58:145–173. https://doi.org/10.1146/annurev.psych.58.110405.085605
Han S, Soleiman MT, Soden ME, Zweifel LS, Palmiter RD (2015) Elucidating an affective pain circuit that creates a threat memory. Cell 162:363–374. https://doi.org/10.1016/j.cell.2015.05.057
Hannon E, Lunnon K, Schalkwyk L, Mill J (2015) Interindividual methylomic variation across blood, cortex, and cerebellum: implications for epigenetic studies of neurological and neuropsychiatric phenotypes. Epigenetics 10(11):1024–1032. https://doi.org/10.1080/15592294.2015.1100786
Harris JL, Pomeranz JL, Lobstein T, Brownell KD (2009) A crisis in the marketplace: how food marketing contributes to childhood obesity and what can be done. Annu Rev Public Health 30:211–225
Hazan C, Shaver P (1987) Romantic love conceptualized as an attachment process. J Pers Soc Psychol 52(3):511–524. http://www.ncbi.nlm.nih.gov/pubmed/3572722
He Y, Yang C, Kirkmire CM, Wang ZJ (2010) Regulation of opioid tolerance by let-7 family MicroRNA targeting the opioid receptor. J Neurosci 30(30):10251–10258. https://doi.org/10.1523/JNEUROSCI.2419-10.2010
Hermans D, Craske MG, Mineka S, Lovibond PF (2006) Extinction in human fear conditioning. Biol Psychiatry. https://doi.org/10.1016/j.biopsych.2005.10.006
Hillemacher T, Frieling H, Moskau S, Muschler MAN, Semmler A, Kornhuber J et al (2008) Global DNA methylation is influenced by smoking behaviour. Eur Neuropsychopharmacol. https://doi.org/10.1016/j.euroneuro.2007.12.005
Horvath S, Raj K (2018) DNA methylation-based biomarkers and the epigenetic clock theory of ageing. Nat Rev Genet 19(6):371–384. https://doi.org/10.1038/s41576-018-0004-3
Horvath S, Garagnani P, Bacalini MG, Pirazzini C, Salvioli S, Gentilini D et al (2015) Accelerated epigenetic aging in Down syndrome. Aging Cell 14(3):491–495. https://doi.org/10.1111/acel.12325
Horvath S, Gurven M, Levine ME, Trumble BC, Kaplan H, Allayee H et al (2016) An epigenetic clock analysis of race/ethnicity, sex, and coronary heart disease. Genome Biol 17:171. https://doi.org/10.1186/s13059-016-1030-0
Hrvatin S, Hochbaum DR, Nagy MA, Cicconet M, Robertson K, Cheadle L et al (2018) Single-cell analysis of experience-dependent transcriptomic states in mouse visual cortex. Nat Neurosci 21(1):120–129. https://doi.org/10.1038/s41593-017-0029-5
Huang EJ, Reichardt LF (2001) Neurotrophins: roles in neuronal development and function. Annu Rev Neurosci 24:677–736. www.annualreviews.org
Huang H, Sabari BR, Garcia BA, Allis DC, Zhao Y (2014) SnapShot: histone modifications. Cell 159:458–458.e1. https://doi.org/10.1016/j.cell.2014.09.037
Hulsebus RC (1974) Operant conditioning of infant behavior: a review. Adv Child Dev Behav 8:111–158. https://doi.org/10.1016/S0065-2407(08)60494-2
Hunter RG, Murakami G, Dewell S, Seligsohn M, Baker ME, Datson NA et al (2012) Acute stress and hippocampal histone H3 lysine 9 trimethylation, a retrotransposon silencing response. Proc Natl Acad Sci U S A 109(43):17657–17662. https://doi.org/10.1073/pnas.1215810109
Husquin LT, Rotival M, Fagny M, Quach H, Zidane N, Mcewen LM et al (2018) Exploring the genetic basis of human population differences in DNA methylation and their causal impact on immune gene regulation. Genome Biol 19(222). https://doi.org/10.1186/s13059-018-1601-3

Hutchinson G, Takei N, Fahy T, Gilvarry C, Hutchinson G, Takei N et al (1997) Morbid risk of schizophrenia in first-degree relatives of White and African-Caribbean patients with psychosis. Br J Psychiatry 169:776–780. https://doi.org/10.1192/bjp.169.6.776

Imai S, Kikuchi R, Kusuhara H, Sugiyama Y (2013) DNA methylation and histone modification profiles of mouse organic anion transporting polypeptides. Drug Metab Dispos 41:72–78. https://doi.org/10.1124/dmd.112.047969

Ingram RE, Luxton DD (2005) Vulnerability-stress models. In: Hankin BL, Abela JRZ (eds) Development of psychopathology: a vulnerability stress perspective. Sage, Thousand Oaks, pp 32–46. https://us.corwin.com/sites/default/files/upm-binaries/5348_Hankin_Final_pdf_Chapter_2.pdf

Islam SA, Goodman SJ, Macisaac JL, Obradović J, Barr RG, Boyce WT, Kobor MS (2019) Integration of DNA methylation patterns and genetic variation in human pediatric tissues help inform EWAS design and interpretation. Epigenetics Chromatin 12(1). https://doi.org/10.1186/s13072-018-0245-6

Janeway CA (2001) How the immune system works to protect the host from infection: a personal view. Proc Natl Acad Sci 98(13):7461–7468. https://doi.org/10.1073/pnas.131202998

Jayaraman S, Fausey CM, Smith LB (2017) Why are faces denser in the visual experiences of younger than older infants? Dev Psychol 53(1):38–49. https://doi.org/10.1037/dev0000230

Jiang Y-H, Bressler J, Beaudet AL (2004) Epigenetics and human disease definitions and historical perspective. Annu Rev Genomics Hum Genet 5:479–510. https://doi.org/10.1146/annurev.genom.5.061903.180014

John S, Sabo PJ, Thurman RE, Sung M-H, Biddie SC, Johnson TA et al (2011) Chromatin accessibility pre-determines glucocorticoid receptor binding patterns. Nat Genet 43 (3):264–268. https://doi.org/10.1038/ng.759

Johnson MH, Dziurawiec S, Ellis H, Morton J (1991) Newborns' preferential tracking of face-like stimuli and its subsequent decline. Cognition. https://doi.org/10.1016/0010-0277(91)90045-6

Johnson MH, Senju A, Tomalski P (2015) The two-process theory of face processing: modifications based on two decades of data from infants and adults. Neurosci Biobehav Rev. https://doi.org/10.1016/j.neubiorev.2014.10.009

Jones MJ, Goodman SJ, Kobor MS (2015) DNA methylation and healthy human aging. Aging Cell 14(6):924–932. https://doi.org/10.1111/acel.12349

Jones MJ, Moore SR, Kobor MS (2018) Annual review of psychology principles and challenges of applying epigenetic epidemiology to psychology. Annu Rev Psychol 69:459–485. https://doi.org/10.1146/annurev-psych-122414

Josselyn SA (2005) What's right with my mouse model? New insights into the molecular and cellular basis of cognition from mouse models of Rubinstein-Taybi syndrome. Learn Mem 12 (2):80–83. https://doi.org/10.1101/lm.93505

Jylhävä J, Jiang M, Foebel AD, Pedersen NL, Hägg S (2019) Can markers of biological age predict dependency in old age? Biogerontology 20:321–329. https://doi.org/10.1007/s10522-019-09795-5

Kaliman P, Párrizas M, Lalanza JF, Camins A, Escorihuela RM, Pallàs M (2011) Neurophysiological and epigenetic effects of physical exercise on the aging process. Ageing Res Rev. https://doi.org/10.1016/j.arr.2011.05.002

Kalkhoven E, Roelfsema JH, Teunissen H, den Boer A, Ariyurek Y, Zantema A et al (2003) Loss of CBP acetyltransferase activity by PHD finger mutations in Rubinstein-Taybi syndrome. Hum Mol Genet 12(4):441–450. https://doi.org/10.1093/hmg/ddg039

Kang HJ, Kim JM, Lee JY, Kim SY, Bae KY, Kim SW et al (2013) BDNF promoter methylation and suicidal behavior in depressive patients. J Affect Disord. https://doi.org/10.1016/j.jad.2013.08.001

Kashimoto RK, Toffoli LV, Manfredo MHF, Volpini VL, Martins-Pinge MC, Pelosi GG, Gomes MV (2016) Physical exercise affects the epigenetic programming of rat brain and modulates the adaptive response evoked by repeated restraint stress. Behav Brain Res. https://doi.org/10.1016/j.bbr.2015.08.038

Kavalali ET, Nelson ED, Monteggia LM, Kavalali ET, Nelson ED, Monteggia LM (2011) Role of MeCP2, DNA methylation, and HDACs in regulating synapse function. J Neurodev Disord 3:250–256. https://doi.org/10.1007/s11689-011-9078-3

Kim VN (2005) MicroRNA biogenesis: coordinated cropping and dicing. Nat Rev Mol Cell Biol 6:376–385. https://doi.org/10.1038/nrm1644

Kim S, Kaang B-K (2017) Epigenetic regulation and chromatin remodeling in learning and memory. Nat Publ Group 49:281. https://doi.org/10.1038/emm.2016.140

Kim VN, Nam JW (2006) Genomics of microRNA. Trends Genet. https://doi.org/10.1016/j.tig.2006.01.003

Kirchner H, Osler ME, Krook A, Zierath JR (2013) Epigenetic flexibility in metabolic regulation: disease cause and prevention? Trends Cell Biol. https://doi.org/10.1016/j.tcb.2012.11.008

Klengel T, Mehta D, Anacker C, Rex-Haffner M, Pruessner JC, Pariante CM et al (2013) Allele-specific FKBP5 DNA demethylation mediates gene-childhood trauma interactions. Nat Neurosci 16(1):33–41. https://doi.org/10.1038/nn.3275

Klengel T, Pape J, Binder EB, Mehta D (2014) The role of DNA methylation in stress-related psychiatric disorders. Neuropharmacology. https://doi.org/10.1016/j.neuropharm.2014.01.013

Koch CM, Wanger W (2011) Epigenetic-aging-signature to determine age in different tissues. Aging 3(10). www.impactaging.com

Koob GF (1996) Drug addition: the yin and yang of hedonic homeostasis. Nature 16:893–896. https://www.cell.com/neuron/pdf/S0896-6273(00)80109-9.pdf

Korzus E, Rosenfeld MG, Mayford M (2004) CBP histone acetyltransferase activity is a critical component of memory consolidation. Neuron. https://doi.org/10.1016/j.neuron.2004.06.002

Koshibu K, Gräff J, Beullens M, Heitz FD, Berchtold D, Russig H et al (2009) Behavioral/systems/cognitive protein phosphatase 1 regulates the histone code for long-term memory. J Neurosci 29(41):13079–13089. https://doi.org/10.1523/JNEUROSCI.3610-09.2009

Kouzarides T (2007) Chromatin modifications and their function. Cell. https://doi.org/10.1016/j.cell.2007.02.005

Kozorovitskiy Y, Gross CG, Kopil C, Battaglia L, Mcbreen M, Stranahan AM, Gould E (2005) Experience induces structural and biochemical changes in the adult primate brain. Proc Natl Acad Sci 102:17478–17482. https://doi.org/10.1073/pnas.0508817102

Kuzumaki N, Ikegami D, Tamura R, Hareyama N, Imai S, Narita M et al (2011) Hippocampal epigenetic modification at the brain-derived neurotrophic factor gene induced by an enriched environment. Hippocampus 21(2):127–132. https://doi.org/10.1002/hipo.20775

LaGraize SC, Borzan J, Peng YB, Fuchs PN (2006) Selective regulation of pain affect following activation of the opioid anterior cingulate cortex system. Exp Neurol. https://doi.org/10.1016/j.expneurol.2005.05.008

Lam LL, Emberly E, Fraser HB, Neumann SM, Chen E, Miller GE, Kobor MS (2012) Factors underlying variable DNA methylation in a human community cohort. Proc Natl Acad Sci U S A 109:17253–17260. https://doi.org/10.1073/pnas.1121249109

Lander SS, Linder-Shacham D, Gaisler-Salomon I (2017) Differential effects of social isolation in adolescent and adult mice on behavior and cortical gene expression. Behav Brain Res. https://doi.org/10.1016/j.bbr.2016.09.005

Laumbach RJ, Kipen HM (2012) Respiratory health effects of air pollution: update on biomass smoke and traffic pollution. J Allergy Clin Immunol 129(1):3–11

LaVeist T, Pollack K, Thorpe R, Fesahazion R, Gaskin D (2011) Place, not race: disparities dissipate in southwest Baltimore when blacks and whites live under similar conditions. Health Aff 30(10):1880–1887

Lee JT (2012) Epigenetic regulation by long noncoding RNAs. Science 338(6113):1435–1439. https://doi.org/10.1126/science.1231776

Leleu A, Rekow D, Poncet F, Rossion B, Durand K, Schaal B, Baudouin J-Y (2019) Maternal odor shapes rapid face categorization in the 4-month-old infant brain. J Vis 18(10):787. https://doi.org/10.1167/18.10.787

Levenson JM, Sweatt JD (2011) Epigenetic mechanisms in memory formation. Epigenet Hum Health 6:287–300. https://doi.org/10.1007/978-3-642-17426-1_14

Levenson JM, O'riordan KJ, Brown KD, Trinh MA, Molfese DL, David Sweatt J (2004) Regulation of histone acetylation during memory formation in the Hippocampus. J Biol Chem 279(39):40545–40559. https://doi.org/10.1074/jbc.M402229200

Levine A, Worrell TR, Zimnisky R, Schmauss C (2012) Early life stress triggers sustained changes in histone deacetylase expression and histone H4 modifications that alter responsiveness to adolescent antidepressant treatment. Neurobiol Dis. https://doi.org/10.1016/j.nbd.2011.09.005

Li H, Zhong X, Fongching Chau K, Cunningham Williams E, Chang Q, Neurosci N (2012) Loss of activity-induced phosphorylation of MeCP2 enhances synaptogenesis, LTP, and spatial memory. Nat Neurosci 14(8):1001–1008. https://doi.org/10.1038/nn.2866

Lienert F, Wirbelauer C, Schübeler D (2011) Identification of genetic elements that autonomously determine DNA methylation states. Nat Genet 43(11):1091. https://doi.org/10.1038/ng.946

Lillycrop KA, Burdge GC (2012) Epigenetic mechanisms linking early nutrition to long term health. Best Pract Res Clin Endocrinol Metab. https://doi.org/10.1016/j.beem.2012.03.009

Ling C, Rönn T (2014) Epigenetic adaptation to regular exercise in humans. Drug Discov Today. https://doi.org/10.1016/j.drudis.2014.03.006

Lipovich L, Dachet F, Cai J, Bagla S, Balan K, Jia H, Loeb JA (2012) Activity-dependent human brain coding/noncoding gene regulatory networks. Genetics 192(3):1133–1148. https://doi.org/10.1534/genetics.112.145128

Lippmann M, Bress A, Nemeroff CB, Plotsky PM, Monteggia LM (2007) Long-term behavioural and molecular alterations associated with maternal separation in rats. Eur J Neurosci 25 (10):3091–3098. https://doi.org/10.1111/j.1460-9568.2007.05522.x

Liu Z, Leung D, Levine M (2019) Comparative analysis of epigenetic aging clocks from CpG characteristics to functional associations. BioRxiv 1:512483. https://doi.org/10.1101/512483

Lubin FD, Roth TL, Sweatt JD (2008) Epigenetic regulation of BDNF gene transcription in the consolidation of fear memory. J Neurosci 28(4):10576–10586. https://doi.org/10.1523/JNEUROSCI.1786-08.2008

Luthar SS (2003) Resilience and vulnerability: adaptation in the context of childhood adversities. Cambridge University Press, New York

Lutz PE, Turecki G (2014) DNA methylation and childhood maltreatment: from animal models to human studies. Neuroscience. https://doi.org/10.1016/j.neuroscience.2013.07.069

Luyten LJ, Saenen ND, Janssen BG, Vrijens K, Plusquin M, Roels HA et al (2018) Air pollution and the fetal origin of disease: a systematic review of the molecular signatures of air pollution exposure in human placenta. Environ Res. https://doi.org/10.1016/j.envres.2018.03.025

Macrì S, Laviola G, Leussis MP, Andersen SL (2010) Abnormal behavioral and neurotrophic development in the younger sibling receiving less maternal care in a communal nursing paradigm in rats. Psychoneuroendocrinology. https://doi.org/10.1016/j.psyneuen.2009.07.016

Mahan AL, Mou L, Shah N, Hu J-H, Worley PF, Ressler KJ (2012) Epigenetic modulation of Homer1a transcription regulation in amygdala and hippocampus with pavlovian fear conditioning. J Neurosci 32(13):4651–4659. https://doi.org/10.1523/JNEUROSCI.3308-11.2012

Mahnke AH, Miranda RC, Homanics GE (2017) Epigenetic mediators and consequences of excessive alcohol consumption. Alcohol. https://doi.org/10.1016/j.alcohol.2017.02.357

Mair C, Roux AVD, Galea S (2008) Are neighborhood characteristics associated with depressive symptoms? A critical review. J Epidemiol Community Health 62(11):940–946

Malenka RC, Bear MF (2004) LTP and LTD: an embarrassment of riches. Neuron. https://doi.org/10.1016/j.neuron.2004.09.012

Malenka RC, Nicoll RA (1999) Long-term potentiation – a decade of progress? Science 285 (5435):1870–1874. http://science.sciencemag.org/

Malenka RC, Kauer JA, Perkel DJ, Mauk MD, Kelly PT, Nicoll RA, Waxham MN (1989) An essential role for postsynaptic calmodulin and protein kinase activity in long-term potentiation. Nature 340(6234):554–557. https://www.nature.com/articles/340554a0.pdf

Malik AN, Vierbuchen T, Hemberg M, Rubin AA, Ling E, Couch CH et al (2014) Genome-wide identification and characterization of functional neuronal activity-dependent enhancers. Nat Neurosci 17(10):1330–1339. https://doi.org/10.1038/nn.3808

Maloney R, Stilling RM, Dinan TG, Cryan JF (2015) Early-life stress-induced visceral hypersensitivity and anxiety behavior is reversed by histone deacetylase inhibition. Neurogastroenterol Motil 27:1831–1836

Mandy M, Nyirenda M (2018) Developmental origins of health and disease: the relevance to developing nations. Int Health 10:66–70. https://doi.org/10.1093/inthealth/ihy006

March D, Hatch SL, Morgan C, Kirkbride JB, Bresnahan M, Fearon P, Susser E (2008) Psychosis and place. Epidemiol Rev 30:84–100. https://doi.org/10.1093/epirev/mxn006

Marioni RE, Shah S, Mcrae AF, Ritchie SJ, Muniz-Terrera G, Harris SE et al (2015) The epigenetic clock is correlated with physical and cognitive fitness in the Lothian Birth Cohort 1936. Int J Epidemiol 44(4):1388–1396. https://doi.org/10.1093/ije/dyu277

Martin E (2008) A dictionary of biology, 6th edn. Oxford University Press, Oxford

Marzi SJ, Sugden K, Arseneault L, Belsky DW, Burrage J, Corcoran DL et al (2018) Analysis of DNA methylation in young people: limited evidence for an association between victimization stress and epigenetic variation in blood. Am J Psychiatr 175:517–529. https://doi.org/10.1176/appi.ajp.2017.17060693

Masataka N (1998) Perception of motherese in Japanese sign language by 6-month-old hearing infants. Dev Psychol 34(2):241

Matas E, Bock J, Braun K (2016) The impact of parent-infant interaction on epigenetic plasticity mediating synaptic adaptations in the infant brain. Psychopathology 49(4):201–210

Mathers JC (2006) Nutritional modulation of ageing: genomic and epigenetic approaches. Mech Ageing Dev. https://doi.org/10.1016/j.mad.2006.01.018

Matthews KA, Gallo LC (2011) Psychological perspectives on pathways linking socioeconomic status and physical health. Annu Rev Psychol 62:501–530. https://doi.org/10.1146/annurev.psych.031809.130711

Matthews KA, Gallo LC, Taylor SE (2010) Are psychosocial factors mediators of socioeconomic status and health connections? Ann N Y Acad Sci 186(1):146–173

Maud C, Ryan J, Mcintosh JE, Olsson CA (2018) The role of oxytocin receptor gene (OXTR) DNA methylation (DNAm) in human social and emotional functioning: a systematic narrative review. BMC Psychiatry 18(154). https://doi.org/10.1186/s12888-018-1740-9

McCartney DL, Stevenson AJ, Hillary RF, Walker RM, Bermingham ML, Morris SW et al (2018) Epigenetic signatures of starting and stopping smoking. EBioMedicine. https://doi.org/10.1016/j.ebiom.2018.10.051

McCrory C, Fiorito G, Ni Cheallaigh C, Polidoro S, Karisola P, Alenius H et al (2019) How does socio-economic position (SEP) get biologically embedded? A comparison of allostatic load and the epigenetic clock(s). Psychoneuroendocrinology. https://doi.org/10.1016/j.psyneuen.2019.02.018

Mcdade TW, Ryan C, Jones MJ, Macisaac JL, Morin AM, Meyer JM et al (2017) Social and physical environments early in development predict DNA methylation of inflammatory genes in young adulthood. Proc Natl Acad Sci U S A 14(29):7611–7618. https://doi.org/10.1073/pnas.1620661114

Mcdade TW, Ryan CP, Jones MJ, Hoke MK, Borja J, Miller GE et al (2019) Genome-wide analysis of DNA methylation in relation to socioeconomic status during development and early adulthood. Am J Phys Anthropol 169:3–11. https://doi.org/10.1002/ajpa.23800

McEwen LM, Goodman SJ, Kobor MS, Jones MJ (2016) The DNA methylome: an interface between the environment, immunity, and ageing. In: Bueno V, Lord J, Jackson T, Lord JM (eds) The ageing immune system and health. Springer, Cham, pp 35–52. https://doi.org/10.1007/978-3-319-43365-3

Mcewen LM, Morin AM, Edgar RD, Macisaac JL, Jones MJ, Dow WH et al (2017) Differential DNA methylation and lymphocyte proportions in a Costa Rican high longevity region. Epigenetics Chromatin:1–14. https://doi.org/10.1186/s13072-017-0128-2

Mcgowan PO, Sasaki A, D'alessio AC, Dymov S, Labonté B, Szyf M et al (2009) Epigenetic regulation of the glucocorticoid receptor in human brain associates with childhood abuse. Nat Neurosci 12(3):342–348. https://doi.org/10.1038/nn.2270

Mcgowan PO, Suderman M, Sasaki A, Huang TCT, Hallett M, Meaney MJ, Szyf M (2011) Broad epigenetic signature of maternal care in the brain of adult rats. PLoS One 6(2):e14739. https://doi.org/10.1371/journal.pone.0014739

Mclaughlin KA, Sheridan MA, Lambert HK, Mclaughlin K (2014) Childhood adversity and neural development: deprivation and threat as distinct dimensions of early experience. Neurosci Biobehav Rev 47:578–591. https://doi.org/10.1016/j.neubiorev.2014.10.012

McNally RJ (2016) The legacy of Seligman's "Phobias and Preparedness" (1971). Behav Ther. https://doi.org/10.1016/j.beth.2015.08.005

Mcvicker G, Van De Geijn B, Degner JF, Cain CE, Banovich NE, Raj A et al (2013) Identification of genetic variants that affect histone modifications in human cells. Science 342(6159):747–749. https://doi.org/10.1126/science.1242463

Meaney MJ (2010) Epigenetics and the biological definition of gene X environment interactions. Child Dev 81(1):41–79. https://doi.org/10.1111/j.1467-8624.2009.01381.x

Mehta D, Klengel T, Conneely KN, Smith AK, Altmann A, Pace TW et al (2013) Childhood maltreatment is associated with distinct genomic and epigenetic profiles in posttraumatic stress disorder. Proc Natl Acad Sci U S A 110(20):8302–8307. https://doi.org/10.1073/pnas.1217750110

Meltzoff AN (1988) Imitation of televised models by infants. Child Dev 59(5):1221–1229. https://www.ncbi.nlm.nih.gov/pmc/articles/PMC3651023/pdf/nihms460666.pdf

Mercer TR, Dinger ME, Mattick JS (2009) Long non-coding RNAs: insights into functions. Nat Rev Genet 10:155–159. www.nature.com/reviews/genetics

Merkin SS, Basurto-Dávila R, Karlamangl A, Bird CE, Lurie N, Escarce J, Seeman T (2009) Neighborhoods and cumulative biological risk profiles by race/ethnicity in a national sample of U.S. adults: NHANES III. Ann Epidemiol 19(3):194–201

Meyer KD, Jaffrey SR (2014) The dynamic epitranscriptome: N 6-methyladenosine and gene expression control. Nat Rev Mol Cell Biol 15(5):313–326. https://doi.org/10.1038/nrm3785

Meyer ML, Williams KD, Eisenberger NI (2015) Why social pain can live on: different neural mechanisms are associated with reliving social and physical pain. PLoS One 10(6):e0128294. https://doi.org/10.1371/journal.pone.0128294

Mikkelsen K, Stojanovska L, Polenakovic M, Bosevski M, Apostolopoulos V (2017) Exercise and mental health. Maturitas. https://doi.org/10.1016/j.maturitas.2017.09.003

Milagro FI, Mansego ML, De Miguel C, Martínez JA (2013) Dietary factors, epigenetic modifications and obesity outcomes: progresses and perspectives. Mol Asp Med. https://doi.org/10.1016/j.mam.2012.06.010

Mill J, Tang T, Kaminsky Z, Khare T, Yazdanpanah S, Bouchard L et al (2008) Epigenomic profiling reveals DNA-methylation changes associated with major psychosis. Am J Hum Genet. https://doi.org/10.1016/j.ajhg.2008.01.008

Miller F, Osbahr H, Boyd E, Thomalla F, Bharwani S, Ziervogel G et al (2010) Resilience and vulnerability: complementary or conflicting concepts? Ecol Soc 15(3). https://www.jstor.org/stable/pdf/26268184.pdf?refreqid=excelsior%3A7a21a88b6284a48d4706ccae657c4dbc

Miller GE, Lachman ME, Chen E, Gruenewald TL, Karlamangla AS, Seeman TE (2011) Pathways to resilience: maternal nurturance as a buffer against the effects of childhood poverty on metabolic syndrome at midlife. Psychol Sci 22(12):1591–1599. https://doi.org/10.1177/0956797611419170

Miller GE, Yu T, Chen E, Brody GH (2015) Self-control forecasts better psychosocial outcomes but faster epigenetic aging in low-SES youth. Proc Natl Acad Sci 112(33):10325–10330. https://doi.org/10.1073/PNAS.1505063112

Minichiello L (2009) Long-term potentiation synaptic plasticity TrkB signalling pathways in LTP and learning. Nat Rev Neurosci 10. https://doi.org/10.1038/nrn2738

Miyashita T, Kubik S, Haghighi N, Steward O, Guzowski JF (2009) Rapid activation of plasticity-associated gene transcription in hippocampal neurons provides a mechanism for encoding of one-trial experience. J Neurosci 29(4):898–906. https://doi.org/10.1523/JNEUROSCI.4588-08.2009

Miyatake M, Narita M, Shibasaki M, Nakamura A, Suzuki T (2005) Glutamatergic neurotransmission and protein kinase C play a role in neuron-glia communication during the development of methamphetamine-induced psychological dependence. Eur J Neurosci 22 (6):1476–1488. https://doi.org/10.1111/j.1460-9568.2005.04325.x

Mohai P, Lantz PM, Morenoff J, House JS, Mero RP (2009) Racial and socioeconomic disparities in residential proximity to polluting industrial facilities: evidence from the Americans' changing lives study. Am J Public Health 99(Suppl 3):S649–S656

Moore SR (2017) Commentary: What is the case for candidate gene approaches in the era of high-throughput genomics? A response to Border and Keller (2017). J Child Psychol Psychiatry 58:331–334. https://doi.org/10.1111/jcpp.12697

Moore SR, Depue RA (2016) Neurobehavioral foundation of environmental reactivity. Psychol Bull 142(2):107–164. https://doi.org/10.1037/bul0000028

Moore SR, Kobor MS (2018) Variability in DNA methylation at the serotonin transporter gene promoter: epigenetic mechanism or cell-type artifact? Mol Psychiatry 1. https://doi.org/10.1038/s41380-018-0121-6

Moore SR, Mcewen LM, Mah SM, Kobor MS (2017) Epigenetic correlates of neonatal contact in humans. Dev Psychopathol 29:1517–1538. https://doi.org/10.1017/S0954579417001213

Morello-Frosch R, Zuk M, Jerrett M, Shamasunder B, Kyle AD (2011) Understanding the cumulative impacts of inequalities in environmental health: implications for policy. Health Aff 30 (5):879–887

Morse S, Butler A, Davis R, Soller I, Lubin F (2015) Environmental enrichment reverses histone methylation changes in the aged hippocampus and restores age-related memory deficits. Biology 4(2):298–313. https://doi.org/10.3390/biology4020298

Morton J, Johnson MH (1991) CONSPEC and CONLERN: a two-process theory of infant face recognition. Psychol Rev 98(2):164–181. https://philpapers.org/archive/MORCAC-15.pdf

Moseley GL, Butler DS (2015) Fifteen years of explaining pain: the past, present, and future. J Pain. https://doi.org/10.1016/j.jpain.2015.05.005

Moszkowski RJ, Stack DM (2007) Infant and child development infant touching behaviour during mother-infant face-to-face interactions. Infant Child Dev 16:307–319. https://doi.org/10.1002/icd.510

Moylan S, Eyre HA, Maes M, Baune BT, Jacka FN, Berk M (2013) Exercising the worry away: how inflammation, oxidative and nitrogen stress mediates the beneficial effect of physical activity on anxiety disorder symptoms and behaviours. Neurosci Biobehav Rev. https://doi.org/10.1016/j.neubiorev.2013.02.003

Mukherjee D, Ignatowska-Jankowska BM, Itskovits E, Gonzales BJ, Turm H, Izakson L et al (2018) Salient experiences are represented by unique transcriptional signatures in the mouse brain. elife 7:1–20. https://doi.org/10.7554/elife.31220

Mustafíc H, Jabre P, Caussin C, Murad MH, Escolano S, Tafflet M et al (2012) Main air pollutants and myocardial infarction. J Am Med Assoc 307(7):713–721

Nair A, Vadodaria K, Banerjee SB, Benekareddy M, Dias BG, Dunman RS, Vaidya VA (2007) Stressor-specific regulation of distinct brain-derived neurotrophic factor transcripts and cyclic AMP response element-binding protein expression in the postnatal and adult rat hippocampus. Neuropsychopharmacology 32(7):1504–1520. https://doi.org/10.1038/sj.npp.1301276

Navarro E, Funtikova AN, Fíto M, Schröder H (2017) Prenatal nutrition and the risk of adult obesity: long-term effects of nutrition on epigenetic mechanisms regulating gene expression. J Nutr Biochem. https://doi.org/10.1016/j.jnutbio.2016.03.012

Niaura R, Shadel WG, Britt DM, Abrams DB (2002) Response to social stress, urge to smoke, and smoking cessation. Addict Behav. https://doi.org/10.1016/S0306-4603(00)00180-5

Odell DW (2018) Epigenetics of pain mediators. Curr Opin Anaesthesiol 31(4):402–406. https://doi.org/10.1097/ACO.0000000000000613
Ohman A, Mineka S (2001) Fears, phobias, and preparedness: toward an evolved module of fear and fear learning. Psychol Rev 108(3):483–522. https://doi.org/10.1037/0033-295X.108.3.483
Oliveira AMM, Wood MA, Mcdonough CB, Abel T (2007) Transgenic mice expressing an inhibitory truncated form of p300 exhibit long-term memory deficits. Learn Mem 14:564–572. https://doi.org/10.1101/lm.656907
Ong M-L, Tuan TA, Poh J, Teh AL, Chen L, Pan H et al (2019) Neonatal amygdalae and hippocampi are influenced by genotype and prenatal environment, and reflected in the neonatal DNA methylome. Genes Brain Behav:e12576. https://doi.org/10.1111/gbb.12576
Ostlund BD, Conradt E, Crowell SE, Tyrka AR, Marsit CJ, Lester BM (2016) Prenatal stress, fearfulness, and the epigenome: exploratory analysis of sex differences in DNA methylation of the glucocorticoid receptor gene. Front Behav Neurosci 10:1–8. https://doi.org/10.3389/fnbeh.2016.00147
Ozbay F, Johnson DC, Dimoulas E, Morgan CA, Charney D, Southwick S (2007) Social support and resilience to stress: from neurobiology to clinical practice. Psychiatry 4(5):35
Ozbay F, Fitterling H, Charney D, Southwick S (2008) Social support and resilience to stress across the life span: a neurobiologic framework. Curr Psychiatry Rep 10(4):304
Paczkowski MM, Galea S (2010) Sociodemographic characteristics of the neighborhood and depressive symptoms. Curr Opin Psychiatry 23(4):337–341
Pang KKL, Sharma M, Sajikumar S (2019) Epigenetics and memory: emerging role of histone lysine methyltransferase G9a/GLP complex as bidirectional regulator of synaptic plasticity. Neurobiol Learn Mem. https://doi.org/10.1016/j.nlm.2019.01.013
Papini MR (2009) Role of opioid receptors in incentive contrast publication date. Int J Comp Psychol 22(3). https://escholarship.org/uc/item/0zk5h8n2
Papini MR, Fuchs PN, Torres C (2015) Behavioral neuroscience of psychological pain. Neurosci Biobehav Rev. https://doi.org/10.1016/j.neubiorev.2014.11.012
Park H, Poo M-M (2013) Neurotrophin regulation of neural circuit development and function. Nat Rev Neurosci 14(7). https://doi.org/10.1038/nrn3379
Peixoto L, Abel T (2012) The role of histone acetylation in memory formation and cognitive impairments. Neuropsychopharmacology 38:62–76. https://doi.org/10.1038/npp.2012.86
Perry BD (2002) Childhood experience and the expression of genetic potential: what childhood neglect tells us about nature and nurture. Brain Mind 3. https://link.springer.com/content/pdf/10.1023/A:1016557824657.pdf
Pickett KE, Pearl M (2001) Multilevel analyses of neighbourhood socioeconomic context and health outcomes: a critical review. J Epidemiol Community Health 55(2):111–122
Pietromonaco PR, Feldman L, Boston College B, Feldman Barrett L (2000) The internal working models concept: what do we really know about the self in relation to others? Rev Gen Psychol 4 (2):155–175. https://doi.org/10.1037/1089-2680.4.2.155
Pietromonaco PR, Feldman L, College BB, Powers SI, Snyder IDK, Simpson JA, Hughes JN (2006) Adult attachment theory and affective reactivity and regulation. In: Snyder DK, Simpson JA, Hughes JN (eds) Emotion regulation in couples and families: pathways to dysfunction and health. American Psychological Association, Washington, pp 57–74. https://www.researchgate.net/profile/Sally_Powers/publication/41528849_Adult_attachment_theory_and_affective_reactivity_and_regulation/links/02e7e524a04d18cb66000000/Adult-attachment-theory-and-affective-reactivity-and-regulation.pdf
Pietromonaco PR, Barrett LF (2000) Attachment theory as an organizing framework: a view from different levels of analysis. Rev Gen Psychol 4(2):107–110. https://doi.org/10.1037/1089-2680.4.2.107
Ponomarev I, Stelly CE, Morikawa H, Blednov YA, Mayfield RD, Harris RA (2017) Mechanistic insights into epigenetic modulation of ethanol consumption. Alcohol. https://doi.org/10.1016/j.alcohol.2017.01.016

Poo M, Pignatelli M, Ryan TJ, Tonegawa S, Bonhoeffer T, Martin KC et al (2016) What is memory? The present state of the engram. BMC Biol 14(1):40. https://doi.org/10.1186/s12915-016-0261-6

Porges S (2011) The polyvagal theory: neurophysiological foundations of emotions, attachment, communication, and self-regulation. Norton, New York

Proag V (2014) The concept of vulnerability and resilience. Procedia Econ Financ. https://doi.org/10.1016/s2212-5671(14)00952-6

Probst AV, Dunleavy E, Almouzni G (2009) Epigenetic inheritance during the cell cycle. Nat Rev Mol Cell Biol 10:192–206. https://doi.org/10.1038/nrm2640

Qi F, Zhou Y, Xiao Y, Tao J, Gu J, Jiang X, Xu GY (2013) Promoter demethylation of cystathionine-β-synthetase gene contributes to inflammatory pain in rats. Pain. https://doi.org/10.1016/j.pain.2012.07.031

Ramírez-Rodríguez G, Ocaña-Fernández MA, Vega-Rivera NM, Torres-Pérez OM, Gómez-Sánchez A, Estrada-Camarena E, Ortiz-López L (2014) Environmental enrichment induces neuroplastic changes in middle age female BalbC mice and increases the hippocampal levels of BDNF, p-Akt and p-MAPK1/2. Neuroscience. https://doi.org/10.1016/j.neuroscience.2013.12.026

Randic M, Jiang MC, Cerne R (1993) Long-term potentiation and long-term depression of primary afferent neurotransmission in the rat spinal cord. J Neurosci 13. http://www.jneurosci.org/content/jneuro/13/12/5228.full.pdf

Reid VM, Dunn K, Young RJ, Amu J, Donovan T, Reissland N (2017) The human fetus preferentially engages with face-like visual stimuli. Curr Biol. https://doi.org/10.1016/j.cub.2017.05.044

Repetti R, Taylor S, Seeman TE (2014) Risky families: family social environments and the mental and physical health of offspring. Psychol Bull 128(2):330–366. https://doi.org/10.1037/0033-2909.128.2.230

Rial Verde EM, Lee-Osbourne J, Worley PFF, Malinow R, Cline HTT (2006) Increased expression of the immediate-early gene arc/Arg3.1 reduces AMPA receptor-mediated synaptic transmission. Neuron. https://doi.org/10.1016/j.neuron.2006.09.031

Rider CF, Yamamoto M, Günther OP, Hirota JA, Singh A, Tebbutt SJ, Carlsten C (2016) Controlled diesel exhaust and allergen coexposure modulates microRNA and gene expression in humans: effects on inflammatory lung markers. J Allergy Clin Immunol. https://doi.org/10.1016/j.jaci.2016.02.038

Roberts AL, Gladish N, Gatev E, Jones MJ, Chen Y, Macisaac JL et al (2018) Exposure to childhood abuse is associated with human sperm DNA methylation. Transl Psychiatry 8:194. https://doi.org/10.1038/s41398-018-0252-1

Rodrigues GM, Toffoli LV, Manfredo MH, Francis-Oliveira J, Silva AS, Raquel HA et al (2015) Acute stress affects the global DNA methylation profile in rat brain: modulation by physical exercise. Behav Brain Res. https://doi.org/10.1016/j.bbr.2014.11.023

Rosenzweig MJ, Renner MR (1986) Social interactions among rats housed in grouped and enriched conditions. Dev Psychobiol 19(4):303–313

Roth TL, Sweatt JD (2011a) Annual research review: epigenetic mechanisms and environmental shaping of the brain during sensitive periods of development. J Child Psychol Psychiatry 52(4):398–408. https://doi.org/10.1111/j.1469-7610.2010.02282.x

Roth TL, Sweatt JD (2011b) Epigenetic marking of the BDNF gene by early-life adverse experiences. Horm Behav. https://doi.org/10.1016/j.yhbeh.2010.05.005

Roth TL, Lubin FD, Funk AJ, Sweatt JD (2009) Lasting epigenetic influence of early-life adversity on the BDNF gene. Biol Psychiatry 65(9):760–769. https://doi.org/10.1016/j.biopsych.2008.11.028

Roth TL, Zoladz PR, Sweatt JD, Diamond DM (2011) Epigenetic modification of hippocampal Bdnf DNA in adult rats in an animal model of post-traumatic stress disorder. J Psychiatr Res. https://doi.org/10.1016/j.jpsychires.2011.01.013

Roux AVD, Mair C (2010) Neighborhoods and health. Ann N Y Acad Sci 186(1):125–145

Rutering J, Ilmer M, Recio A, Coleman M, Vykoukal J, Alt E, Orleans N (2016) Traumatic stress and accelerated DNA methylation age: a meta-analysis. Nat Rev Drug Discov 5(6):1–8. https://doi.org/10.4172/2157-7633.1000305.Improved

Rye M, Sandve GK, Daub CO, Kawaji H, Carninci P, Forrest A, Drabløs F (2014) Chromatin states reveal functional associations for globally defined transcription start sites in four human cell lines. BMC Genomics 15:120. https://doi.org/10.1186/1471-2164-15-120

Saletore Y, Meyer K, Korlach J, Vilfan ID, Jaffrey S, Mason CE (2012) The birth of the epitranscriptome: deciphering the function of RNA modifications. Genome Biol 13(175). https://doi.org/10.1186/gb-2012-13-10-175

Sallis JF, Glanz K (2006) The role of built environments in physical activity, eating, and obesity in childhood. Future Child 16(1):89–108

Sandkühler J, Gruber-Schoffnegger D (2012) Hyperalgesia by synaptic long-term potentiation (LTP): an update. Curr Opin Pharmacol. https://doi.org/10.1016/j.coph.2011.10.018

Saul MC, Seward CH, Troy JM, Zhang H, Sloofman LG, Lu X et al (2017) Transcriptional regulatory dynamics drive coordinated metabolic and neural response to social challenge in mice. Genome Res 27(6):959–972. https://doi.org/10.1101/gr.214221.116

Sayette MA (2017) The effects of alcohol on emotion in social drinkers. Behav Res Ther. https://doi.org/10.1016/j.brat.2016.06.005

Schaefer A, Sampath SC, Intrator A, Min A, Gertler TS, Surmeier DJ et al (2009) Control of cognition and adaptive behavior by the GLP/G9a epigenetic suppressor complex. Neuron. https://doi.org/10.1016/j.neuron.2009.11.019

Schwartz S (2016) Cracking the epitranscriptome. RNA 22:169–174. https://doi.org/10.1261/rna.054502.115

Schwarz JM, Hutchinson MR, Bilbo SD (2011) Early-life experience decreases drug-induced reinstatement of morphine CPP in adulthood via microglial-specific epigenetic programming of anti-inflammatory IL-10 expression. J Neurosci 32(49):17835–17847. https://doi.org/10.1523/JNEUROSCI.3297-11.2011

Selten J-P (2005) Schizophrenia and migration: a meta-analysis and review. Am J Psychiatr 162:12–24. https://doi.org/10.1176/appi.ajp.162.1.12

Sen P, Shah PP, Nativio R, Berger SL (2016) Epigenetic mechanisms of longevity and aging. Cell. https://doi.org/10.1016/j.cell.2016.07.050

Seo MK, Ly NN, Lee CH, Cho HY, Choi CM, Nhu LH et al (2016) Early life stress increases stress vulnerability through BDNF gene epigenetic changes in the rat hippocampus. Neuropharmacology. https://doi.org/10.1016/j.neuropharm.2016.02.009

Shahbazian MD, Grunstein M (2007) Functions of site-specific histone acetylation and deacetylation. Annu Rev Biochem 76:75–100. https://doi.org/10.1146/annurev.biochem.76.052705.162114

Shahbazian MD, Young JI, Yuva-Paylor LA, Spencer CM, Antalffy BA, Noebels JL et al (2002) Mice with truncated MeCP2 recapitulate many Rett syndrome features and display hyperacetylation of histone H3. Neuron. https://doi.org/10.1016/S0896-6273(02)00768-7

Sharma M, Sajikumar S (2018) G9a/GLP complex acts as a bidirectional switch to regulate metabotropic glutamate receptor-dependent plasticity in hippocampal CA1 pyramidal neurons. Cereb Cortex:1–15. https://doi.org/10.1093/cercor/bhy161

Shen X, Dong Y, Xu Z, Wang H, Miao C, Soriano SG et al (2013) Selective anesthesia-induced Neuroinflammation in developing mouse brain and cognitive impairment. Anesthesiology 118:502–515. www.anesthesiology.org

Shih J, May LDV, Gonzalez HE, Lee EW, Alvi RS, Sall JW et al (2012) Delayed environmental enrichment reverses sevoflurane-induced memory impairment in rats. Anesthesiology 116:586–602. www.anesthesiology.org

Shpigler HY, Saul MC, Murdoch EE, Cash-Ahmed AC, Seward CH, Sloofman L et al (2017) Behavioral, transcriptomic and epigenetic responses to social challenge in honey bees. Genes Brain Behav 16:579–591. https://doi.org/10.1111/gbb.12379

Simons LE, Elman I, Borsook D (2014) Psychological processing in chronic pain: a neural systems approach. Neurosci Biobehav Rev. https://doi.org/10.1016/j.neubiorev.2013.12.006
Simons RL, Lei MK, Beach SRH, Philibert RA, Cutrona CE, Gibbons FX, Barr A (2016) Economic hardship and biological weathering: the epigenetics of aging in a U.S. sample of black women. Soc Sci Med 150:192–200. https://doi.org/10.1016/j.socscimed.2015.12.001
Skene PJ, Illingworth RS, Webb S, Kerr ARW, James KD, Turner DJ et al (2010) Neuronal MeCP2 is expressed at near histone-octamer levels and globally alters the chromatin state. Mol Cell. https://doi.org/10.1016/j.molcel.2010.01.030
Somineni HK, Zhang X, Biagini Myers JM, Kovacic MB, Ulm A, Jurcak N et al (2016) Ten-eleven translocation 1 (TET1) methylation is associated with childhood asthma and traffic-related air pollution. J Allergy Clin Immunol. https://doi.org/10.1016/j.jaci.2015.10.021
Song Y, Ren H, Pei-Yuan Z (2017) Collaborations between CpG sites in DNA methylation. DNAMethy Int J Mod Phys B 6(7). https://doi.org/10.1101/149815
Stefanelli G, Azam AB, Walters BJ, Brimble MA, Gettens CP, Bouchard-Cannon P et al (2018) Learning and age-related changes in genome-wide H2A.Z binding in the mouse hippocampus. Cell Rep. https://doi.org/10.1016/j.celrep.2018.01.020
Suderman M, Borghol N, Pappas JJ, Pinto Pereira SM, Pembrey M, Hertzman C et al (2014) Childhood abuse is associated with methylation of multiple loci in adult DNA. BMC Med Genet 7(13):1–12. https://doi.org/10.1186/1755-8794-7-13
Suzuki K (2018) The developing world of DOHaD. J Dev Orig Health Dis 9(3):266–269. https://doi.org/10.1017/S2040174417000691
Swain JE, Lorberbaum JP, Kose S, Strathearn L (2007) Brain basis of early parent-infant interactions: psychology, physiology, and in vivo functional neuroimaging studies. J Child Psychol Psychiatry 48(3–4):262–287. https://doi.org/10.1111/j.1469-7610.2007.01731.x
Sweeting H, Hunt K (2015) Adolescent socioeconomic and school-based social status, smoking, and drinking. J Adolesc Health. https://doi.org/10.1016/j.jadohealth.2015.03.020
Tajerian M, Alvarado S, Millecamps M, Vachon P, Crosby C (2013) Peripheral nerve injury is associated with chronic, reversible changes in global DNA methylation in the mouse prefrontal cortex. PLoS One 8(1):55259. https://doi.org/10.1371/journal.pone.0055259
Takei N, Inamura N, Kawamura M, Namba H, Hara K, Yonezawa K, Nawa H (2004) Brain-derived neurotrophic factor induces mammalian target of rapamycin-dependent local activation of translation machinery and protein synthesis in neuronal dendrites. J Neurosci 24 (44):9760–9769. https://doi.org/10.1523/JNEUROSCI.1427-04.2004
Talge NM, Neal C, Glover V (2007) Antenatal maternal stress and long-term effects on child neurodevelopment: how and why? J Child Psychol Psychiatry Allied Discip 48(3–4):245–261. https://doi.org/10.1111/j.1469-7610.2006.01714.x
Tehranifar P, Wu H-C, Fan X, Flom JD, Ferris JS, Cho H et al (2013) Early life socioeconomic factors and genomic DNA methylation in mid-life. Epigenetics 8(1):23–27. https://doi.org/10.4161/epi.22989
Tessarz P, Kouzarides T (2014) Histone core modifications regulating nucleosome structure and dynamics. Nat Rev Mol Cell Biol. https://doi.org/10.1038/nrm3890
Thompson R (2000) The legacy of early attachments. Child Dev 71(1):145–152. https://onlinelibrary.wiley.com/doi/pdf/10.1111/1467-8624.00128
Thompson LA, Fagan JF, Fulker DW (1991) Longitudinal prediction of specific cognitive abilities from infant novelty preference. Child Dev 62(3):530–538. https://www.jstor.org/stable/pdf/1131128.pdf?casa_token=8L-b_hycWOsAAAAA:epxEIbY04JVtj-SYzGfR68HzIsPSZq13_bsbc9YTP_X21WbYEqMc_aBNG-sm5jJyIVD3jNAQ-uHI_Ha2IpxOIoscAru03l3XSFc7AS5VPLwslQtigw
Trehub SE, Trainor LJ, Unyk AM (1993) Music and speech processing in the first year of life. Adv Child Dev Behav. https://doi.org/10.1016/S0065-2407(08)60298-0
Troisi A, Frazzetto G, Carola V, Lorenzo GD (2011) Social neuroscience social headonic capacity is associated with the A118G polymorphism of the mu-opioid receptor gene (OPRM1) in adult

healthy volunteers and psychiatric patients. Soc Neurosci 6(1):88–97. https://doi.org/10.1080/17470919.2010.482786
Tulisiak CT, Harris RA, Ponomarev I (2017) DNA modifications in models of alcohol use disorders. Alcohol. https://doi.org/10.1016/j.alcohol.2016.11.004
Turati C, Simion F, Milani I, Umiltà C (2002) Newborns' preference for faces: what is crucial? Dev Psychol 38(6):875–882. https://doi.org/10.1037/0012-1649.38.6.875
Turecki G, Meaney MJ (2016) Effects of the social environment and stress on glucocorticoid receptor gene methylation: a systematic review. Biol Psychiatry 79:87–96. https://doi.org/10.1016/j.biopsych.2014.11.022
Turinsky AL, Butcher DT, Choufani S, Weksberg R, Brudno M (2019) Don't brush off buccal data heterogeneity. Epigenetics 14(2):109–117. https://doi.org/10.1080/15592294.2019.1581592
Tyssowski KM, DeStefino NR, Cho JH, Dunn CJ, Poston RG, Carty CE et al (2018) Different neuronal activity patterns induce different gene expression programs. Neuron. https://doi.org/10.1016/j.neuron.2018.04.001
Tzivian L (2011) Outdoor air pollution and asthma in children. J Asthma 48(5):470–481
Van Oers JAM, Bongers IMB, Van de Goor LAM, Garretsen HFL (1999) Alcohol consumption, and alcohol-related problems, problem drinking, and socioeconomic status. Alcohol Alcohol 34(1):78–88. https://watermark.silverchair.com/78.pdf?token=AQECAHi208BE49Ooan9kkhW_Ercy7Dm3ZL_9Cf3qfKAc485ysgAAAmIwggJeBgkqhkiG9w0BBwagggJPMIICSwIBADCCAkQGCSqGSIb3DQEHATAeBglghkgBZQMEAS4wEQQMJi-xPZN5VN26EWO3AgEQgIICFZilbhfvErPOA6vDopqm_HGA89b5lPjiUY5_VRqhxgB4Cp6gj879
Van Os J, Kenis G, Rutten BPF (2010) The environment and schizophrenia. Nature 468(7321):203. https://doi.org/10.1038/nature09563
Wadhwa PD, Buss C, Entringer S, Swanson JM (2009) Developmental origins of health and disease: brief history of the approach and current focus on epigenetic mechanisms. Semin Reprod Med 27(5):358–368. https://doi.org/10.1055/s-0029-1237424
Wagner W (2017) Epigenetic aging clocks in mice and men. Genome Biol 18(107). https://doi.org/10.1186/s13059-017-1203-5
Wagner M, Essica Steinbacher J, Kraus TFJ, Michalakis S, Pfaffeneder T, Perera A et al (2015) Age-dependent levels of 5-methyl-, 5-hydroxymethyl-, and 5-formylcytosine in human and mouse brain tissues. Angew Chem 54:12511–12514. https://doi.org/10.1002/ange.201502722
Wang J, Weaver ICG, Gauthier-Fisher A, Wang H, He L, Yeomans J et al (2010) CBP histone acetyltransferase activity regulates embryonic neural differentiation in the normal and Rubinstein-Taybi syndrome brain. Dev Cell. https://doi.org/10.1016/j.devcel.2009.10.023
Weaver ICG, Cervoni N, Champagne FA, D'alessio AC, Sharma S, Seckl JR et al (2004) Epigenetic programming by maternal behavior. Nat Neurosci 7(8). https://doi.org/10.1038/nn1276
Weder N, Zhang H, Jensen K, Yang BZ, Simen A, Jackowski A et al (2014) Child abuse, depression, and methylation in genes involved with stress, neural plasticity, and brain circuitry. J Am Acad Child Adolesc Psychiatry. https://doi.org/10.1016/j.jaac.2013.12.025
Wu J, Bie B, Naguib M (2016a) Epigenetic manipulation of brain-derived neurotrophic factor improves memory deficiency induced by neonatal anesthesia in rats. Anesthesiology 124:624–640. https://doi.org/10.1097/ALN.0000000000000981
Wu TP, Wang T, Seetin MG, Lai Y, Zhu S, Lin K et al (2016b) DNA methylation on N 6-adenine in mammalian embryonic stem cells. Nature 532(7599):329–333. https://doi.org/10.1038/nature17640
Yao B, Cheng Y, Wang Z, Li Y, Chen L, Huang L et al (2017) DNA N6-methyladenine is dynamically regulated in the mouse brain following environmental stress. Nat Commun 8(1):1122. https://doi.org/10.1038/s41467-017-01195-y

Younger ST, Corey DR (2011) Transcriptional gene silencing in mammalian cells by miRNA mimics that target gene promoters. Nucleic Acids Res 39(13):5682–5691. https://doi.org/10.1093/nar/gkr155

Yousefi P, Huen K, Davé V, Barcellos L, Eskenazi B, Holland N (2015) Sex differences in DNA methylation assessed by 450 K BeadChip in newborns. BMC Genomics 16(911). https://doi.org/10.1186/s12864-015-2034-y

Zammit S, Lewis G, Gustafsson J-E (2010) Individuals, schools, and neighborhood: a multilevel longitudinal study of variation in incidence of psychotic disorders. Arch Gen Psychiatry 67:914–922. https://doi.org/10.1001/archgenpsychiatry.2010.101

Zannas AS, Arloth J, Carrillo-Roa T, Iurato S, Röh S, Ressler KJ et al (2015a) Lifetime stress accelerates epigenetic aging in an urban, African American cohort: relevance of glucocorticoid signaling. Genome Biol 16(266). https://doi.org/10.1186/s13059-015-0828-5

Zannas AS, Provençal N, Binder EB (2015b) Epigenetics of posttraumatic stress disorder: current evidence, challenges, and future directions. Biol Psychiatry. https://doi.org/10.1016/j.biopsych.2015.04.003

Zayas V, Hazan C (2015) Bases of adult attachment: linking brain, mind and behavior. Springer, New York, pp 1–196. https://doi.org/10.1007/978-1-4614-9622-9

Zayas V, Shoda Y (2005) Do automatic reactions elicited by thoughts of romantic partner, mother, and self relate to adult romantic attachment? Personal Soc Psychol Bull 31(8):1011–1025. https://doi.org/10.1177/0146167204274100

Zayas V, Shoda Y (2015) Love you? Hate you? Maybe it's both: evidence that significant others trigger bivalent-priming. Soc Psychol Personal Sci 65(1):56–64. https://doi.org/10.1177/1948550614541297

Zayas V, Shoda Y, Mischel W, Osterhout L, Takahashi M (2009) Neural responses to partner rejection cues. Psychol Sci 20(7):813–821. https://journals.sagepub.com/doi/pdf/10.1111/j.1467-9280.2009.02373.x

Zayas V, Mischel W, Shoda Y, Aber JL (2011) Roots of adult attachment: maternal caregiving at 18 months predicts adult peer and partner attachment. Soc Psychol Personal Sci 2(3):289–297. https://doi.org/10.1177/1948550610389822

Zhang N (2015) Epigenetic modulation of DNA methylation by nutrition and its mechanisms in animals. Anim Nutr. https://doi.org/10.1016/j.aninu.2015.09.002

Zhao J, Lee M-C, Momin A, Cendan C-M, Shepherd ST, Baker MD et al (2010) Small RNAs control sodium channel expression, nociceptor excitability, and pain thresholds. J Neurosci 30 (32):10860–10871. https://doi.org/10.1523/JNEUROSCI.1980-10.2010

Zheleznyakova GY, Cao H, Schiöth HB (2016) BDNF DNA methylation changes as a biomarker of psychiatric disorders: literature review and open access database analysis. Behav Brain Funct 12:17. https://doi.org/10.1186/s12993-016-0101-4

Zheng Y-C, Ma J, Wang Z, Li J, Jiang B, Zhou W et al (2015) A systematic review of histone lysine-specific demethylase 1 and its inhibitors. Med Res Rev 35(5):1032–1071. https://doi.org/10.1002/med.21350

Zhou X-L, Yu L-N, Wang Y, Tang L-H, Peng Y-N, Cao J-L, Yan M (2014) Increased methylation of the MOR gene proximal promoter in primary sensory neurons plays a crucial role in the decreased analgesic effect of opioids in neuropathic pain. Mol Pain 10:51–65. https://doi.org/10.1186/1744-8069-10-51

Ziller MJ, Gu H, Müller F, Donaghey J, Tsai LT-Y, Kohlbacher O et al (2013) Charting a dynamic DNA methylation landscape of the human genome. Nature 500(7463):477–481. https://doi.org/10.1038/nature12433

Zimmer P, Schenk A, Bloch W (2016) Epigenetics in exercise science and sports medicine. In: Medical epigenetics. https://doi.org/10.1016/B978-0-12-803239-8.00029-6

Zimmerberg B, Foote HE, Van Kempen TA (2009) Olfactory association learning and brain-derived neurotrophic factor in an animal model of early deprivation. Dev Psychobiol 51 (4):333–344. https://doi.org/10.1002/dev.20373

The Role of Dynamic Histone Modifications in Learning Behavior

Andre Fischer

Contents

Abstract Epigenetic mechanisms have been linked to memory formation under physiological and pathological conditions. Therapeutic strategies that target the epigenetic machinery have been successfully used in preclinical studies investigating cognitive phenotypes linked to neuropsychiatric and neurodegenerative diseases. This chapter will specifically discuss the role of histone modification in the adult brain with a focus on learning and memory processes in the healthy and diseased brain. Data on dynamic changes in histone modification during memory processes as well as the most current knowledge on the corresponding enzymatic machinery in the adult brain will be summarized and discussed in the context of potential therapeutic opportunities to treat brain diseases.

Keywords Epigenetic therapies · Histone methylation · Learning · Memory · Neurodegenerative diseases · Neuroepigenetics · Neuropsychiatric diseases

A. Fischer (✉)
Department for Psychiatry and Psychotherapy, University Medical Center Göttingen, Göttingen, Germany

Department for Systems Medicine and Brain Diseases, German Center for Neurodegenerative Diseases (DZNE) site Göttingen, Göttingen, Germany
e-mail: andre.fischer@dzne.de

© Springer Nature Switzerland AG 2019

Curr Topics Behav Neurosci (2019) 42: 127–158
DOI 10.1007/7854_2019_108
Published Online: 27 August 2019

1 Introduction: Neuro-epigenetics and Histone Modifications

The tight control of gene expression programs is essential for cellular function. In turn, deregulated gene expression has been implicated with the pathogenesis of various diseases. In addition to the activity of transcription factors, epigenetic mechanisms are key processes that control gene expression at a systems level (Allis et al. 2007a). The term epigenetics has been originally introduced by Conrad Waddington to describe heritable changes of a phenotype that do not depend on altered DNA sequence (Waddington 1953). In addition to DNA methylation and the action of noncoding RNAs, a central epigenetic process is the modifications of histone proteins. The four core histones (H) build the nucleosome around which 147 bp of DNA is wrapped. The histone tails are subjected to posttranslational modifications such as acetylation, methylation, phosphorylation, ubiquitination, sumoylation, ADP-ribosylation, etc. (Vaquero et al. 2003) which are mediated by the counteracting activities of enzymes that add or remove histone modifications and are thus called "writers" and "erasers." For example, histone acetylation is mediated by histone acetyltransferases (HAT, writers) and histone deacetylases (HDAC, erasers). The activity of such enzymes is believed to give rise to a combinatorial pattern of chromatin modifications, the so-called histone code (Strahl and Allis 2000). Such histone modifications are recognized by proteins that subsequently affect cellular processes such as gene expression and are referred to as chromatin "readers" (Fischer 2014a). The field of neuro-epigenetics investigates the abovementioned epigenetic processes in the context of neuronal plasticity, memory function, and brain diseases (Fischer 2014a; Day and Sweatt 2011; Jakovcevski et al. 2013b).

Various cellular mechanisms have been linked to memory function, and tightly controlled dynamic changes in gene expression are one requirement to enable the consolidation of long-term memories (Flexner et al. 1962, 1963; Davis 1984; Igaz et al. 2002; Frey and Morris 1998; Moncada et al. 2011; Fischer 2014a). In addition, deregulated gene expression is seen in brain diseases linked to memory impairment (Lu et al. 2004; Liang et al. 2010; Ginsberg et al. 2010; Twine et al. 2011; Blalock et al. 2011; Kim et al. 2012a; Arefin et al. 2012; Caldeira et al. 2013; Mills et al. 2013). Understanding the mechanisms that orchestrate gene expression programs in brain cells is thus of utmost importance, and epigenetic processes have emerged as key players that enable synaptic plasticity and memory formation.

A number of excellent review articles have discussed the role of epigenetic processes such as DNA methylation and noncoding RNAs in synaptic plasticity, memory function, and brain diseases (Day and Sweatt 2010; Fiore et al. 2011; Fischer 2014a; Jakovcevski and Akbarian 2012; Campbell and Wood 2019). This chapter will focus on recent findings on the role of histone modifications in memory formation and cognitive diseases.

1.1 Histone Modifications and Memory Consolidation in Health and Disease

First evidence that histone acetylation might play a role in memory consolidation stems already from 1979. Using C14-labeled acetate, Schmitt and Matthies could show that acetylation of histones is altered when rats are subjected to a learning paradigm that leads to the formation of long-term memories (Schmitt and Matthies 1979). Only at the beginning of this century, such studies were followed up, and a number of labs could demonstrate that memory formation in rodents is correlated to increased activity of HATs (Swank and Sweatt 2001) and changes in histone acetylation that were measured by semiquantitative immunoblot analysis using antibodies specific to acetylated lysine residues of histones (Levenson et al. 2004; Fischer et al. 2007; Bousiges et al. 2010). These methods are able to detect changes in bulk levels of histone acetylation but do not allow for any deeper insight to genes regulated by changes in histone acetylation. Such issues were circumvented when researchers started to perform chromatin immunoprecipitation (ChIP) followed by either qPCR or later next-generation sequencing approaches (ChIP-seq) to study histone acetylation at the single gene or genome-wide level (Peleg et al. 2010). An important consideration in this context is the complexity of cell types in the brain. Technological advances now also allow the analysis of histone modifications via ChIP-seq at the single-cell level, but these methods have not been applied yet to the neurosciences (Grosselin et al. 2019). Most researchers nowadays perform cell type-specific analysis of histone acetylation via ChIP-seq using FACS-sorted nuclei that distinguish at least between neuronal and non-neuronal cells (Shulha et al. 2012; Benito et al. 2015; Halder et al. 2016), which is of utmost importance for the analysis of brain diseases characterized by neuronal cell death in combination with inflammatory processes.

Thus, there is now substantial evidence confirming the initial observation by Schmitt and Matthies that changes in histone acetylation but also other histone modifications are observed in response to learning stimuli or during the pathogenesis of brain diseases. Moreover, there is now convincing evidence from the analysis of mutant mice or other model systems that these epigenetic processes play a key role in cognition. In the following we will discuss these data in detail.

1.2 Histone Acetylation

As discussed above, early findings reported altered histone acetylation in response to memory training, and in most cases, these changes were linked to active gene expression, hence an increase of histone marks linked to active gene expression such as histone acetylation at various lysine residues of H3 and H4 (Levenson et al. 2004; Alarcon et al. 2004; Fischer et al. 2007; Miller et al. 2008; Federman et al. 2009; Peleg et al. 2010; Dagnas and Mons 2013; Fischer 2014a). These data led to the hypothesis that inhibition of HDAC proteins might increase memory formation, which was confirmed in multiple studies (Alarcon et al. 2004; Levenson et al. 2004; Wood et al. 2005; Fischer et al. 2007; Fischer 2014b). As a consequence HDAC inhibitors were tested in models for cognitive diseases such as Alzheimer's disease but also other cognitive diseases, and indeed administration of HDAC inhibitors was found to ameliorate memory impairments and cognitive phenotypes in numerous studies (Fischer 2014a, b). Some of the HDAC inhibitors such as vorinostat are already used in patients for other indications, and it should be mentioned that oral administration of vorinostat could ameliorate memory impairment and deregulated transcriptome plasticity in various mouse models for age-associated memory impairment and Alzheimer's disease (Kilgore et al. 2010; Benito et al. 2015). Thus, this drug is currently tested in Alzheimer's disease patients (https://clinicaltrials.gov/ct2/show/NCT03056495).

Histone acetylation is generally linked to active gene expression, and thus initially not much attention was paid to potential difference of, for example, learning-induced increases in H3K9ac, H4K14ac, or the H4 acetylation sites H4K5, H4K8, H4K12, and H4K12, and often histone acetylation was assayed using pan-antibodies that would detect all of the abovementioned H3 and H4 acetylation sites. While this is still a valid approach, there is also evidence that H3 and H4 acetylation sites might serve specific cellular functions. For example, the onset of cognitive decline during aging was linked to deregulated acetylation of H4K12 (H4K12ac). While memory training induced a transient increase in various histone acetylation marks in the hippocampus of young and aged mice, H4K12ac did not respond anymore to such a stimulus in old animals. When ChIP-Seq was combined with transcriptome analysis, the data revealed that H4K12ac is critical for the initiation of a learning-induced gene expression program via a mechanism that involved transcriptional elongation (Peleg et al. 2010). These data are interesting since the chromatin reader proteins BRD2 and BRD4, which belong to the BET (Bromodomain Extraterminal) subfamily of chromatin readers, were found to specifically recognize H4K12ac. Similar to HDAC inhibitors, the BRD2/BRD4 inhibitor JQ1 was found to improve memory function in wild-type mice and in a mouse model for Alzheimer's disease (Benito et al. 2017).

All of these data are summarized in a number of excellent review articles (Fischer 2014a, b; Deussing and Jakovcevski 2013) and inspire the interested to better understand the enzymatic machinery that controls histone acetylation.

The mammalian genome encodes for at least 18 HATs that are subdivided into the GNAT (*G*cn5 *N*-*a*cetyl*t*ransferases) family, the MYST (*M*OZ, *Y*bf2/Sas3, *S*as2, *T*IP60) family, the p300/CBP family, the SCR family nuclear receptor coactivators, and the several other HATs that cannot be grouped into a certain family (Lee and Workman 2007; Allis et al. 2007a). Most of these proteins are expressed in the mouse and human brain (Table 1). The best-studied HATs in the context of mammalian brain function belong to the p300/CBP family. Multiple studies demonstrated a role for CBP/*KAT3A* in memory consolidation (Alarcon et al. 2004; Korzus et al. 2004; Wood et al. 2005, 2006; Chen et al. 2010; Barrett et al. 2011). It should be mentioned at this point that memory function is routinely analyzed in rodents via a

Table 1 Regulators of histone acetylation and methylation in learning and memory

Enzyme	Catalytic action	Linked to memory-related processes
Histone acetyltransferases		
KAT1A (HAT1)	Histone acetylation	No conclusive data
KAT2A (GCN5)	Histone acetylation	Yes, loss of KAT2A impairs memory
KAT2B (PCAF)	Histone acetylation	Yes, loss of KAT2B impairs memory
KAT3A (CBP)	Histone acetylation	Yes, loss of KAT3A impairs memory
KAT3B (P300)	Histone acetylation	Yes, loss of KAT3B impairs memory
KAT4 (TAF1)	Histone acetylation	No conclusive data
KAT5A (TIP60)	Histone acetylation	Yes, loss of KAT5A impairs memory
KAT6a (MYST3)	Histone acetylation	No conclusive data
KAT6B (MYST4)	Histone acetylation	No conclusive data, linked to neurogenesis
KAT7 (MYST2)	Histone acetylation	No conclusive data
KAT8 (MYST1)	Histone acetylation	No conclusive data
KAT9 (ELP3)	Histone acetylation	No conclusive data
KAT12 (GTF3C4)	Histone acetylation	No conclusive data
KAT13A (NCOA1)	Histone acetylation	Yes, loss of KAT2B impairs memory
KAT13B (NCOA3)	Histone acetylation	No conclusive data
KAT13C (NCOA2)	Histone acetylation	No conclusive data
KAT13D (CLOCK)	Histone acetylation	No conclusive data, circadian rhythm
Histone deacetylases (zinc-dependent, excluding sirtuins)		
HDAC1	Histone deacetylation	Yes, HDAC1 affects memory
HDAC2	Histone deacetylation	Yes, loss of HDAC2 improves memory
HDAC3	Histone deacetylation	Yes, loss of HDAC3 improves memory
HDAC4	Histone deacetylation	Yes, HDAC4 affects memory
HDAC5	Histone deacetylation	Yes, loss of HDAC5 impairs memory
HDAC6	Histone deacetylation	Yes, loss of HDAC6 improves memory
HDAC7	Histone deacetylation	No conclusive data
HDAC8	Histone deacetylation	No conclusive data
HDAC9	Histone deacetylation	No conclusive data
HDAC10	Histone deacetylation	No conclusive data
HDAC11	Histone deacetylation	No conclusive data

(continued)

Table 1 (continued)

Enzyme	Catalytic action	Linked to memory-related processes
Histone methyltransferases		
KMT2A (Mll1)	H3K4 mono-, di-, and tri-methylation	Yes, loss of KMT2A impairs memory
KMT2B (Mll2)	H3K4 mono-, di-, and tri-methylation	Yes, loss of KMT2A impairs memory
KMT2C (Mll3)	H3K4 mono-, di-, and tri-methylation	No conclusive data
KMT2D (Mll4)	H3K4 mono-, di-, and tri-methylation	No conclusive data
SETD1A	H3K4 mono-, di-, and tri-methylation	No conclusive data
SETD1B	H3K4 mono-, di-, and tri-methylation	No conclusive data
KMT1A (SUV39H)	H3K9 di- and tri-methylation	No conclusive data
KMT1B (SUV39H2)	H3K9 di- and tri-methylation	No conclusive data
KMT1C (G9A/EHMT2)	H3K9 mono- and di-methylation	Yes, loss of KMT1C impairs memory
KMT1D (GLP/EHMT1)	H3K9 mono- and di-methylation	Yes, inhibition of KMT1D affected memory
KMT1E (SETDB1)	H3K9 mono- and di-methylation	No conclusive data
KMT8A (PRDM2)	H3K4 mono-, di-, and tri-methylation	No conclusive data
KMT6A (EZH2)	H3K27 mono-, di-, and tri-methylation	Yes, loss of KMT6A impairs memory
KMT6B (EZH1)	H3K27 mono, di-, and tri-methylation	No conclusive data
KMT3F	H3K4 and H3K27 mono-, di-, and tri-methylation	No conclusive data
KMT3A (SET2)	H3K36 mono- di- and tri-methylation	No conclusive data
KMT3B (NDS1)	H3K36 di- and tri-methylation	No conclusive data
KMT3C (SMYD2)	H3K36 di- and tri-methylation	No conclusive data
KMT4 (Dot1)	H3K79 mono-, di-, and tri-methylation	No conclusive data
KMT5A (SET8)	H4K20 mono-methylation	No conclusive data
KMT5B (SUV4-20H1)	H4K20 mono-, di-, and tri-methylation	No conclusive data
KMT5C (SUV4-20H2)	H4K20 mono-, di-, and tri-methylation	No conclusive data
Histone demethylases		
KDM5A (JARID1A)	H3K4 demethylation	No conclusive data
KDM5B (JARID1B)	H3K4 demethylation	No conclusive data
KDM5C (JARIC1C)	H3K4 demethylation	Yes, loss of KDM5C affects memory

(continued)

Table 1 (continued)

Enzyme	Catalytic action	Linked to memory-related processes
KDM5D (JARID1C)	H3K4 demethylation	No conclusive data
KDM2A (JHDM2A)	H3K9 demethylation	No conclusive data
KDM4A (JMJD2A)	H3K9/H3K36 demethylation	No conclusive data
KDM4B (JMJD2B)	H3K9/H3K36 demethylation	Yes, loss of KDM5C impairs memory
KDM4C (JMJD2C)	H3K9/H3K36 demethylation	No conclusive data
KDM4D (JMJD2D)	H3K9/H3K36 demethylation	No conclusive data
KDM4E (JMJD2E)	H3K9/H3K36 demethylation	No conclusive data
KDM1 (LDS1)	H3K4 and H3K9 demethylation	Yes, inhibition of KDM1 impairs memory
KMD6A (UTX)	H3K27 demethylation	Yes, loss of KDM6A impairs memory
KDM6B (JMJD3)	H3K27 demethylation	No conclusive data
KDM7B	H3K27 demethylation	No conclusive data
KDM2A (JHDM1A)	H3K36 mono- and di-demethylation	No conclusive data
KDM2B (JHDM1B)	H3K36/79 mono- and di-demethylation	No conclusive data
KDM8 (JMJD5)	H3K36 mono- and di-demethylation	No conclusive data

number of well-established memory tests that assay different cognitive domains such as spatial reference memory or associative fear learning. Loss of CBP/KAT3A does not affect all types of memory, and specifically spatial reference learning in the hippocampus-dependent Morris water maze is only medley affected in mutant mice (Josselyn 2005). In line with these observations, experiments in which CBP/KAT3A agonists were administered to mice during spatial reference learning did not improve the consolidation of memories that was measured 48 h after training but improved the consolidation of remote memories when analyzed up to 16 days after the training (Chatterjee et al. 2013). The role of CBP/KAT3A in neuronal plasticity has been linked to its potential to regulate the expression of specific genes via the modulation of histone acetylation. For example, its effect on memory consolidation was linked its ability to induce the expression of nuclear receptor 4a1 (Nr4a1) (Vecsey et al. 2007; McNulty et al. 2012). However, detailed genome-wide analysis of transcriptional networks associated with CBP function in the adult brain is sparse, and gene array approaches thus far linked CBP function to calcium signaling, transcription and synaptic plasticity (Chen et al. 2010), and deregulated gene expression in response to environmental enrichment training (Lopez-Atalaya et al. 2011). Evidence for a role of P300/KAT3B in memory function stems for work showing that genetic reduction of P300/KAT3B activity in the mouse brain impairs memory consolidation (Viosca et al. 2010; Oliveira et al. 2007, 2011; Lipinski et al. 2019). Moreover, CBP/KAT3A and P300/KAT3B are genetically linked to Rubinstein-

Taybi syndrome, an autosomal dominantly inherited form of mental retardation that affects 1 in 125,000 individuals (Petrij et al. 1995; Oike et al. 1999). It is therefore important to note that administration of an HDAC inhibitor was able to ameliorate the phenotypes in KAT3A/CBP mutant mice (Alarcon et al. 2004). KAT3A/CBP has also been linked to Alzheimer's disease. For example, it was found that phospho-CBP levels are downregulated in the 3xAD mice and that viral-mediated overexpression of CBP was able to ameliorate deficits in phospho-CBP levels and memory function (Caccamo et al. 2010, CBP gene transfer increases BDNF levels and ameliorates learning and memory deficits in a mouse model of Alzheimer's disease). Moreover, deletion of presenilins 1 and 2 in mice affected CBP-mediated gene expression (Saura et al. 2004). There is also evidence PCAF/KAT2B is required for memory formation (Maurice et al. 2008; Duclot et al. 2010; Wei et al. 2012; Mitchnick et al. 2016). PCAF has also been linked to AD pathogenesis. One study found that mice which lack PCAF (KAT2*B*) are resistant to the detrimental effects of amyloid beta peptides that are directly injected into the lateral ventricles of mice (Duclot et al. 2010). This data is somewhat in conflict with the fact that the same PCAF mutant mice were found to show impaired memory function (Maurice et al. 2008), and it remains to be seen if reducing PCAF levels would have similar effects in more established models for amyloid pathology. Nevertheless, such findings remind us that the role of histone and protein acetylation in AD is likely very complex and should not be exclusively reduced to the hypothesis that inhibition of HDACs and activation of HATs would ameliorate cognitive impairment.

Studies from our laboratory used an unbiased RNA-sequencing approach to compare the expression of the 18 HATs in the hippocampal CA1 region of adult mice, a region that is intimately linked to memory formation (Stilling et al. 2014) (Table 1). These data revealed that GCN5/KAT2A and TIP60/KAT5 are the highest expressed HATs in this brain region, at least at the mRNA level. Consequently, deletion of GCN5/KAT2A from excitatory forebrain neurons of the adult brain leads to severely impaired spatial reference and associative memory consolidation, as well as impaired hippocampal long-term potentiation (LTP), that is considered to be a molecular correlate of memory formation (Stilling et al. 2014). Loss of GCN5/KAT2A also had a profound impact on hippocampal gene expression that was linked to the acetylation of histone 4 at lysine 12 and histone 3 at lysine 14 and 18 as well as the acetylation of the p65, a co-regulation of the transcription factor NFkappaB (Stilling et al. 2014). Similarly, unpublished data from our laboratory suggest that deletion of the other highly expressed HAT in the hippocampus, namely, KAT5/TIP60, leads to severe memory impairment and a massive deregulation of neuronal gene expression when deleted from excitatory forebrain neurons. In line with these data, pharmacological inhibition of KAT5A/TIP60 impaired remote memory retrieval, a finding that was linked to the role of KAT5A/TIP60 in the acetylation of the histone variant H2A.Z (Narkaj et al. 2018). KAT5A7TIP60 has also been implicated with Alzheimer's disease. To this end it was shown that increased TIP60 activity can rescue amyloid beta-induced neurotoxicity (Pirooznia et al. 2012) and axonal transport deficits in a *Drosophila* model (Johnson et al. 2013). In line with the data that KMTs do not exclusively act on histones, loss of

TIP60 activity in drosophila causes reduced microtubule acetylation (Sarthi and Elefant 2011). Nevertheless, the protective effect of increasing TIP60 activity is – at least in part – also linked to the expression of genes that regulate apoptotic cell death (Pirooznia et al. 2012) and axonal transport (Johnson et al. 2013). A potential role for TIP60 in Alzheimer's disease has been furthermore suggested even earlier since it was found that TIP60 regulates gene expression in a complex with the amyloid precursor protein intracellular domain (AICD) (Cao et al. 2004; Müller et al. 2013). Moreover TIP60 plays a critical role in orchestrating the cellular response to DNA damage (Kaidi and Jackson 2013) which is emerging as another critical player in AD pathogenesis (Mao and Reddy 2011; Herrup et al. 2013).

Comparatively little is known about the role of the other HATs in the adult brain and memory function. HAT1/KAT1 expression increases in mice treated with antidepressants, but the functional consequences are not understood (Kocki et al. 2018). These data are however in line with the observation that neuronal histone modifications are altered in depression (Sun et al. 2013). Myst4/KAT6B has (Merson et al. 2006) been linked to adult neurogenesis, a process that has been linked to cognitive function (Ernst and Frisén 2015), while MYST1/KAT8 has been linked to Alzheimer's disease by a recent GWAS study (Marioni et al. 2018). RNAi-mediated knockdown of NCOA1/KAT13A in the hippocampal CA1 region of mice impairs spatial reference memory and LTP which was accompanied by reduced synapse density (Bian et al. 2018). The other HATs have not been linked to memory function yet, although mutations in TAF1/KAT4 are linked to X-linked dystonia-parkinsonism (Bragg et al. 2019). It is important to mention that HATs act as part of larger protein complexes and proteins with HAT activity are also part of the RNA-pol II complex (Lee and Workman 2007). This is of particular interest since acetyl-CoA is the main donor of acetyl groups essential for histone acetylation. It was recently shown that acetyl-CoA synthase 2 (ACSS2) is the main enzyme that provides acetyl-CoA to histone acetylation in the adult brain and that reduced ACSS2 levels lead to memory impairment (Mews et al. 2017). This is in line with previous data suggesting that age-associated memory decline is linked to reduced hippocampal citrate levels (Peleg et al. 2010), a main donor for acetyl-CoA synthesis.

The activity of HATs is counteracted by the histone deacetylases (HDAC) that are grouped into the zinc-dependent HDACs and the non-zinc-dependent sirtuins. A number of excellent reviews discuss the role of sirtuins and HDACs in various cellular systems (Haigis and Sinclair 2010; Donmez 2013; Fischer et al. 2010; Fischer 2014b). Here we will specifically focus on the role of the zinc-dependent HDACs in memory function. In line with the data that have generally shown that loss of HAT activity in linked to memory impairment, it was found that pharmacological inhibition of HDACs enhances the consolidation of memories in rodents (Levenson et al. 2004; Fischer et al. 2007; Stefanko et al. 2009; Federman et al. 2009; Benito et al. 2015; Agís-Balboa et al. 2017).

The 11 mammalian zinc-dependent HDACs belong to an ancient protein family and require a Zn^{2+} ion as cofactor (Gregoretti et al. 2004; de Ruijter et al. 2003). Under naïve conditions, all HDAC genes are expressed within the adult rodent brain

(Broide et al. 2007). Mice that lack HDAC1 or overexpress it in all neurons from early developmental stages show no impairments in contextual fear learning or spatial memory formation (Guan et al. 2009), suggesting that HDAC1 has no obvious role at least in the abovementioned types of memory function. HDAC1 was however found to be essential for fear extinction learning, a process that is important for the treatment of neuropsychiatric diseases associated with aversive behaviors as they occur in post-traumatic stress disorder (Bahari-Javan et al. 2012). Here, the induction of immediate early genes was suppressed via HDAC1-mediated deacetylation of H3K9 and subsequent H3K9 tri-methylation (Bahari-Javan et al. 2012). Although HDAC1 and HDAC2 are close homologues that derived from gene duplications, their roles in memory function differ. Overexpression of HDAC2 in neurons impaired contextual fear learning and spatial memory formation in mice, while deletion of HDAC2 in neurons from early developmental stages improved memory function and synaptic plasticity (Guan et al. 2009). Notably, enhanced learning behavior in HDAC2 knockout mice correlated with increased hippocampal H4K12 acetylation, while H3K14 acetylation was not affected. Although the authors measured bulk changes, these data are interesting taking into account that genome-wide analysis of chromatin in the aging hippocampus suggested a key role for H4K12 acetylation in age-associated memory impairment (Peleg et al. 2010). Later studies suggest that the memory enhancing effect of HDAC2 reduction is linked to the activity of the transcription factor SP1 (Yamakawa et al. 2017). On the structural level, loss of HDAC2 increased the number of synapses (Guan et al. 2009) which is in line with a role of HDAC2 – but also HDAC1 – in synapse formation during development (Akhtar et al. 2009). Loss of HDAC2 was also found to improve fear extinction learning (Morris et al. 2013) which is opposite to the function of HDAC1 (Bahari-Javan et al. 2012). HDAC2 was found to bind promoter regions of genes linked to memory formation, but the precise mechanisms by which HDAC2 acts as a memory repressor are not well understood (Guan et al. 2009). One study showed that HDAC2 is essential for the survival of adult born neurons in the dentate gyrus (Jawerka et al. 2010). Since adult neurogenesis has been linked to memory function, it is clear that the role of HDAC2 in the adult brain awaits a more detailed analysis. An important next step would be to understand the role of HDAC1 and 2 in a cell type-specific manner. A recent study suggests that HDAC2 is produced in many neuronal cell types and in oligodendrocytes but not in astro- or microglia (Yao et al. 2013), a view that is however not undisputed since recent data point to an important role of HDAC1 and 2 in glia cells (Wendeln et al. 2018). HDAC1 has been also linked to neuropsychiatric diseases (Jakovcevski et al. 2013a). For example, a number of studies have demonstrated a role for HDAC1 in schizophrenia. Thus, HDAC1 was upregulated in postmortem brain tissue of schizophrenia patients (Dong et al. 2007; Sharma et al. 2008; Bahari-Javan et al. 2017). In a recent study, our laboratory was able to show that HDAC1 is increased specifically in schizophrenia patients that encountered early life stress (Bahari-Javan et al. 2017). Interestingly, this upregulation of HDAC1 was linked to DNA methylation-dependent regulation of the HPA axis and was also prominent in corresponding human blood samples (Bahari-Javan et al. 2017), suggesting the possibility that HDAC1 might serve as a biomarker for stratified therapy of patients with an unfavorable course of the disease (Bahari-Javan et al. 2017). Similar to the data available for HDAC2, mice that lack HDAC3 in the adult hippocampus show enhanced memory function (McQuown et al. 2011) which is mechanistically linked to the regulation of the Nr4a gene (Kwapis et al. 2019). The function of HDAC8 in adult brain has not been addressed in detail, but a recent study found that an HDAC inhibitor with some selectivity toward HDAC8 improves memory function in rats (Yang et al. 2013). In conclusion, it appears that the class I HDACs act as molecular inhibitors of memory formation and are thus potential therapeutic targets to ameliorate memory impairment.

As for the class II HDACs (Table 1), HDAC4 is known to shuttle between the cytoplasm and the nucleus of cultured hippocampal neurons in response to calcium signaling and CamKII activity (Chawla et al. 2003; Backs et al. 2006). In a *C. elegans* model, deletion of HDAC4 gene increases long-term memory for thermosensation in a CamKII-dependent manner (Wang et al. 2011). Specific expression of mammalian HDAC4 in the nucleus was able to revert this phenotype suggesting that nuclear export of HDAC4 is a critical process for memory formation. In line with this data, cytoplasmatic expression of HDAC4 increased memory formation in wild-type worms (Wang et al. 2011), suggesting that during learning HDAC4 regulates counteracting molecular processes in the nucleus and the cytoplasm. In line with these data, another study demonstrated that overexpression of HDAC4 impairs long-term memory in a *Drosophila* model (Fitzsimons et al. 2013). In contrast to such findings, a recent study suggests that HDAC4 is essential for memory function in mammalian systems. Mice that lack HDAC4 in the adult forebrain exhibit impaired hippocampus-dependent memory formation and plasticity (Kim et al. 2012b). This data is in line with findings showing that haploinsufficiency of HDAC4 is linked to mental retardation in humans (Williams et al. 2010). Another study confirmed that lack of HDAC4 in the adult brain results in impaired memory function and synaptic plasticity in mice and could identify synaptic plasticity genes that are regulated by nuclear HDAC4 (Sando et al. 2012). There is also evidence that HDAC4 controls genes essential for adult neurogenesis (Saha et al. 2019) that contribute to the development of psychiatric phenotypes in response to stressful events (Maddox et al. 2018).

Regarding HDAC5, there is a substantial amount of evidence implicating this HDAC with the mechanisms linked to drug abuse (Smith and Kenny 2018). Loss of HDAC5 in the nucleus accumbens renders mice hypersensitive to chronic cocaine (Renthal et al. 2007) and regulates cocaine-conditioned behaviors (Taniguchi et al. 2017), while 2-month-old mice that lack HDAC5 from the adult forebrain show no changes in hippocampus-dependent memory formation (Kim et al. 2012b). A role of HDAC5 in memory formation may not only be brain-region but also age-related. To this end 10-month-old mice that lack HDAC5 show hippocampus-dependent memory disturbances (Agis-Balboa et al. 2013).

HDAC6 is best known for its acetylation of tubulin or heat shock protein 90 (Govindarajan et al. 2013). Loss of HDAC6 has mild impact on memory function in wild-type mice but improves cognition in mouse models for Alzheimer's disease-related protein aggregation (Govindarajan et al. 2013; Fan et al. 2018; Selenica et al.

2014) or Fragile X syndrome (Kozikowski et al. 2019). HDAC7 was found to be downregulated in response to memory training in mice, which was linked to the expression of genes essential for long-term memory consolidation (Jing et al. 2017). Little is known on the role of HDAc9 in the adult brain, but a recent study points to a role of an HDAC9-dependent circRNA in dementia (Lu et al. 2019). Moreover, HDAC9 has been linked to dendritic plasticity (Sugo et al. 2010) and is associated with stroke and schizophrenia (Lang et al. 2011; Markus et al. 2013). Almost nothing is known about the role of HDAC10 and 11 in the adult brain. In conclusion, several HATs and HDACs have emerged as key players in memory function. Future research will be essential to understand how these epigenetic enzymes are regulated and which gene expression pathways they control. Moreover, it is important to mention that this enzymatic machinery not only regulates the acetylation state of specific histone proteins but also affects non-histone proteins such as transcription factors but also proteins generally linked, for example, to metabolic processes (Choudhary et al. 2009).

1.3 Histone Methylation: H3K4

In addition to acetylation, another well-studied histone modification is methylation (Vaquero et al. 2003). Similar to histone acetylation, it is regulated by the counteracting activity of histone methyltransferases (HMTs) and histone demethylases (HMDs). However, in contrast to acetylation, the lysine residues of histones can be either mono-, di-, or tri-methylated which is catalyzed by specific enzymes. The functional consequences of histone methylation are thus also more complex and have been either linked to active euchromatin or the formation of facultative and constitutive heterochromatin that is associated with gene silencing. As such, there are more HMTs and HDMs than there are HATs and HDACs, and their general role has been discussed in a number of review articles (Shi 2007; Shi and Whetstine 2007; Badeaux and Shi 2013). In the context of memory formation, most data has been generated on the role of histone 3 methylation (H3K4me3) that is enriched around transcription start site (TSS) regions of actively transcribed and/or poised genes (Guenther et al. 2006), whereas histone 3 lysine 4 monomethylation (H3K4me1) is enriched at enhancers (Heintzman et al. 2009). H3K4 methylation is mediated by SET proteins. Set1 is the only H3K4 methylase in yeast (Roguev et al. 2001), whereas in *Drosophila*, three different SET proteins – Set1, trithorax (trx), and trithorax-related (trr) – are expressed (Mohan et al. 2011). Mammals, in turn, possess six SET-related H3K4 methyltransferases – Kmt2a (Mll1), Kmt2b (Mll2), Kmt2c (Mll3), Kmt2d (Mll4), Setd1a, and Setd1b – that have been linked to mono-, di-, and tri-methylation. These proteins are subdivided into three groups, based on their homology to each other and to the corresponding *Drosophila* homologues. Kmt2a/Kmt2b are related to trx, Kmt2c/Kmt2d to trr, and Setd1a/Setd1b to Set1 (Shilatifard 2012). The existence of several homologous histone-modifying enzymes mediating the same modifications raises the question of redundancy vs. specificity of their actions. This is especially true for brain regions such as the hippocampus, which is intimately linked to memory function and cognitive diseases, where all six H3K4-KMTs are strongly expressed (Kerimoglu et al. 2017}. It is also important to note that mutations in any of the six H3K4-KMTs are linked to various rare diseases that include cognitive phenotypes and intellectual disability (Kleefstra et al. 2014).

In line with this, a number of studies implicated the regulation of H3K4me3 with memory function. For example, it was found that H3K4me3 correlates with the expression of glutamate receptors in the human brain (Stadler et al. 2005). Moreover, bulk levels of hippocampal H3K4me3 were found to increase in response to fear learning (Gupta et al. 2010), and mice that constitutively lack one allele of the H3K4-KMT KMT2A display impaired memory formation (Gupta et al. 2010). In line with this, conditional deletion of KMT2A from neurons of the adult brain impaired synaptic plasticity and working memory (Jakovcevski et al. 2015). Another study also demonstrated a role for the H3K4-specific HMT Mll2/*KMT2B* in memory function. Mice that lack KMT2B in the dorsal dentate gyrus of hippocampal region show memory impairment that is linked to deregulation of learning-relevant genes (Kerimoglu et al. 2013). Loss of KMT2B not only affected H3K4me3 at the promoter regions of learning-regulated genes but also reduced H3K9 acetylation, while H4K16 acetylation and H3K4me1 were unaffected at the same gene promoters (Kerimoglu et al. 2013). Such data further confirm the view that histone methylation and histone acetylation are tightly linked and also demonstrate the need to better understand the protein complexes that regulate chromatin plasticity in the adult brain. Especially the latter is in utmost importance since recent data compared the role of two closely related H3K4me3 KMTs – KMT2A and KMT2B – in excitatory neurons of the adult brain (Kerimoglu et al. 2017). These data show that loss of either enzyme leads to the impairment of hippocampus-dependent memories, but it was somewhat surprising to see that the gene expression pathways controlled by KMT2A and KMT2B in the hippocampus were entirely different. In fact, loss of neuronal KTM2A mainly affected gene expression pathways linked to neuronal identity and synaptic plasticity, while loss of KMT2B was linked to more general metabolic functions (Kerimoglu et al. 2013, 2017). Preliminary data from our lab suggest that similar observations are made for other KMTs that regulate H3K4me3. Mechanistically this phenomenon is presently not understood, but multiple explanations can be considered. Different KMTs could simply occupy different genomic regions, a hypothesis that is at present difficult to be tested since antibodies that allow the reliable analysis of KMTs in ChIP-seq experiments are necessary for this approach. Some evidence for this hypothesis stems however from the finding that in the neurons of KMT2A mutant mice, the deregulated genes contain different transcription factor binding sites when compared to KMT2B (Kerimoglu et al. 2017). KMTs may however also occupy similar genomic regions but might be regulated via different posttranslational modifications that are associated with distinct protein complexes. This could have different function outcomes. For example, KMT2A has been also linked to the regulation of higher-order chromatin structure and long-distance interactions (Bharadwaj et al. 2014). Mechanistic data on the other H3K4 HMTs in memory formation are comparatively rare. Loss of KMT2D in mice leads to phenotypes that partially recapitulate the Kabuki syndrome including hippocampus-dependent memory impairment, which can be rescued by the inhibition of HDAC inhibitors (Bjornsson et al. 2014) or a ketogenic diet (Benjamin et al. 2017). These data are highly interesting, since they suggest that defects in neural H3K4me3 can be attenuated by HDAC inhibitors that increase histone acetylation. While this is in line with other finding suggesting that also in neurons H3K4me3 is functionally coupled to histone acetylation (Kerimoglu et al. 2013, 2017), data from T-cells had suggested that genes which do not carry H3K4me3 cannot be regulated by HDAC inhibitors (Wang et al. 2009). Virtually nothing is known on the role of SETD1A in the adult brain, but it has been genetically linked to autism and schizophrenia (Singh et al. 2016; Takata et al. 2016). Also on the role of SET1B in the adult brain, there is so far little data available. Mutations in the corresponding genes are also linked to intellectual disability and autism (Labonne et al. 2016; Hiraide et al. 2018).

Four histone demethylases (KDMs) counteract the activity of the H3K4-KMTs. These are KDM5A (JARID1A), KDM5B (JARID1B), KDM5C (JARIC1C), and KDM5D (JARID1C). Little is known on the role of KDM5A and KDM5D. There is in vitro evidence that reducing the levels of KDM5B might increase neurogenesis in the subventricular zone of the adult brain (Zhou et al. 2016). Mutations in KDM5C have been linked to mental retardation (Jensen et al. 2005; Rujirabanjerd et al. 2012) and to short-term memory deficits in female humans (Simensen et al. 2012). A recent study analyzed mutant mice that lack one KDM5C allele either constitutively from early developmental stages and in addition tested mice that lack KDM5C in excitatory neurons of the adult brain (Scandaglia et al. 2017). While loss of KMD5C from developmental stages leads to memory impairment, this effect is less pronounced when KDM5C is deleted at adult stages. Similar findings are observed when H3K4me3-dependent gene expression was analyzed. The finding that deletion of KDM5C from the adult brain has no severe phenotype is interesting taking into account that reduced H3K4me3 has been linked to adult onset memory impairment and cognitive disease such as Alzheimer's disease. These data suggest that inhibition of KMD5 proteins may ameliorate corresponding disease phenotypes. In line with this, a study available via bioRxiv shows that the phenotypes including memory impairment which are linked to the loss of KMT2A are ameliorated when these mutant mice are crossed with mice that lack KDM5C (Vallianatos et al. 2019). In conclusion, the current data point to a key role of H3K4 methylation in cognitive function and suggest that loss of H3K4 methylation in neurons is linked to memory impairment. In turn, drugs that increase H3K4 methylation could be suitable avenues for the treatment of cognitive diseases. Whether this could be achieved by the inhibition of HDAC or KDMs or both needs to be investigated. Also, clearly more research on the role of the H3K4 regulating enzymes in brain circuitries linked to memory function is needed.

1.4 Histone Methylation: H3K9

In addition to H3K4 methylation, there is a substantial amount of data on H3K9 methylation in the adult brain that is a mark for heterochromatin-mediating gene silencing. Similar to the situation described for H3K4 methylation, a number of different enzymes that are all expressed in the adult brain mediate H3K9 methylation. These are KMT1A (SUV39H), KMT1B (SUV39H2), KMT1C (G9A/EHMT2), KMT1D (GLP/EHMT1), KMT1E (SETDB1), and KMT8A (PRDM2).

While little is known on the role of H3K9 tri-methylase KMT1A in the adult brain, KMT1B was found to be upregulated in the hippocampus of rats in response to restraint stress, a procedure that affects memory function (Hunter et al. 2012). KMT1C and KMT1D have been linked to H3K9 mono- and di-methylation and act mainly as a complex. Pharmacological compounds affect in most cases both proteins. KMT1C and KMT1D were linked to intellectual disability and autism spectrum disorder (Koemans et al. 2018). Mice that lack KMT1C in the adult forebrain develop mental retardation-like phenotypes (Schaefer et al. 2009). However, loss of KMT1C in the nucleus accumbens was shown to increase cocaine-induced neuronal plasticity (Maze et al. 2010). In *Drosophila*, loss of KMT1C impaired memory formation (Kramer et al. 2011). Another study found that pharmacological inhibition of KMT1C/KMT1D in the entorhinal cortex facilitated memory function, while administration of the same inhibitor into the hippocampus resulted in impaired memory function in mice (Gupta-Agarwal et al. 2012). Another KMT1C/KMT1D inhibitor was found to enhance hippocampal LTP (Sharma et al. 2018), while yet another compound reduced anxiety when administered to adult animals, whereas administration during development increased anxiety when the animals were adult (Wang et al. 2018). Another recent study reported that pharmacological inhibition of KMT1C/KMT1D rescues memory impairment and synaptic function in a mouse model for amyloid deposition (Zheng et al. 2019). In sum, these data do not leave a clear picture on the role of KMT1C/KMT1D in the adult brain, but they clearly demonstrate an important role for these enzymes. One likely explanation to this could be the different roles of KMT1C/KMT1D in the developing and adult brain. Thus, timing of KMT1C/KMT1D manipulation appears to be essential for the interpretation of the resulting phenotypes.

Increased expression of KMT1E in the forebrain of mice reduced depressive-like behavior via a mechanism that involved regulation of NMDA receptor subunit 2B (Jiang et al. 2010). KMT1E is linked to H3K9-tri-methylation and was also found to be an essential regulator chromosomal conformation in mouse neurons of the adult brain (Bharadwaj et al. 2014; Jiang et al. 2017). KMT8A has been linked to mono-methylation of H3K9. It has been implicated with the detrimental effect of alcohol abuse (Barbier et al. 2017).

H3K9 methylation is reversed by the action of H3K9 KDMs that are however also known to affect other methylation sites such as H3K36. These are KDM2A (JHDM2A), KDM4A (JMJD2A), KDM4B (JMJD2B), KDM4C (JMJD2C),

KDM4D (JMJD2D), and KDM4E (JMJD2E). KDM4A was found in a complex with HDAC1 and protein phosphatase 1 (PP1) which was required for memory formation (Koshibu et al. 2009). When KMT4B was deleted from adult forebrain neurons, the authors observed altered morphology of dendritic spines. Moreover mutant mice exhibited impaired working memory (Fujiwara et al. 2016). Another KMD that acts on H3K4 and K3K9 is KMD1 (LSD1). It was shown that pharmacological inhibition of LSD1 impairs memory function in mice (Neelamegam et al. 2012). In line with this data, recent data show that deletion of KDM1 from adult forebrain neurons in mice leads to neurodegeneration and memory impairment and induces gene expression changes linked to Alzheimer's disease and frontotemporal dementia (Christopher et al. 2017).

In conclusion, the current data suggest that H3K4 methylation, as well as the activity of the corresponding HMTs and HDMs play a role in memory formation, are heavily linked to intellectual disability disorders and that targeting this machinery could a suitable approach for the treatment of cognitive diseases.

1.5 Histone Methylation: H3K27

Tri-methylation of H3K27 has been linked to facultative heterochromatin and gene silencing, while its mono- and di-methylation has been linked to active gene expression. H3K27me3 counteracts the role of H3K4me3, and both marks are often present at the same gene promoter which is then called bivalent promoter since they carry an active and repressive mark. Such genes are often induced by specific stimuli; hence they are repressed but can be induced rapidly by decreasing, for example, H3K27me3. Thus, H3K27me3 is associated with facultative heterochromatin. Interestingly, large portions of non-facultative heterochromatin are also marked by histone H3K9me3 and are thereby tethered to the nuclear lamina and associate with it in what is called lamina-associated domains (LADs) (van Steensel and Belmont 2017). Interestingly a recent publication elegantly showed that mechanical forces derived from tensile loading in human epithelial progenitor cells resulted in a switch from the constitutive heterochromatin marker H3K9me3 to the facultative H3K27me3 (Le et al. 2016). Deregulation of H3K27 methylation has been linked to neurodegenerative diseases. For example, changes in H3K27me3 a specific form of frontotemporal dementia is caused by mutations in the C9ORF72 genes that result in GGGGCC expanded repeats. It was found that mutant H3K27me3 level is increased in the mutant C9ORF72 gene and contributes to its downregulation (Belzil et al. 2013). Mutant C9ORF72 variants are also linked to translation leading to the generation of dipeptide repeat proteins (DRPs). A recent study showed that one of these DRPs, namely, proline-arginine (PR)-DRP, causes heterochromatin abnormalities linked to altered H3K27 methylation (Zheng et al. 2019). H3K27 methylation is mediated by the polycomb repression complex 2 (PRC2) with the essential subunits KMT6A (EZH2) and KMT6B (EZH1). KMT3F methylates H3K4 and H3K27. Neuronal deletion of KMT6A in mice from early developmental stages resulted in hippocampal memory impairment and reduced adult neurogenesis (Zhang et al. 2014). Another study found that in response to memory training, KMT6A regulated H3K27me3-dependent expression of the PTEN gene (Jarome et al. 2018). Ataxia-telangiectasia (A-T), also named Louis-Bar syndrome, is a rare progressive neurodegenerative disease that leads to cerebellar ataxia and caused by mutations of the ATM kinase. ATM was found to act on KMT6A, and the corresponding changes in H3K27 methylation have been linked to the disease progression (Li et al. 2013). Regarding KMT6B, it was shown that its expression is regulated by microRNA-132 (Johnstone et al. 2018), a microRNA intimately linked to memory formation (Fischer 2014a). The same study reports the microRNA-132-dependent downregulation of KMT6B in mice that were chronically exposed to antipsychotics (Johnstone et al. 2018). Consequently, knockdown of KMT6B in the prefrontal cortex affected motivational behavior. Another H3K27 KMT is KMT3F (NSD3) that can however also methylate H3K4. Nothing is known on the role of this enzyme in the adult brain. Among the H3K27 KDMs, KMD6A (UTX) and KDM6B (JMJD3) are rather specific to the demethylation of H3K27me2 and HeK27me3. KMD6A is encoded on the X chromosome, and there is indeed evidence for a sex-specific expression of KMD6A in neurons; more specifically KMD6A levels are higher in females (Xu et al. 2008). KMD6A has also been linked to the Kabuki syndrome, and deleting KDM6A from neurons from early developmental stages in mice results in impaired hippocampus-dependent memory formation and reduced LTP (Tang et al. 2017). Gene expression analysis revealed that KDM6A control genes linked to serotonergic signaling (Tang et al. 2017). Knockdown of KDM6B in hippocampal neurons impaired the stimulus-dependent expression of candidate genes (Wijayatunge et al. 2014). There is also evidence that KMD6B function is essential to keep neurogenesis genes in a poised state and thereby regulates neurogenesis in the developing and neurogenic niches of the adult brain (Park et al. 2014). Moreover, pharmacological inhibition of KMD6A in mice was shown to impair reinstatement of cocaine reward memory (Zhang et al. 2018). Nothing is known on the role of KDM7B on memory function in the adult brain.

In sum, although comparatively less studied than H3K4, the current data link H3K27 methylation and the corresponding enzymatic machinery to cognitive function and brain diseases.

1.6 Other Histone Methylation Sites

Little is known about the dynamic regulation of H3K36 methylation, a mark linked to complex regulation of gene expression depending on its mono-, di-, or tri-methylation, in the adult brain. This histone lysine residue can be methylated by KMT3A (SET2) that mediates H3K36me3, while KMT3B (NDS1) and KMT3C (SMYD2) mainly regulate H3K36 mono- and di-methylation. H3K36 demethylation is catalyzed by KDM2A (JHDM1A), KDM2B (JHDM1B), and KDM8 (JMJD5) that act on H3K36 mono- and di-methylation. Besides the observation that KDM2B has been genetically linked to intellectual disability and neuronal progenitor dysfunction leading to exencephaly (Fukuda et al. 2011; Labonne et al. 2016), nothing is known on the role of these enzymes in the adult brain. Demethylation of H3K36me3 is mediated by KDM4A–E that have been discussed above. Another histone modification is H3K79 methylation which has been linked to active gene expression but is also implicated with more complex regulations depending on the cellular context. Dynamic changes in H3K79 methylation are involved with neurodevelopmental processes (Büttner et al. 2010) and are reduced in FTLD/ALS patients that carry mutations in the C9ORF72 gene (Belzil et al. 2013). H3K79 mono-, di-, and tri-methylation is mediated by KMT4 (Dot1) that play a role in cortical development (Roidl et al. 2016). Demethylation of H3K79 has been linked to the activity of KDM2B that also acts on H3K36 and has been discussed above. H4K20 is so far the main methylation site on histone 4. H4K20me1 has been linked to active transcription, while H4K20me2 occurs during specific cellular processes such as cell cycle control. In contrast H4K20me3 is linked to repression of gene expression. Reduced H4K20me3 was observed in FTLD/ALS patients with the C9ORF72 mutation (Belzil et al. 2013) and in a mouse model for accelerated aging (Wang et al. 2010). H4K20 methylation is mediated by KMT5A (SET8) that mediated mono-methylation, while KMT5B (SUV4-20 h1) and KMT5C (SUV4-20H2) control di- and tri-methylation. While nothing is known on the role of KMT5A and C in the brain, KMT5B is associated with autism and developmental-disability biases (Stessman et al. 2017). A very interesting finding was that LSD1n is a H4K20 demethylase. LSD1n is a neuronal-specific splice variant of KMT1 (LSD1). Interestingly, while LSD1 affect H3K4 and H3K9, LSD1n de-methylates H4K20. Genetic ablation of LSD1n caused memory impairment and deregulation of neuronal gene expression (Wang et al. 2015).

1.7 Other Histone Modifications

Histones can also be reversibly ubiquitylated, which is probably the most severe form of a histone modification considering that a 76-amino-acid polypeptide is added to a histone tail. Histone ubiquitylation occurs mainly at lysine 119 of H2A (H2AK119ub) and lysine 120 of histone H2B (H2BK120ub). Both modifications occur as mono-ubiquitylation are thus not priming histones for degradation but serve signaling function that are diverse and have been linked to activation and repression of gene expression depending on the cellular context (Allis et al. 2007b). Early studies already observed increased H2A and H2B ubiquitylation in the aging mouse brain (Morimoto et al. 1993) which is in line with more recent data showing that elevated H2A ubiquitylation is a general marker for aging in a *Drosophila* model which is conserved in rodents, primates, and humans and that reduction of H2AK119ub in *Drosophila* increases life span (Yang et al. 2019). H2AK119ub not only increases in aging but also in neurodegenerative disease such as Huntington chorea (McFarland et al. 2013). Ubiquitination of histones is, for example, mediated by the E3 ubiquitin ligases RNF20 (BRE1A) and RNF40 (BRE1B) which have been linked to astrocyte differentiation (Liang et al. 2018). Histones can also be phosphorylated, and a prime example is the phosphorylation of the noncanonical histone gamma-H2AX that plays an important role in the DNA-damage response (Kinner et al. 2008). Interestingly, activity-induced DNA damage has been implicated with synaptic plasticity, memory consolidation, and neurodegenerative disease (Suberbielle et al. 2013; Madabhushi et al. 2015). Many other histone modifications have been reported (Vaquero et al. 2003), but their role in the adult brain remains to be studied. It has also been mentioned that in addition to the comparatively well-studied acetylation and methylation modifications of histone 3 and 4, proteome analysis suggests that more sites may exist, which still await to be studied in the brain (Brunner et al. 2012).

In addition to histone modifications, it is interesting to mention that under certain conditions, the canonical histones can be replaced by histone variants. A prime example is the replacement of H2A with H2A.Z that is considered represent a mechanisms for a transcriptional memory (Buschbeck and Hake 2017). Other variants are, for example, H3.3 or macroH2A. While in the brain H3 is constantly replaced by H3.3 along aging, so that older individuals contain almost exclusively to the replication-independent H3.3, the amount of H2A.Z was initially found to be rather constant (Piña and Suau 1987). Interestingly, it was shown that memory training in mice induced the eviction of H2A.Z at selected genes in the hippocampus and cortex of mice. Moreover H2A.Z levels were found to increase in the hippocampus of aging mice, and viral-mediated knockdown of H2A.Z improved memory function in mice (Zovkic et al. 2013; Stefanelli et al. 2018). Interestingly KAT5/TIP60 has been implicated with the deposition of H2A.Z, and more recent data suggest that inhibition of this process can improve memory consolidation in mice (Narkaj et al. 2018). While H3.3 has been linked to the general aging process (Bano et al. 2017), little is known on it and other histone variants in the adult brain, and also the role of reader proteins is only beginning to emerge.

2 Conclusion

Histone modifications are dynamically regulated during memory formation and in cognitive diseases. Histone acetylation appears to be a prerequisite for neuronal plasticity, and loss of the corresponding KMTs leads to memory impairments. In turn, the inhibition of class I HDAC has emerged as a suitable and promising approach to improve memory function in cognitive diseases. Histone methylation is also critical for proper neuronal plasticity, but the picture is more complex. Mutations in many of the enzymes machinery that control the various histone methylation events are linked to intellectual disability disorders. Best studied is the role of H3K4 and H3K9. Future research is needed to explore the full spectrum of histone modifications in the adult brain and understand how these modifications build a combinatorial code that controls transcriptome plasticity in neuronal circuitries.

References

Agis-Balboa RC, Pavelka Z, Kerimoglu C, Fischer A (2013) Loss of HDAC5 impairs memory function: implications for Alzheimer's disease. J Alzheimers Dis 33:35–44

Agís-Balboa RC, Pinhero P, Rebola N, Kerimoglu C, Benito E, Gertig M, Bahari-Javan S, Jain G, Burkhardt S, Delalle I, Jatzko A, Dettenhofer M, Zunszain PA, Schmitt A, Falkai P, Pape JC, Binder EB, Mulle C, Fischer A, Sananbenesi F (2017) Formin 2 links neuropsychiatric phenotypes at young age to an increased risk for dementia. EMBO J 36:2815–2828

Akhtar MW, Raingo J, Nelson ED, Montgomery RL, Olson EN, Kavalali ET, Monteggia LM (2009) Histone deacetylases 1 and 2 form a developmental switch that controls excitatory synapse maturation and function. J Neurosci 29(25):8288–8297

Alarcon JM, Malleret G, Touzani K, Vronskaya S, Ishii S, Kandel ER, Barco A (2004) Chromatin acetylation, memory, and LTP are impaired in CBP+/− mice: a model for the cognitive deficit in Rubinstein-Taybi syndrome and its amelioration. Neuron 42:947–959

Allis CD, Berger SL, Cote J, Dent S, Jenuwein T, Kouzarides T, Pillus L, Reinberg D, Shi Y, Shiekhattar R, Shilatifard A, Workman J, Zhang Y (2007a) New nomenclature for chromatin-modifying enzymes. Cell 131:633–636

Allis CD, Jenuwein T, Reinberg D, Caparros ML (2007b) Epigenetics. Cold Spring Habor Laboratory Press, Cold Spring Habor

Arefin AS, Mathieson L, Johnstone D, Berretta R, Moscato P (2012) Unveiling clusters of RNA transcript pairs associated with markers of Alzheimer's disease progression. PLoS One 7: e45535

Backs J, Song K, Bezprozvannaya S, Chang S, Olson EN (2006) CaM kinase II selectively signals to histone deacetylase 4 during cardiomyocyte hypertrophy. J Clin Invest 116:1853–1864

Badeaux AI, Shi Y (2013) Emerging roles for chromatin as a signal integration and storage platform. Nat Rev Mol Cell Biol 14:211–224

Bahari-Javan S, Maddalena A, Kerimoglu C, Wittnam J, Held T, Bähr M, Burkhardt S, Delalle I, Kügler S, Fischer A, Sananbenesi F (2012) HDAC1 Regulates Fear Extinction in Mice. J Neurosci 32:5062–5073

Bahari-Javan S, Varbanov H, Halder R, Benito E, Kaurani L, Burkhardt S, Anderson-Schmidt H, Anghelescu I, Budde M, Stilling RM, Costa J, Medina J, Dietrich DE, Folkerts H, Gade K, Heilbronner U, Koller M, Konrad C, Nussbeck SY, Scherk H, Spitze C, Stierl S, Stöckel J, Thiel J, Hagen M, Zimmermann J, Zitzelsberger A, Schulz A, Schmitt A, Delalle I, Falkai P, Schulze TG, Dityatev A, Sananbenesi F, Fischer A (2017) HDAC1 links early life stress to schizophrenia-like phenotypes. Proc Natl Acad Sci U S A 114:E4686–E4694

Bano D, Piazzesi A, Salomoni P, Nicotera P (2017) The histone variant H3.3 claims its place in the crowded scene of epigenetics. Aging (Albany NY) 9:602–614

Barbier E, Johnstone AL, Khomtchouk BB, Tapocik JD, Pitcairn C, Rehman F, Augier E, Borich A, Schank JR, Rienas CA, Van Booven DJ, Sun H, Nätt D, Wahlestedt C, Heilig M (2017) Dependence-induced increase of alcohol self-administration and compulsive drinking mediated by the histone methyltransferase PRDM2. Mol Psychiatry 22:1746–1758

Barrett RM, Malvaez M, Kramar E, Matheos DP, Arrizon A, Cabrera SM, Lynch G, Greene RW, Wood MA (2011) Hippocampal focal knockout of CBP affects specific histone modifications, long-term potentiation, and long-term memory. Neuropsychopharmacology 36:1545–1556

Belzil VV, Bauer PO, Prudencio M, Gendron TF, Stetler CT, Yan IK, Pregent L, Daughrity L, Baker MC, Rademakers R, Boylan K, Patel TC, Dickson DW, Petrucelli L (2013) Reduced C9orf72 gene expression in c9FTD/ALS is caused by histone trimethylation, an epigenetic event detectable in blood. Acta Neuropathol 126:895–905

Benito E, Urbanke E, Barth J, Halder R, Capece V, Jain G, Burkhardt S, Navarro M, Schutz AL, Bonn S, Fischer A (2015) Reinstating transcriptome plasticity and memory function in mouse models for cognitive decline. J Clin Invest 125:3572–3584

Benito E, Ramachandran B, Schroeder H, Schmidt G, Urbanke H, Burkhardt S, Capece V, Dean C, Fischer A (2017) The BET/BRD inhibitor JQ1 improves brain plasticity in WT and APP mice. Transl Psychiatry 7:e1239

Benjamin JS, Pilarowski GO, Carosso GA, Zhang L, Huso DL, Goff LA, Vernon HJ, Hansen KD, Bjornsson HT (2017) A ketogenic diet rescues hippocampal memory defects in a mouse model of Kabuki syndrome. Proc Natl Acad Sci U S A 114:125–130

Bharadwaj R, Peter CJ, Jiang Y, Roussos P, Vogel-Ciernia A, Shen EY, Mitchell AC, Mao W, Whittle C, Dincer A, Jakovcevski M, Pothula V, Rasmussen TP, Giakoumaki SG, Bitsios P, Sherif A, Gardner PD, Ernst P, Ghose S, Sklar P, Haroutunian V, Tamminga C, Myers RH, Futai K, Wood MA, Akbarian S (2014) Conserved higher-order chromatin regulates NMDA receptor gene expression and cognition. Neuron 84:997–1008

Bian C, Huang Y, Zhu H, Zhao Y, Zhao J, Zhang J (2018) Steroid receptor coactivator-1 knockdown decreases synaptic plasticity and impairs spatial memory in the hippocampus of mice. Neuroscience 1:114–125

Bjornsson HT, Benjamin JS, Zhang L, Weissman J, Gerber EE, Chen YC, Vaurio RG, Potter MC, Hansen KD, Dietz HC (2014) Histone deacetylase inhibition rescues structural and functional brain deficits in a mouse model of Kabuki syndrome. Sci Transl Med 6:256ra135

Blalock EM, Buechel HM, Popovic J, Geddes JW, Landfield PW (2011) Microarray analyses of laser-captured hippocampus reveal distinct gray and white matter signatures associated with incipient Alzheimer's disease. J Chem Meuroanat 42:118–126

Bousiges O, Vasconcelos AP, Neidl R, Cosquer B, Herbeaux K, Panteleeva I, Loeffler JP, Cassel JC, Boutillier AL (2010) Spatial memory consolidation is associated with induction of several lysine-acetyltransferase (histone acetyltransferase) expression levels and H2B/H4 acetylation-dependent transcriptional events in the rat hippocampus. Neuropsychopharmacology 35:2521–2537

Bragg DC, Sharma N, Ozelius LJ (2019) X-linked dystonia-parkinsonism: recent advances. Curr Opin Neurol 32(4):604–609

Broide RS, Redwine JM, Aftahi N, Young W, Bloom FE, Winrow CJ (2007) Distribution of histone deacetylases 1-11 in the rat brain. J Mol Neurosci 31:47–58

Brunner AM, Tweedie-Cullen RY, Mansuy IM (2012) Epigenetic modifications of the neuroproteome. Proteomics 12:2404–2420

Buschbeck M, Hake SB (2017) Variants of core histones and their roles in cell fate decisions, development and cancer. Nat Rev Mol Cell Biol 18:299–314

Büttner N, Johnsen SA, Kügler S, Vogel T (2010) Af9/Mllt3 interferes with Tbr1 expression through epigenetic modification of histone H3K79 during development of the cerebral cortex. Proc Natl Acad Sci U S A 107:7042–7047

Caccamo A, Maldonado MA, Bokov AF, Majumder S, Oddo S (2010) CBP gene transfer increases BDNF levels and ameliorates learning and memory deficits in a mouse model of Alzheimer's disease. Proc Natl Acad Sci U S A 107:22687–22692

Caldeira GL, Ferreira IL, Rego AC (2013) Impaired transcription in Alzheimer's disease: key role in mitochondrial dysfunction and oxidative stress. J Alzheimers Dis 34:115–131

Campbell RR, Wood MA (2019) How the epigenome integrates information and reshapes the synapse. Nat Rev Neurosci 20:133–147

Cao L, Jiao X, Zuzga DS, Liu Y, Fong DM, Young D, During MJ (2004) VEGF links hippocampal activity with neurogenesis, learning and memory. Nat Genet 36:827–835

Chatterjee S, Mizar P, Cassel R, Neidl R, Selvi BR, Mohankrishna DV, Vedamurthy BM, Schneider A, Bousiges O, Mathis C, Cassel JC, Eswaramoorthy M, Kundu TK, Boutillier AL (2013) A novel activator of CBP/p300 acetyltransferases promotes neurogenesis and extends memory duration in adult mice. J Neurosci 33:10698–10712

Chawla S, Vanhoutte P, Arnold FJ, Huang CL, Bading H (2003) Neuronal activity-dependent nucleocytoplasmic shuttling of HDAC4 and HDAC5. J Neurochem 85:151–159

Chen G, Zou X, Watanabe H, van Deursen JM, Shen J (2010) CREB binding protein is required for both short-term and long-term memory formation. J Neurosci 30:13066–13070

Choudhary C, Kumar C, Gnad F, Nielsen ML, Rehman M, Walther TC, Olsen JV, Mann M (2009) Lysine acetylation targets protein complexes and co-regulates major cellular functions. Science 325:834–840
Christopher MA, Myrick DA, Barwick BG, Engstrom AK, Porter-Stransky KA, Boss JM, Weinshenker D, Levey AI, Katz DJ (2017) LSD1 protects against hippocampal and cortical neurodegeneration. Nat Commun 8:805
Dagnas M, Mons N (2013) Region- and age-specific patterns of histone acetylation related to spatial and cued learning in the water maze. Hippocampus 23:581–591
Davis HS, Squire LR (1984) Protein synthesis and memory: a review. Psychol Bull 96:518–559
Day JJ, Sweatt JD (2010) DNA methylation and memory formation. Nat Neurosci 13:1319–1323
Day JJ, Sweatt JD (2011) Epigenetic mechanisms in cognition. Neuron 70:813–829
de Ruijter AJ, van Gennip AH, Caron HN, Kemp S, van Kuilenburg AB (2003) Histone deacetylases (HDACs): characterization of the classical HDAC family. Biochem J 370:737–749
Deussing JM, Jakovcevski M (2013) Histone modifications in major depressive disorder and related rodent models. Adv Exp Med Biol 978:169–183
Dong E, Guidotti A, Grayson DR, Costa E (2007) Histone hyperacetylation induces demethylation of reelin and 67-kDa glutamic acid decarboxylase promoters. Proc Natl Acad Sci U S A 104:4676–4681
Donmez G (2013) Sirtuins as possible targets in neurodegenerative diseases. Curr Drug Targets 14:644–647
Duclot F, Meffre J, Jacquet C, Gongora C, Maurice T (2010) Mice knock out for the histone acetyltransferase p300/CREB binding protein-associated factor develop a resistance to amyloid toxicity. Neuroscience 167:850–863
Ernst A, Frisén J (2015) Adult neurogenesis in humans- common and unique traits in mammals. PLoS Biol 13:e1002045
Fan SJ, Huang FI, Liou JP, Yang CR (2018) The novel histone de acetylase 6 inhibitor, MPT0G211, ameliorates tau phosphorylation and cognitive deficits in an Alzheimer's disease model. Cell Death Dis 9:655
Federman N, Fustiñana MS, Romano A (2009) Histone acetylation is recruited in consolidation as a molecular feature of stronger memories. Learn Mem 16:600–606
Fiore R, Khudayberdiev S, Saba R, Schratt G (2011) MicroRNA function in the nervous system. Prog Mol Biol Transl Sci 102:47–100
Fischer A (2014a) Epigenetic memory: the Lamarckian brain. EMBO J 33:945–967
Fischer A (2014b) Targeting histone-modifications in Alzheimer's disease. What is the evidence that this is a promising therapeutic avenue? Neuropsychopharmacology 80:95–012
Fischer A, Sananbenesi F, Wang X, Dobbin M, Tsai LH (2007) Recovery of learning & memory after neuronal loss is associated with chromatin remodeling. Nature 447:178–182
Fischer A, Sananbenesi F, Mungenast A, Tsai LH (2010) Targeting the right HDAC(s) to treat cognitive diseases. Trends Pharmacol Sci 31:605–617
Fitzsimons HL, Schwartz S, Given FM, Scott MJ (2013) The histone deacetylase HDAC4 regulates long-term memory in Drosophila. PLoS One 8:e83903
Flexner JB, Flexner LB, Stellar E, De La Haba G, Roberts RB (1962) Inhibition of protein synthesis in brain and learning and memory following puromycin. J Neurochem 9:595–605
Flexner JB, Flexner LB, Stellar E (1963) Memory in mice as affected by intracerebral puromycin. Science 141:57–59
Frey U, Morris RG (1998) Synaptic tagging: implications for late maintenance of hippocampal long-term potentiation. Trends Neurosci 21:181–188
Fujiwara K, Fujita Y, Kasai A, Onaka Y, Hashimoto H, Okada H, Yamashita T (2016) Deletion of JMJD2B in neurons leads to defective spine maturation, hyperactive behavior and memory deficits in mouse. Transl Psychiatry 6:e766
Fukuda T, Tokunaga A, Sakamoto R, Yoshida N (2011) Fbxl10/Kdm2b deficiency accelerates neural progenitor cell death and leads to exencephaly. Mol Cell 46:614–624

Ginsberg SD, Alldred MJ, Counts SE, Cataldo AM, Neve RL, Jiang Y, Wuu J, Chao MV, Mufson EJ, Nixon RA, Che S (2010) Microarray analysis of hippocampal CA1 neurons implicates early endosomal dysfunction during Alzheimer's disease progression. Biol Psychiatry 68:885–893

Govindarajan N, Rao P, Burkhardt S, Sananbenesi F, Schlüter OM, Bradke F, Lu J, Fischer A (2013) Reducing HDAC6 ameliorates cognitive deficits in a mouse model for Alzheimer's disease. EMBO Mol Med 5:52–63

Gregoretti IV, Lee YM, Goodson HV (2004) Molecular evolution of the histone deacetylase family: functional implications of phylogenetic analysis. J Mol Biol 338:17–31

Grosselin K, Durand A, Marsolier J, Poitou A, Marangoni E, Nemati F, Dahmani A, Lameiras S, Reyal F, Frenoy O, Pousse Y, Reichen M, Woolfe A, Brenan C, Griffiths AD, Vallot C, Gérard A (2019) High-throughput single-cell ChIP-seq identifies heterogeneity of chromatin states in breast cancer. Nat Genet 51:1060–1066

Guan JS, Haggarty SJ, Giacometti E, Dannenberg JH, Joseph N, Gao J, Nieland TJ, Zhou Y, Wang X, Mazitschek R, Bradner JE, DePinho RA, Jaenisch R, Tsai LH (2009) HDAC2 negatively regulates memory formation and synaptic plasticity. Nature 459:55–60

Guenther MG, Jenner RG, Chevalier B, Nakamura T, Croce CM, Canaani E, Young RA (2006) Global and Hox-specific roles for the MLL1 methyltransferase. Proc Natl Acad Sci U S A 102:8603–8608

Gupta S, Kim SY, Artis S, Molfese DL, Schumacher A, Sweatt JD, Paylor RE, Lubin FD (2010) Histone methylation regulates memory formation. J Neurosci 30:3589–3599

Gupta-Agarwal S, Franklinm AV, Deramusm T, Wheelockm M, Davism RL, McMahonm LL, Lubin FD (2012) G9a/GLP histone lysine dimethyltransferase complex activity in the hippocampus and the entorhinal cortex is required for gene activation and silencing during memory consolidation. J Neurosci 32:5440–5453

Haigis MC, Sinclair DA (2010) Mammalian sirtuins: biological insights and disease relevance. Annu Rev Pathol 5:253–295

Halder R, Hennion M, Vidal RO, Shomroni O, Rahman RU, Rajput A, Centeno TP, van Bebber F, Capece V, Vizcaino JC, Schuetz AL, Burkhardt S, Benito E, Sala MN, Javan SB, Haass C, Schmid B, Fischer A, Bonn S (2016) DNA methylation changes in plasticity genes accompany the formation and maintenance of memory. Nat Neurosci 19:102–110

Heintzman ND, Hon GC, Hawkins RD, Kheradpour P, Stark A, Harp LF, Ye Z, Lee LK, Stuart RK, Ching CW, Ching KA, Antosiewicz-Bourget JE, Liu H, Zhang X, Green RD, Lobanenkov VV, Stewart R, Thomson JA, Crawford GE, Kellis M, Ren B (2009) Histone modifications at human enhancers reflect global cell-type-specific gene expression. Nature 459:108–112

Herrup K, Li J, Chen J (2013) The role of ATM and DNA damage in neurons: upstream and downstream connections. DNA Repair 12:600–604

Hiraide T, Nakashima M, Yamoto K, Fukuda T, Kato M, Ikeda H, Sugie Y, Aoto K, Kaname T, Nakabayashi K, Ogata T, Matsumoto N, Saitsu H (2018) De novo variants in SETD1B are associated with intellectual disability, epilepsy and autism. Hum Genet 137:95–104

Hunter RG, Murakami G, Dewell S, Seligsohn M, Baker ME, Datson NA, McEwen BS, Pfaff DW (2012) Acute stress and hippocampal histone H3 lysine 9 trimethylation, a retrotransposon silencing response. Proc Natl Acad Sci U S A 109:17657–17662

Igaz LM, Vianna MR, Medina JH, Izquierdo I (2002) Two time periods of hippocampal mRNA synthesis are required for memory consolidation of fear-motivated learning. J Neurosci 22:6781–6791

Jakovcevski M, Akbarian S (2012) Epigenetic mechanisms in neurological disease. Nat Med 18:1194–1204

Jakovcevski M, Bharadwaj R, Straubhaar J, Gao G, Gavin DO, Jakovcevski I, Mitchell AC, Akbarian S (2013a) Prefrontal cortical dysfunction after overexpression of histone deacetylase 1. Biol Psychiatry 74:696–705

Jakovcevski M, Bharadwaj R, Straubhaar J, Gao G, Gavin DP, Jakovcevski I, Mitchell AC, Akbarian S (2013b) Prefrontal cortical dysfunction after overexpression of histone deacetylase 1. Biol Psychiatry 74:696–705

Jakovcevski M, Ruan H, Shen EY, Dincer A, Javidfar B, Ma Q, Peter CJ, Cheung I, Mitchell AC, Jiang Y, Lin CL, Pothula V, Stewart AF, Ernst P, Yao WD, Akbarian S, Impey S (2015) Neuronal Kmt2a/Mll1 histone methyltransferase is essential for prefrontal synaptic plasticity and working memory. J Neurosci 35:5097–5108

Jarome TJ, Perez GA, Hauser RM, Hatch KM, Lubin FD (2018) EZH2 methyltransferase activity controls Pten expression and mTOR signaling during fear memory reconsolidation. J Neurosci 38:7635–7648

Jawerka M, Colak D, Dimou L, Spiller C, Lagger S, Montgomery RL, Olson EN, Wurst W, Göttlicher M, Götz M (2010) The specific role of histone deacetylase 2 in adult neurogenesis. Neuron Glia Biol 14:1–15

Jensen LR, Amende M, Gurok U, Moser B, Gimmel V, Tzschach A, Janecke AR, Tariverdian G, Chelly J, Fryns JP, Van Esch H, Kleefstra T, Hamel B, Moraine C, Gecz J, Turner G, Reinhardt R, Kalscheuer VM, Ropers HH, Lenzner S (2005) Mutations in the JARID1C gene, which is involved in transcriptional regulation and chromatin remodeling, cause X-linked mental retardation. Am J Hum Genet 76:227–236

Jiang Y, Jakovcevski M, Bharadwaj R, Connor C, Schroeder FA, Lin CL, Straubhaar J, Martin G, Akbarian S (2010) Setdb1 histone methyltransferase regulates mood-related behaviors and expression of the NMDA receptor subunit NR2B. J Neurosci 30:7152–7267

Jiang Y, Loh YE, Rajarajan P, Hirayama T, Liao W, Kassim BS, Javidfar B, Hartley BJ, Kleofas L, Park RB, Labonte B, Ho SM, Chandrasekaran S, Do C, Ramirez BR, Peter CJ, Safaie BM, Morishita H, Roussos P, Nestler EJ, Schaefer A, Tycko B, Brennand KJ, Yagi T, Shen L, Akbarian S (2017) The methyltransferase SETDB1 regulates a large neuron-specific topological chromatin domain. Nat Genet 49:1239–1250

Jing X, Sui WH, Wang S, Xu XF, Yuan RR, Chen XR, Ma HX, Zhu YX, Sun JK, Yi F, Chen ZY, Wang Y (2017) HDAC7 ubiquitination by the E3 ligase CBX4 is involved in contextual fear conditioning memory formation. J Neurosci 37:3848–3863

Johnson AA, Sarthi J, Pirooznia SK, Reube W, Elefant F (2013) Increasing Tip60 HAT levels rescues axonal transport defects and associated behavioral phenotypes in a Drosophila Alzheimer's disease model. J Neurosci 33:7535–7547

Johnstone AL, O'Reilly JJ, Patel AJ, Guo Z, Andrade NS, Magistri M, Nathanson L, Esanov R, Miller BH, Turecki G, Brothers SP, Zeier Z, Wahlestedt C (2018) EZH1 is an antipsychotic-sensitive epigenetic modulator of social and motivational behavior that is dysregulated in schizophrenia. Neurobiol Dis 119:149–158

Josselyn SA (2005) What's right with my mouse model? New insights into the molecular and cellular basis of cognition from mouse models of Rubinstein-Taybi Syndrome. Learn Mem 12:80–83

Kaidi A, Jackson SP (2013) KAT5 tyrosine phosphorylation couples chromatin sensing to ATM signalling. Nature 498:70–74

Kerimoglu C, Agis-Balboa RC, Kranz A, Stilling R, Bahari-Javan S, Benito-Garagorri E, Halder R, Burkhardt S, Stewart AF, Fischer A (2013) Histone-methyltransferase mll2 (kmt2b) is required for memory formation in mice. J Neurosci 33:3452–3464

Kerimoglu C, Sakib MS, Jain G, Benito E, Burkhardt S, Capece V, Kaurani L, Halder R, Agís-Balboa RC, Stilling R, Urbanke H, Kranz A, Stewart AF, Fischer A (2017) KMT2A and KMT2B mediate memory function by affecting distinct genomic regions. Cell Rep 20:538–548

Kilgore M, Miller CA, Fass DM, Hennig KM, Haggarty SJ, Sweatt JD, Rumbaugh G (2010) Inhibitors of class 1 histone deacetylases reverse contextual memory deficits in a mouse model of Alzheimer's disease. Neuropsychopharmacology 35:870–880

Kim KH, Moon M, Yu SB, Mook-Jung I, Kim JI (2012a) RNA-Seq analysis of frontal cortex and cerebellum from 5XFAD mice at early stage of disease pathology. J Alzheimers Dis 29:793–808

Kim MS, Akhtar MW, Adachi M, Mahgoub M, Bassel-Duby R, Kavalali ET, Olson EN, Monteggia LM (2012b) An essential role for histone deacetylase 4 in synaptic plasticity and memory formation. J Neurosci 32:10879–10886

Kinner A, Wu W, Staudt C, Iliakis G (2008) Gamma-H2AX in recognition and signaling of DNA double-strand breaks in the context of chromatin. Nucleic Acids Res 36:5678–5694

Kleefstra T, Schenck A, Kramer JM, van Bokhoven H (2014) The genetics of cognitive epigenetics. Neuropharmacology 80:83–94

Kocki T, Urbańska EM, Kocki J, Kloc R, Kocka K, Olajossy M, Owe-Larsson B (2018) Prolonged therapy with antidepressants increases hippocampal level of kynurenic acid and expression of Kat1 and Kat2 genes. Pharmacol Rep 70:737–745

Koemans TS, Kleefstra T, Chubak MC, Stone MH, Reijnders MRF, de Munnik S, Willemsen MH, Fenckova M, Stumpel CTRM, Bok LA, Sifuentes B, Saenz M, Byerly KA, Baughn LB, Stegmann APA, Pfundt R, Zhou H, van Bokhoven H, Schenck A, Kramer JM (2018) Functional convergence of histone methyltransferases EHMT1 and KMT2C involved in intellectual disability and autism spectrum disorder. PLoS Genet 13(10):e1006864

Korzus E, Rosenfeld MG, Mayford M (2004) CBP histone acetyltransferase activity is a critical component of memory consolidation. Neuron 42:961–972

Koshibu K, Gräff J, Beullens M, Heitz FD, Berchtold D, Russig H, Farinelli M, Bollen M, Mansuy IM (2009) Protein phosphatase 1 regulates the histone code for long-term memory. J Neurosci 29:13079–13089

Kozikowski AP, Shen S, Pardo M, Tavares MT, Szarics D, Benoy V, Zimprich CA, Kutil Z, Zhang G, Bařinka C, Robers MB, Van Den Bosch L, Eubanks JH, Jope RS (2019) Brain penetrable histone deacetylase 6 inhibitor SW-100 ameliorates memory and learning impairments in a mouse model of fragile X syndrome. ACS Chem Nerosci 10:1679–1695

Kramer JM, Kochinke K, Oortveld MA, Marks H, Kramer D, de Jong EK, Asztalos Z, Westwood JT, Stunnenberg HG, Sokolowski MB, Keleman K, Zhou H, van Bokhoven H, Schenck A (2011) Epigenetic regulation of learning and memory by Drosophila EHMT/G9a. PLoS Biol 9: e1000569

Kwapis JL, Alaghband Y, López AJ, Long JM, Li X, Shu G, Bodinayake KK, Matheos DP, Rapp PR, Wood MA (2019) HDAC3-mediated repression of the Nr4a family contributes to age-related impairments in long-term memory. J Neurosci 39:4999–5009

Labonne JD, Lee KH, Iwase S, Kong IK, Diamond MP, Layman LC, Kim CH, Kim HG (2016) An atypical 12q24.31 microdeletion implicates six genes including a histone demethylase KDM2B and a histone methyltransferase SETD1B in syndromic intellectual disability. Hum Genet 135:757–771

Lang B, Alrahbeni TM, Clair DS, Blackwood DH, International Schizophrenia Consortium, McCaig CD, Shen S (2011) HDAC9 is implicated in schizophrenia and expressed specifically in post-mitotic neurons but not in adult neural stem cells. Am J Stem Cells 18:31–41

Le HQ, Ghatak S, Yeung CY, Tellkamp F, Günschmann C, Dieterich C, Yeroslaviz A, Habermann B, Pombo A, Niessen CM, Wickström SA (2016) Mechanical regulation of transcription controls polycomb-mediated gene silencing during lineage commitment. Nat Cell Biol 18:864–875

Lee KK, Workman JL (2007) Histone acetyltransferase complexes: one size doesn't fit all. Nat Rev Mol Cell Biol 8:285–295

Levenson JM, O'Riordan KJ, Brown KD, Trinh MA, Molfese DL, Sweatt JD (2004) Regulation of histone acetylation during memory formation in the hippocampus. J Biol Chem 279:40545–40559

Li J, Hart RP, Mallimo EM, Swerdel MR, Kusnecov AW, Herrup K (2013) EZH2-mediated H3K27 trimethylation mediates neurodegeneration in ataxia-telangiectasia. Nat Neurosci 16:1745–1753

Liang WS, Dunckley T, Beach TG, Grover A, Mastroeni D, Ramsey K, Caselli RJ, Kukull WA, McKeel D, Morris JC, Hulette CM, Schmechel D, Reiman EM, Rogers J, Stephan DA (2010) Neuronal gene expression in non-demented individuals with intermediate Alzheimer's disease neuropathology. Neurobiol Aging 31:549–566

Liang Q, Xia W, Li W, Jiao J (2018) RNF20 controls astrocytic differentiation through epigenetic regulation of STAT3 in the developing brain. Cell Death Differ 25:294–306

Lipinski M, Del Blanco B, Barco A (2019) CBP/p300 in brain development and plasticity: disentangling the KAT's cradle. Curr Opin Neurobiol 8:1–8

Lopez-Atalaya JP, Ciccarelli A, Viosca J, Valor LM, Jimenez-Minchan M, Canals S, Giustetto M, Barco A (2011) CBP is required for environmental enrichment-induced neurogenesis and cognitive enhancement. EMBO J 30:4287–4298

Lu T, Pan Y, Kao SY, Li C, Kohane I, Chan J, Yankner BA (2004) Gene regulation and DNA damage in the ageing human brain. Nature 429:883–891

Lu Y, Tan L, Wang X (2019) Circular HDAC9/microRNA-138/Sirtuin-1 pathway mediates synaptic and amyloid precursor protein processing deficits in Alzheimer's disease. Neurosci Bull 18. https://doi.org/10.1007/s12264-019-00361-0

Madabhushi R, Gao F, Pfenning AR, Pan L, Yamakawa S, Seo J, Rueda R, Phan TX, Yamakawa H, Pao PC, Stott RT, Gjoneska E, Nott A, Cho S, Kellis M, Tsai LH (2015) Activity-induced DNA breaks govern the expression of neuronal early-response genes. Cell 161:1592–1605

Maddox SA, Kilaru V, Shin J, Jovanovic T, Almli LM, Dias B, Norrholm SD, Fani N, Michopoulos V, Ding Z, Conneely KN, Binder EB, Ressler KJ, Smith AK (2018) Estrogen-dependent association of HDAC4 with fear in female mice and women with PTSD. Mol Psychiatry 23:658–665

Mao P, Reddy PH (2011) Aging and amyloid beta-induced oxidative DNA damage and mitochondrial dysfunction in Alzheimer's disease: implications for early intervention and therapeutics. Biochim Biophys Acta 1812:1359–1370

Marioni RE, Harris SE, Zhang Q, McRae AF, Hagenaars SP, Hill WD, Davies G, Ritchie CW, Gale CR, Starr JM, Goate AM, Porteous DJ, Yang J, Evans KL, Deary IJ, Wray NR, Visscher PM (2018) GWAS on family history of Alzheimer's disease. Transl Psychiatry 8:161

Markus HS, Mäkelä KM, Bevan S, Raitoharju E, Oksala N, Bis JC, O'Donnell C, Hainsworth A, Lehtimäki T (2013) Evidence HDAC9 genetic variant associated with ischemic stroke increases risk via promoting carotid atherosclerosis. Stroke 44:1220–1225

Maurice T, Duclot F, Meunier J, Naert G, Givalois L, Meffre J, Célérier A, Jacquet C, Copois V, Mechti N, Ozato K, Gongora C (2008) Altered memory capacities and response to stress in p300/CBP-associated factor (PCAF) histone acetylase knockout mice. Neuropsychopharmacology 33:1584–1602

Maze I, Covington HE 3rd, Dietz DM, LaPlant Q, Renthal W, Russo SJ, Mechanic M, Mouzon E, Neve RL, Haggarty SJ, Ren Y, Sampath SC, Hurd YL, Greengard P, Tarakhovsky A, Schaefer A, Nestler EJ (2010) Essential role of the histone methyltransferase G9a in cocaine-induced plasticity. Science 327:213–216

McFarland KN, Das S, Sun TT, Leyfer D, Kim MO, Xia E, Sangrey GR, Kuhn A, Luthi-Carter R, Clark TW, Sadri-Vakili G, Cha JH (2013) Genome-wide increase in histone H2A ubiquitylation in a mouse model of Huntington's disease. J Huntingtons Dis 2:263–277

McNulty SE, Barrett RM, Vogel-Ciernia A, Malvaez M, Hernandez N, Davatolhagh MF, Matheos DP, Schiffman A, Wood MA (2012) Differential roles for Nr4a1 and Nr4a2 in object location vs. object recognition long-term memory. Learn Mem 19:588–592

McQuown SC, Barrett RM, Matheos DP, Post RJ, Rogge GA, Alenghat T, Mullican SE, Jones S, Rusche JR, Lazar MA, Wood MA (2011) HDAC3 is a critical negative regulator of long-term memory formation. J Neurosci 31:764–774

Merson TD, Dixon MP, Collin C, Rietze RL, Bartlett PF, Thomas T, Voss AK (2006) The transcriptional coactivator Querkopf controls adult neurogenesis. J Neurosci 26:11359–11370

Mews P, Donahuem G, Drakem AM, Luczakm V, Abelm T, Bergerm SL (2017) Acetyl-CoA synthetase regulates histone acetylation and hippocampal memory. Nature 546(7658):381–386

Miller CA, Campbell SL, Sweatt JD (2008) DNA methylation and histone acetylation work in concert to regulate memory formation and synaptic plasticity. Neurobiol Learn Mem 89:599–603

Mills JD, Nalpathamkalam T, Jacobs HI, Janitz C, Merico D, Hu P, Janitz M (2013) RNA-Seq analysis of the parietal cortex in Alzheimer's disease reveals alternatively spliced isoforms related to lipid metabolism. Neurosci Lett 1:90–95

Mitchnick KA, Creighton SD, Cloke JM, Wolter M, Zaika O, Christen B, Van Tiggelen M, Kalisch BE, Winters BD (2016) Dissociable roles for histone acetyltransferases p300 and PCAF in hippocampus and perirhinal cortex-mediated object memory. Genes Brain Behav 2016:6

Mohan M, Herz HM, Smith ER, Zhang Y, Jackson J, Washburn MP, Florens L, Eissenberg JC, Shilatifard A (2011) The COMPASS family of H3K4 methylases in Drosophila. Mol Cell Biol 31:4310–4318

Moncada D, Ballarini F, Martinez MC, Frey JU, Viola H (2011) Identification of transmitter systems and learning tag molecules involved in behavioral tagging during memory formation. Proc Natl Acad Sci U S A 108:12931–12936

Morimoto S, Komatsu S, Takahashi R, Matsuo M, Goto S (1993) Age-related change in the amount of ubiquitinated histones in the mouse brain. Arch Gerontol Geriatr 16:217–224

Morris MJ, Mahgoub M, Na ES, Pranav H, Monteggia LM (2013) Loss of histone deacetylase 2 improves working memory and accelerates extinction learning. J Neurosci 33:6401–6411

Müller T, Schrötter A, Loosse C, Pfeiffer K, Theiss C, Kauth M, Meyer HE, Marcus K (2013) A ternary complex consisting of AICD, FE65, and TIP60 down-regulates Stathmin1. Biochim Biophys Acta 1834:387–394

Narkaj K, Stefanelli G, Wahdan M, Azam AB, Ramzan F, Steininger CFDJ, Walters BJ, Zovkic IB (2018) Blocking H2A.Z incorporation via Tip60 inhibition promotes systems consolidation of fear memory in mice. eNeuro 5(5). pii: ENEURO.0378-18.2018

Neelamegam R, Ricq EL, Malvaez M, Patnaik D, Norton S, Carlin SM, Hill IT, Wood MA, Haggarty SJ, Hooker JM (2012) Brain-penetrant LSD1 inhibitors can block memory consolidation. ACS Chem Nerosci 3:120–128

Oike Y, Hata A, Mamiya T, Kaname T, Noda Y, Suzuki M, Yasue H, Nabeshima T, Araki K, Yamamura K (1999) Truncated CBP protein leads to classical Rubinstein-Taybi syndrome phenotypes in mice: implications for a dominant-negative mechanism. Hum Mol Genet 8:387–396

Oliveira AM, Wood MA, McDonough CB, Abel T (2007) Transgenic mice expressing an inhibitory truncated form of p300 exhibit long-term memory deficits. Learn Mem 14:564–572

Oliveira AM, Estévez MA, Hawk JD, Grimes S, Brindle PK, Abel T (2011) Subregion-specific p300 conditional knock-out mice exhibit long-term memory impairments. Learn Mem 18:161–169

Park DH, Hong SJ, Salinas RD, Liu SJ, Sun SW, Sgualdino J, Testa G, Matzuk MM, Iwamori N, Lim DA (2014) Activation of neuronal gene expression by the JMJD3 demethylase is required for postnatal and adult brain neurogenesis. Cell Rep 8:1290–1299

Peleg S, Sananbenesi F, Zovoilis A, Burkhardt S, Bahari-Java S, Agis-Balboa RC, Cota P, Wittnam J, Gogul-Doering A, Opitz L, Salinas-Riester G, Dettenhofer M, KAng H, Farinelli L, Chen W, Fischer A (2010) Altered histone acetylation is associated with age-dependent memory impairment in mice. Science 328:753–756

Petrij F, Giles RH, Dauwerse HG, Saris JJ, Hennekam RC, Masuno M, Tommerup N, van Ommen GJ, Goodman RH, Peters DJ (1995) Rubinstein-Taybi syndrome caused by mutations in the transcriptional co-activator CBP. Nature 376:348–351

Piña B, Suau P (1987) Changes in histones H2A and H3 variant composition in differentiating and mature rat brain cortical neurons. Dev Biol 123:51–58

Pirooznia SK, Sarthi J, Johnson AA, Toth MS, Chiu K, Koduri S, Elefant F (2012) Tip60 HAT activity mediates APP induced lethality and apoptotic cell death in the CNS of a Drosophila Alzheimer's disease model. PLoS One 7:e41776

Renthal W, Maze I, Krishnan V, Covington HE 3rd, Xiao G, Kumar A, Russo SJ, Graham A, Tsankova N, Kippin TE, Kerstetter KA, Neve RL, Haggarty SJ, TA MK, Bassel-Duby R, Olson EN, Nestler EJ (2007) Histone deacetylase 5 epigenetically controls behavioral adaptations to chronic emotional stimuli. Neuron 56:517–529

Roguev A, Schaft D, Shevchenko A, Pijnappel WW, Wilm M, Aasland R, Stewart AF (2001) The Saccharomyces cerevisiae Set1 complex includes an Ash2 homologue and methylates histone 3 lysine 4. EMBO J 20:7137–7143

Roidl D, Hellbach N, Bovio PP, Villarreal A, Heidrich S, Nestel S, Grüning BA, Boenisch U, Vogel T (2016) DOT1L activity promotes proliferation and protects cortical neural stem cells from activation of ATF4-DDIT3-mediated ER stress in vitro. Stem Cells 34:233–245

Rujirabanjerd S, Nelson J, Tarpey PS, Hackett A, Edkins S, Raymond FL, Schwartz CE, Turner G, Iwase S, Shi Y, Futreal PA, Stratton MR, Gecz J (2012) Identification and characterization of two novel JARID1C mutations: suggestion of an emerging genotype-phenotype correlation. Eur J Hum Genet 18:330–335

Saha P, Gupta RK, Sen T, Sen N (2019) Histone deacetylase 4 downregulation elicits post-traumatic psychiatric disorders through impairment of neurogenesis. J Neurotrauma. https://doi.org/10.1089/neu.2019.6373

Sando R 3rd, Gounko N, Pieraut S, Liao L, Yates J 3rd, Maximov A (2012) HDAC4 governs a transcriptional program essential for synaptic plasticity and memory. Cell 151:821–834

Sarthi J, Elefant F (2011) dTip60 HAT activity controls synaptic bouton expansion at the Drosophila neuromuscular junction. PLoS One 6:e26202

Saura CA, Choi SY, Beglopoulos V, Malkani S, Zhang D, Shankaranarayana Rao BS, Chattarji S, Kelleher RJ 3rd, Kandel ER, Duff K, Kirkwood A, Shen J (2004) Loss of presenilin function causes impairments of memory and synaptic plasticity followed by age-dependent neurodegeneration. Neuron 42:23–36

Scandaglia M, Lopez-Atalaya JP, Medrano-Fernandez A, Lopez-Cascales MT, Del Blanco B, Lipinski M, Benito E, Olivares R, Iwase S, Shi Y, Barco A (2017) Loss of Kdm5c causes spurious transcription and prevents the fine-tuning of activity-regulated enhancers in neurons. Cell Rep 21:47–59

Schaefer A, Sampath SC, Intrator A, Min A, Gertler TS, Surmeier DJ, Tarakhovsky A, Greengard P (2009) Control of cognition and adaptive behavior by the GLP/G9a epigenetic suppressor complex. Neuron 64:678–691

Schmitt M, Matthies H (1979) Biochemical studies on histones of the central nervous system. III. Incorporation of [14C]-acetate into the histones of different rat brain regions during a learning experiment. Acta Biol Med Ger 38:683–689

Selenica ML, Benner L, Housley SB, Manchec B, Lee DC, Nash KR, Kalin J, Bergman JA, Kozikowski A, Gordon MN, Morgan D (2014) Histone deacetylase 6 inhibition improves memory and reduces total tau levels in a mouse model of tau deposition. Alzheimers Res Ther 6. https://doi.org/10.1186/alzrt1124

Sharma RP, Grayson DR, Gavin DP (2008) Histone deacetylase 1 expression is increased in the prefrontal cortex of schizophrenia subjects: analysis of the National Brain Databank microarray collection. Schizophr Bull 98:111–117

Sharma M, Razali NB, Sajikumar S (2018) Inhibition of G9a/GLP complex promotes long-term potentiation and synaptic tagging/capture in hippocampal CA1 pyramidal neurons. Cereb Cortex 27:3161–3171

Shi Y (2007) Histone lysine demethylases: emerging roles in development, physiology and disease. Nat Rev Genet 8:829–833

Shi Y, Whetstine JR (2007) Dynamic regulation of histone lysine methylation by demethylases. Mol Cell 25:1–14

Shilatifard A (2012) The COMPASS family of histone H3K4 methylases: mechanisms of regulation in development and disease pathogenesis. Annu Rev Biochem 81:65–95

Shulha HP, Crisci JL, Reshetov D, Tushir JS, Cheung I, Bharadwaj R, Chou HJ, Houston IB, Peter CJ, Mitchell AC, Yao WD, Myers RH, Chen JF, Preuss TM, Rogaev EI, Jensen JD, Weng Z, Akbarian S (2012) Human-specific histone methylation signatures at transcription start sites in prefrontal neurons. PLoS Biol 10:e1001427

Simensen RJ, Rogers RC, Collins JS, Abidi F, Schwartz CE, Stevenson RE (2012) Short-term memory deficits in carrier females with KDM5C mutations. Genet Couns 23:31–40

Singh T, Kurki MI, Curtis D, Purcell SM, Crooks L, McRae J, Suvisaari J, Chheda H, Blackwood D, Breen G, Pietiläinen O, Gerety SS, Ayub M, Blyth M, Cole T, Collier D, Coomber EL, Craddock N, Daly MJ, Danesh J, DiForti M, Foster A, Freimer NB,

Geschwind D, Johnstone M, Joss S, Kirov G, Körkkö J, Kuismin O, Holmans P, Hultman CM, Iyegbe C, Lönnqvist J, Männikkö M, McCarroll SA, McGuffin P, McIntosh AM, McQuillin A, Moilanen JS, Moore C, Murray RM, Newbury-Ecob R, Ouwehand W, Paunio T, Prigmore E, Rees E, Roberts D, Sambrook J, Sklar P, St Clair D, Veijola J, Walters JT, Williams H, Swedish Schizophrenia Study; INTERVAL Study; DDD Study; UK10 K Consortium, Sullivan PF, Hurles ME, O'Donovan MC, Palotie A, Owen MJ, Barrett JC (2016) Rare loss-of-function variants in SETD1A are associated with schizophrenia and developmental disorders. Nat Neurosci 19:571–577

Smith AC, Kenny PJ (2018) HDAC5 regulates the formation of drug memories. Trends Mol Med 24:106–108

Stadler F, Kolb G, Rubusch L, Baker SP, Jones EG, Akbarian S (2005) Histone methylation at gene promoters is associated with developmental regulation and region-specific expression of ionotropic and metabotropic glutamate receptors in human brain. J Neurochem 94:324–336

Stefanelli G, Azam AB, Walters BJ, Brimble MA, Gettens CP, Bouchard-Cannon P, Cheng HM, Davidoff AM, Narkaj K, Day JJ, Kennedy AJ, Zovkic IB (2018) Learning and age-related changes in genome-wide H2A.Z binding in the mouse Hippocampus. Cell Rep 22:1124–1131

Stefanko DP, Barrett RM, Ly AR, Reolon GK, Wood MA (2009) Modulation of long-term memory for object recognition via HDAC inhibition. Proc Natl Acad Sci U S A 106:9447–9452

Stessman HA, Xiong B, Coe BP, Wang T, Hoekzema K, Fenckova M, Kvarnung M, Gerdts JTS, Cosemans N, Vives L, Lin J, Turner TN, Santen G, Ruivenkamp C, Kriek M, van Haeringen A, Aten E, Friend K, Liebelt J, Barnett C, Haan E, Shaw M, Gecz J, Anderlid BM, Nordgren A, Lindstrand A, Schwartz C, Kooy RF, Vandeweyer G, Helsmoortel C, Romano C, Alberti A, Vinci M, Avola E, Giusto S, Courchesne E, Pramparo T, Pierce K, Nalabolu S, Amaral DG, Scheffer IE, Delatycki MB, Lockhart PJ, Hormozdiari F, Harich B, Castells-Nobau A, Xia K, Peeters H, Nordenskjöld M, Schenck A, Bernier RA, Eichler EE (2017) Targeted sequencing identifies 91 neurodevelopmental-disorder risk genes with autism and developmental-disability biases. Nat Genet 49:515–526

Stilling R, Rönicke R, Benito-Garagorri E, Urbanke H, Capece V, Burckhard S, Bahari-Javan S, Barth J, Sananbenesi F, Schütz AL, Dyczkowski J, Martinez-Hernandez A, Kerimoglu C, Dent SR, Bonn S, Reymann KG, Fischer A (2014) K-Lysine acetyltransferase 2A regulates a hippocampal gene-expression network linked to memory formation. EMBO J 33:1912–1927

Strahl BD, Allis CD (2000) The language of covalent histone modifications. Nature 403:41–45

Suberbielle E, Sanchez PE, Kravitz AV, Wang X, Ho K, Eilertson K, Devidze N, Kreitzer AC, Mucke L (2013) Physiologic brain activity causes DNA double-strand breaks in neurons, with exacerbation by amyloid-β. Nat Neurosci 16:613–621

Sugo N, Oshiro H, Takemura M, Kobayashi T, Kohno Y, Uesaka N, Song WJ, Yamamoto N (2010) Nucleocytoplasmic translocation of HDAC9 regulates gene expression and dendritic growth in developing cortical neurons. Eur J Neurosci 31:152–1532

Sun H, Kennedy PJ, Nestler EJ (2013) Epigenetics of the depressed brain: role of histone acetylation and methylation. Neuropsychopharmacology 38:124–137

Swank MW, Sweatt JD (2001) Increased histone acetyltransferase and lysine acetyltransferase activity and biphasic activation of the ERK/RSK cascade in insular cortex during novel taste learning. J Neurosci 21:3383–3391

Takata A, Ionita-Laza I, Gogos JA, Xu B, Karayiorgou M (2016) De novo synonymous mutations in regulatory elements contribute to the genetic etiology of autism and schizophrenia. Neuron 89:940–947

Tang GB, Zeng YQ, Liu PP, Mi TW, Zhang SF, Dai K, Tang QY, Yang L, Xu YJ, Yan HL, Du HZ, Teng ZQ, Zhou FQ, Liu CM (2017) The histone H3K27 demethylase UTX regulates synaptic plasticity and cognitive behaviors in mice. Front Mol Neurosci 10:267

Taniguchi M, Carreira MB, Cooper YA, Bobadilla AC, Heinsbroek JA, Koike N, Larson EB, Balmuth EA, Hughes BW, Penrod RD, Kumar J, Smith LN, Guzman D, Takahashi JS, Kim TK, Kalivas PW, Self DW, Lin Y, Cowan CW (2017) HDAC5 and its target gene, Npas4, function in the nucleus accumbens to regulate cocaine-conditioned behaviors. Neuron 96:130–144

Twine NA, Janitz K, Wilkins MR, Janitz M (2011) Whole transcriptome sequencing reveals gene expression and splicing differences in brain regions affected by Alzheimer's disease. PLoS One 6:e16266

Vallianatos CN, Raines B, Porter RS, Wu MC, Garay PM, Collette KM, Seo YA, Dou Y, Keegan CE, Tronson N, Iwase S (2019) Amelioration of brain histone methylopathies by balancing a writer-eraser duo KMT2A-KDM5C. bioRxiv. doi: https://doi.org/10.1101/567917

van Steensel B, Belmont AS (2017) Lamina-associated domains: links with chromosome architecture, heterochromatin, and gene repression. Cell 169(5):780–791

Vaquero A, Loyola A, Reinberg D (2003) The constantly changing face of chromatin. Sci Aging Knowledge Environ 14:RE4

Vecsey CG, Hawk JD, Lattal KM, Stein JM, Fabian SA, Attner MA, Cabrera SM, McDonough CB, Brindle PK, Abel T, Wood MA (2007) Histone deacetylase inhibitors enhance memory and synaptic plasticity via CREB:CBP-dependent transcriptional activation. J Neurosci 27:6128–6140

Viosca J, Lopez-Atalaya JP, Olivares R, Eckner R, Barco A (2010) Syndromic features and mild cognitive impairment in mice with genetic reduction on p300 activity: differential contribution of p300 and CBP to Rubinstein-Taybi syndrome etiology. Neurobiol Dis 37:186–194

Waddington CH (1953) Epigenetics and evolution. Symp Soc Exp Biol 7:186–199

Wang Z, Zang C, Cui K, Schones DE, Barski A, Peng W, Zhao K (2009) Genome-wide mapping of HATs and HDACs reveals distinct functions in active and inactive genes. Cell 138:1019–1031

Wang CM, Tsai SN, Yew TW, Kwan YW, Ngai SM (2010) Identification of histone methylation multiplicities patterns in the brain of senescence-accelerated prone mouse 8. Biogerontology 11:87–102

Wang WH, Cheng LC, Pan FY, Xue B, Wang DY, Chen Z, Li CJ (2011) Intracellular trafficking of histone deacetylase 4 regulates long-term memory formation. Anat Rec (Hoboken) 294:1025–1034

Wang J, Telese F, Tan Y, Li W, Jin C, He X, Basnet H, Ma Q, Merkurjev D, Zhu X, Liu Z, Zhang J, Ohgi K, Taylor H, White RR, Tazearslan C, Suh Y, Macfarlan TS, Pfaff SL, Rosenfeld MG (2015) LSD1n is an H4K20 demethylase regulating memory formation via transcriptional elongation control. Nat Neurosci 18:1256–1264

Wang DY, Kosowan J, Samsom J, Leung L, Zhang KL, Li YX, Xiong Y, Jin J, Petronis A, Oh G, Wong AHC (2018) Inhibition of the G9a/GLP histone methyltransferase complex modulates anxiety-related behavior in mice. Acta Pharmacol Sin 39:866–874

Wei W, Coelho CM, Li X, Marek R, Yan S, Anderson S, Meyers D, Mukherjee C, Sbardella G, Castellano S, Milite C, Rotili D, Mai A, Cole PA, Sah P, Kobor MS, Bredy TW (2012) p300/CBP-associated factor selectively regulates the extinction of conditioned fear. J Neurosci 32:11930–11941

Wendeln AC, Degenhardt K, Kaurani L, Gertig M, Ulas T, Jain G, Wagner J, Häsler LMWK, Skodras A, Blank T, Staszewski O, Datta M, Centeno TP, Capece VIM, Kerimoglu C, Staufenbiel M, Schultze JL, Beyer M, Prinz M, Jucker M, Fischer ANJ (2018) Innate immune memory in the brain shapes neurological disease hallmarks. Nature 556(7701):332–338

Wijayatunge R, Chen LF, Cha YM, Zannas AS, Frank CL, West AE (2014) The histone lysine demethylase Kdm6b is required for activity-dependent preconditioning of hippocampal neuronal survival. Mol Cell Neurosci 61:187–200

Williams SR, Aldred MA, Der Kaloustian VM, Halal F, Gowans G, McLeod DR, Zondag S, Toriello HV, Magenis RE, Elsea SH (2010) Haploinsufficiency of HDAC4 causes brachydactyly mental retardation syndrome, with brachydactyly type E, developmental delays, and behavioral problems. Am J Hum Genet 87:219–228

Wood MA, Kaplan MP, Park A, Blanchard EJ, Oliveira AM, Lombardi TL, Abel T (2005) Transgenic mice expressing a truncated form of CREB-binding protein (CBP) exhibit deficits in hippocampal synaptic plasticity and memory storage. Learn Mem 12:111–119

Wood MA, Attner MA, Oliveira AM, Brindle PK, Abel T (2006) A transcription factor-binding domain of the coactivator CBP is essential for long-term memory and the expression of specific target genes. Learn Mem 13:609–617

Xu J, Deng X, Watkins R, Disteche CM (2008) Sex-specific differences in expression of histone demethylases Utx and Uty in mouse brain and neurons. J Neurosci 28:4521–4527

Yamakawa H, Cheng J, Penney J, Gao F, Rueda R, Wang J, Yamakawa S, Kritskiy O, Gjoneska E, Tsai LH (2017) The transcription factor Sp3 cooperates with HDAC2 to regulate synaptic function and plasticity in neurons. Cell Rep 20:1319–1334

Yang YC, Chen CN, Wu CI, Huang WJ, Kuo TY, Kuan MC, Tsai TH, Huang JS, Huang CY (2013) NBM-T-L-BMX-OS01, semisynthesized from osthole, is a novel inhibitor of histone deacetylase and enhances learning and memory in rats. Evid Based Complement Alternat Med 2013:514908. https://doi.org/10.1155/2013/514908

Yang L, Ma Z, Wang H, Niu K, Cao Y, Sun L, Geng Y, Yang B, Gao F, Chen Z, Wu Z, Li Q, Shen Y, Zhang X, Jiang H, Chen Y, Liu R, Liu N, Zhang Y (2019) Ubiquitylome study identifies increased histone 2A ubiquitylation as an evolutionarily conserved aging biomarker. Nat Commun 10(1):2191

Yao ZG, Zhang L, Huang L, Zhu H, Liu Y, Ma CM, Sheng SL, Qin C (2013) Regional and cell-type specific distribution of HDAC2 in the adult mouse brain. Brain Strct Funct 218:563–573

Zhang J, Ji F, Liu Y, Lei X, Li H, Ji G, Yuan Z, Jiao J (2014) Ezh2 regulates adult hippocampal neurogenesis and memory. J Neurosci 34:5184–5199

Zhang YX, Akumuo RC, España RA, Yan CX, Gao WJ, Li YC (2018) The histone demethylase KDM6B in the medial prefrontal cortex epigenetically regulates cocaine reward memory. Neuropharmacology 141:113–125

Zheng Y, Liu A, Wang ZJ, Cao Q, Wang W, Lin L, Ma K, Zhang F, Wei J, Matas E, Cheng J, Chen GJ, Wang X, Yan Z (2019) Inhibition of EHMT1/2 rescues synaptic and cognitive functions for Alzheimer's disease. Brain 142:787–807

Zhou Q, Obana EA, Radomski KL, Sukumar G, Wynder C, Dalgard CL, Doughty ML (2016) Inhibition of the histone demethylase Kdm5b promotes neurogenesis and derepresses Reln (reelin) in neural stem cells from the adult subventricular zone of mice. Mol Biol Cell 27:627–639

Zovkic IB, Meadows JP, Kaas GA, Sweatt JD (2013) Interindividual variability in stress susceptibility: a role for epigenetic mechanisms in PTSD. Front Psych 4. https://doi.org/10.3389/fpsyt.2013.00060

Disease Modeling of Neuropsychiatric Brain Disorders Using Human Stem Cell-Based Neural Models

Johanna Kaindl and Beate Winner

Contents

Abstract Human pluripotent stem (PS) cells are a relevant platform to model human-specific neurological disorders. In this chapter, we focus on human stem cell models for neuropsychiatric disorders including induced pluripotent stem (iPS) cell-derived neural precursor cells (NPCs), neurons and cerebral organoids. We discuss crucial steps for planning human disease modeling experiments. We introduce the different strategies of human disease modeling including transdifferentiation, human embryonic stem (ES) cell-based models, iPS cell-based models and genome editing options. Analysis of disease-relevant phenotypes is discussed. In more detail, we provide exemplary insight into modeling of the neurodevelopmental defects in autism spectrum disorder (ASD) and the process of neurodegeneration in Alzheimer's disease (AD). Besides monogenic diseases, iPS cell-derived models also generated data from idiopathic and sporadic cases.

Keywords Disease modeling · hiPSCs · hiPSC-derived neurons · Neuropsychiatric disorders · Organoids

J. Kaindl and B. Winner (✉)
Department of Stem Cell Biology, Friedrich-Alexander-University (FAU) Erlangen-Nürnberg, Erlangen, Germany
e-mail: beate.winner@fau.de

© Springer Nature Switzerland AG 2019

Curr Topics Behav Neurosci (2019) 42: 159–184
DOI 10.1007/7854_2019_111
Published Online: 13 August 2019

1 Introduction

Neuropsychiatric disorders are associated with heavy social and economic burden, contributing to the leading cause of disabilities worldwide (Whiteford et al. 2017; WHO 2014). Psychiatric diseases, such as autism spectrum disorder (ASD), schizophrenia (SCZ), and bipolar disorder (BPD), have heterogeneous etiologies. Symptoms are variable and complex, including abnormal social interaction and communication, restricted behavior, and deficits in cognition, combined with the lack of objective biomarkers. This heterogeneity has been challenging for experimental studies (Soliman et al. 2017). Genetically engineered mice frequently carry single mutations in relevant genes. The limitation of these models is that they only recapitulate certain aspects of the human disease phenotypes. However, due to the differences in brain size and architecture and development of the rodent and the complex human brain, studying human-specific neurological disorders in other species has certain limits. Therefore, a wide range of studies considering the pathology of neuropsychiatric disorders have been performed using brain imaging, genomic studies, and analysis of postmortem tissue. Genome-wide association studies (GWAS) reported heterogeneous combinations of numerous alleles of small effect, besides the rare and highly penetrant mutations (Gratten et al. 2014). However, the identification of underlying mechanisms and pathways beyond the genetic risk associations in cellular disease models has been challenging. Since many neuropsychiatric diseases appear sporadic, affect complex high-order brain functions, and involve multiple cell types across different brain regions, new human-specific disease models are needed for preclinical studies and understanding of disease-causing phenotypes. Human stem cell-based models for neuropsychiatric disorders have continuously emerged over the last decade. Here we review the recent progress of disease modeling of neuropsychiatric disorders using human stem cell-based models. Since over 500 studies covering human pluripotent stem cells and neurosciences have been published over the last 10 years, our review comprises a selection of two disease entities and exemplary studies on neuropsychiatric disease modeling.

2 Disease Modeling of Neuropsychiatric Brain Disorders Using Human Stem Cell-Based Neural Models

The first and crucial step when planning human disease modeling for neuropsychiatric diseases is to decide for the specific cell type/types needed for the experiment. This decision tree includes various experimental levels: (1) starting material, (2) necessary steps of reprogramming or genome editing, (3) the technique of generation of neural progenitors and neurons, and (4) neurotransmitter phenotypes. Depending on the hypothesis of the experiment, these are the first crucial decisions to be made. Therefore, within this chapter, we will describe the basic principles to generate neural models and neurons and discuss respective selection criteria.

2.1 *First Step to Plan Human Disease Modeling Experiments*

This is one of the most critical steps when setting out to plan a human disease modeling experiment. Different starting materials will be necessary, depending on whether the question can be best answered by performing genome editing in human pluripotent stem cells or patient-derived material. In the latter, somatic cells (fibroblasts, blood, others) will be needed for reprogramming into induced pluripotent stem cells or transdifferentiation/generation of induced neurons (for further details see Sect. 2.2.1). A check for respective rules of the Institutional Review Board and obtaining the necessary permissions according to the governmental regulations are a must.

2.2 *Human Stem Cell-Based Disease Models*

2.2.1 Transdifferentiation

Transdifferentiation, or lineage reprogramming, is the conversion of somatic cells into a cell of a different somatic lineage without passing the intermediate stage of pluripotency (Fig. 1). The force to change identity is achieved by ectopic expression

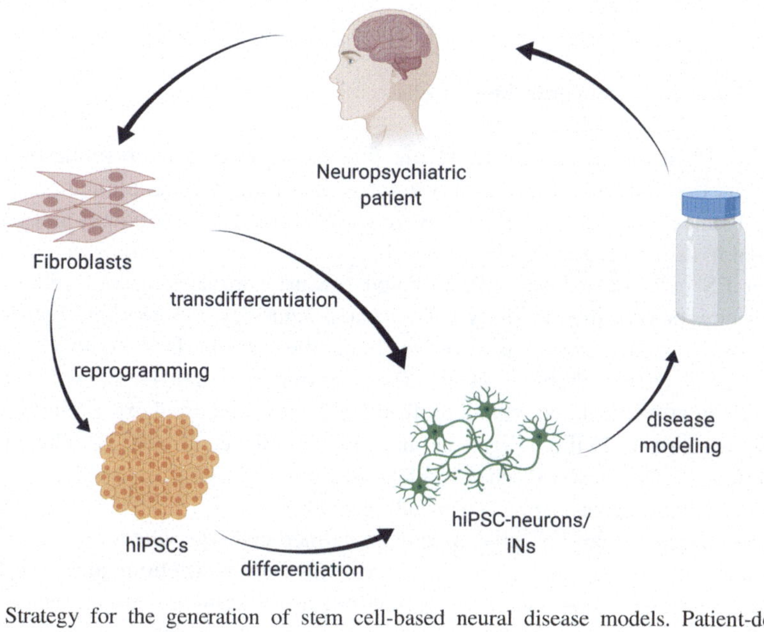

Fig. 1 Strategy for the generation of stem cell-based neural disease models. Patient-derived somatic cells are transdifferentiated into iNs or reprogrammed into hiPSCs and subsequently differentiated into disease-relevant cell types. Based on in vitro disease modeling, treatment of patients can be personalized in the future. Created with Biorender.com

of tissue-specific transcription factors (Davis et al. 1987). This switch to another cell type is considered to have great promise for regenerative medicine. For neuropsychiatric disease, especially disease modeling and compound testing are potential applications. Human fibroblasts, due to accessability and easy in vitro expansion, are frequently the cell type of choice for direct conversion into functional induced neurons (iNs), though hepatocytes or adipocytes have been successfully reprogrammed (Colasante et al. 2015; Vierbuchen et al. 2010). For example, a combination of the transcription factors achaete-scute homolog 1 (ASCL1), neurogenin 2 (NGN2), neurodifferentiation D1 (NEUROD1), POU domain class 3 transcription factor 2 (POU3F2), and myelin transcription factor 1-like protein (MYT1L) are required for efficient conversion into iNs (Mertens et al. 2016; Pang et al. 2011; Pfisterer et al. 2011; Son et al. 2011; Vadodaria et al. 2016). Moreover, computational approaches are made to predict transcription factors for developing transdifferentiation protocols (Rackham et al. 2016). The time-saving technique of lineage reprogramming into mature cells compared to pluripotent reprogramming could be applied faster for compound screening (Mertens et al. 2016). The applicability of direct lineage reprogramming of patient-derived cells for disease modeling has been shown for Alzheimer's disease (AD) (Hu et al. 2015). The major advantages of transdifferentiation are the fast speed of generating a human neuron and that the epigenetic landscape and aging signatures are preserved (Mertens et al. 2015a, b). However, limiting factors of iNs are the huge amount of needed patient material, the genetic mosaicism, and maturity of iN cultures (Gascón et al. 2017; Mertens et al. 2016).

2.2.2 Human Embryonic Stem Cells

In 1998, Thomson et al. described the first isolation of human embryonic stem (ES) cells derived from the blastocyst stage of a human preimplantation embryo. The self-renewable human ES cells were characterized as pluripotent and have the potential to differentiate into all three embryonic germ layers (Thomson et al. 1998). Human ES cells have been differentiated toward various neuronal cell fate and opened new possibilities to study early human neurodevelopment and pathological mechanisms causing neuronal disorders (Chambers et al. 2009; Kriks et al. 2011; Maroof et al. 2013; Nicholas et al. 2013; Perrier et al. 2004). ES cells carrying specific mutations can be obtained from embryos identified by preimplantation genetic diagnosis (PGD) (Eiges et al. 2007; Niclis et al. 2013). The classical generation of ESC-based disease models was done via viral expression or knockdown of specific genes (Urbach et al. 2004, 2009).

Even though human ES cells represent a defined cell type where huge experience exists also for differentiation protocols, governmental restrictions and ethical controversies need to be considered when planning an experiment with human ES cells. However, with the establishment of new genome editing methods, disease modeling based on human ES cells became more popular in the past years (Avior et al. 2017; Dever et al. 2016; Fogarty et al. 2017; Li et al. 2013; Liao et al. 2015; Soldner et al. 2016).

2.2.3 Human-Induced Pluripotent Stem Cells

The difficulty of limited availability and ethical controversy using human embryos was resolved by the discovery to jump-start pluripotency by overexpression of the transcription factors OCT3/4, SOX2, KLF4, and MYC (Takahashi et al. 2007). Human somatic cells can be reprogrammed to induced pluripotent stem (iPS) cells by transient expression of the pluripotency-specific transcription factors. For the first time, fibroblasts directly isolated from patients were reprogrammed and characterized for induced pluripotency and differentiation potential (Park et al. 2008). The patient-specific iPS cell lines offer a great opportunity for disease modeling of human brain disorders in vitro (Fig. 1), supported by the abundant availability of reliable differentiation protocols for various neuronal subtypes. Major discoveries for neuropsychiatric diseases were made through modeling of complex psychiatric disorders SCZ, ASD, and BPD using reprogrammed patient iPS cells (Brennand et al. 2011; Mertens et al. 2015b; Marchetto et al. 2010; Robicsek et al. 2013). Prior to planning an iPS cell-based experiment, one needs to think of the minimal number of patients and cell lines per patient, for example, including at least three diseased and three probands. The major advantage of iPS cells is the opportunity to generate patient-specific stem cells and develop personalized medicine for both genetic and sporadic diseases (Fig. 1). On the other hand, high variability, low reproducibility, and the choice of respective controls challenge stem cell-based experiments. Isogenic controls of patient-specific iPS cells could be a profitable approach enabled by genome editing.

2.2.4 Genome Editing

With the focus on monogenic disorders, genome editing can either be used to induce disease-causing mutations in healthy control cells or correct patient-specific genetic variants and generate isogenic control cell lines. Transcription activator-like effector nucleases (TALENs) and zinc-finger nucleases (ZFNs) have been utilized to target specific DNA sequences in human pluripotent stem cells (Hockemeyer et al. 2017; Soldner et al. 2011). With the discovery of the clustered regularly interspaced short palindromic repeats (CRISPR) and CRISPR-associated 9 (Cas9) system, a revolutionary opportunity for disease modeling has emerged (Jinek et al. 2012). The CRISPR/Cas9 system efficiently targets and cleaves desired DNA sequences, previously also shown for human iPS cells (Mali et al. 2013). Because of the remaining risk of potential indel formations at off-target sites in the genome, due to imperfect complementary of the guide RNA, recent studies used Cas9 mutants to increase the specificity of genome editing (Slaymaker et al. 2016; Chen et al. 2017). Unwanted and undetected off-targets in iPS cells endanger the reliability of new genome editing methods and still need to be optimized for preclinical studies. However, the major advantage of the CRISPR/Cas9 system is the possibility to perform rescue experiments and generate isogenic control cell lines that reduce the heterogeneity of iPS cell cultures and increase the reproducibility of stem cell-based experiments.

2.3 *Analysis of Disease-Relevant Phenotypes*

2.3.1 The "-OMIC" Tools for Disease Modeling

GWAS, whole exome sequencing (WES) studies, and copy number variation (CNV) studies of large patient cohorts have identified a range of genetic risk factors and biological pathways associated with mental disorders, such as ASD, BPD, SCZ, major depressive disorder, and attention-deficit hyperactivity disorder (ADHD) (Lee et al. 2013; Purcell et al. 2014; Ripke et al. 2013, 2014). Only a small percentage of psychiatric disorders are caused by single-gene mutations, for instance, Rett syndrome, Timothy syndrome, and familial AD (fAD). Hence, patient-derived iPS cells represent an efficient way to study the disease-relevant phenotypes of sporadic psychiatric disorders on the cellular and molecular level. Next-generation sequencing technologies identified altered gene expression in patient-derived iPS, neural progenitor cells (NPCs), and neurons (Brennand et al. 2011; Lin et al. 2014; Madison et al. 2015; Paşca et al. 2011; Wang et al. 2015; Wen et al. 2014). RNA-Seq analysis can also be achieved on single-cell level and could be used to identify molecularly distinct cell types (Bardy et al. 2016; Camp et al. 2015; Quadrato et al. 2017).

However, disease-causing expression profiles might as well result from epigenetic changes. Since the reprogramming process is thought to remove many epigenetic modifications, diseases with solely epigenetic causes might be difficult to study in human iPS cells. Previous studies have shown that iPS cells retain an epigenetic memory from their somatic origin, suggesting hot spots in the genome with unknown function that escape from epigenetic reprogramming (Kim et al. 2010; Lister et al. 2011; Nishino et al. 2011). The influences of epigenetic profiles of DNA methylation or histone modifications on the pathology of psychiatric disorders can be analyzed among others by chromatin immunoprecipitation sequencing (ChIP-Seq) (Brennand et al. 2015; Sheridan et al. 2011; Sugathan et al. 2014; Urbach et al. 2010). Nonetheless, the discussion about the epigenetic landscape after reprogramming and transdifferentiation is still ongoing. During reprogramming, closed and compacted chromatin of a somatic cells acts as a barrier for the complete establishment of pluripotency and an epigenetic landscape that is similar to ES cells. On the other hand, transdifferentiation combines downregulation of donor cell-specific genes and upregulation of target cell-specific genes and preserves certain epigenetic modifications (Qin et al. 2016). For disease modeling, one needs to keep in mind that direct conversion of somatic cells skips developmental precursor cell stages, and therefore one might miss associated phenotypes. Therefore, neurodevelopmental diseases such as ASD and Rett syndrome should be studied using iPSC-derived NPCs and neurons (Fig. 2). On the other hand, neurodegenerative diseases in certain circumstances might be better off with transdifferentiated iNs, because of the preserved age-related epigenetic signatures (Mertens et al. 2015a, b).

Lastly, quantitative proteomic mass spectrometry using stable isotope labeling by amino acids in cell culture (SILAC) of disease-relevant cell types and subsequent

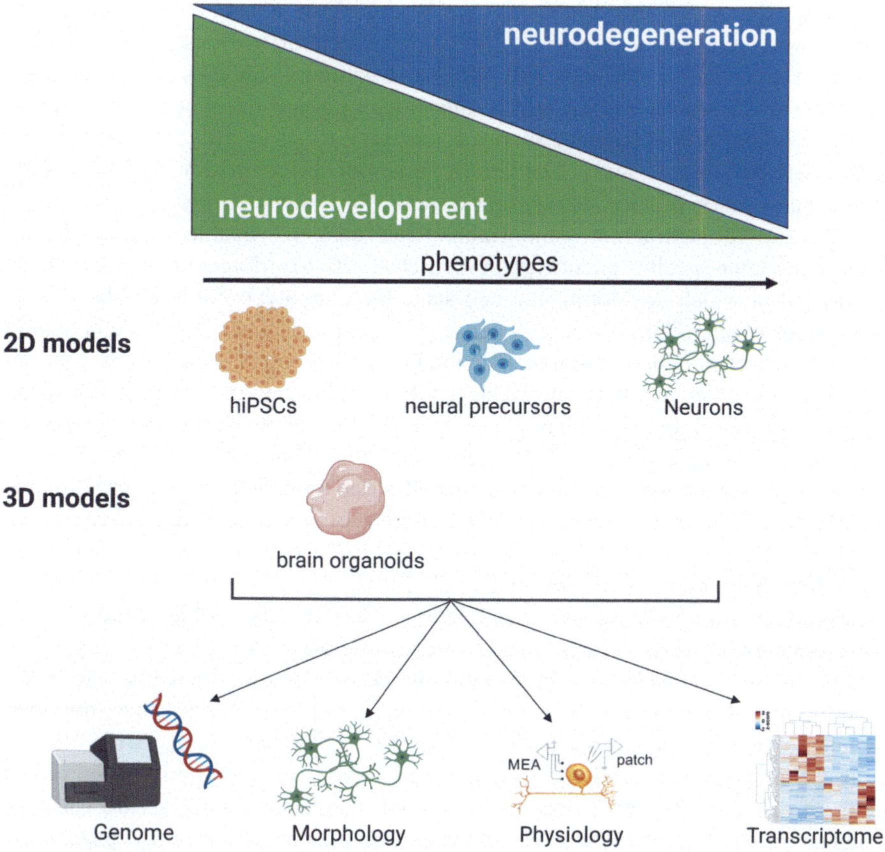

Fig. 2 Current approaches of stem cell-based neural disease models for neuropsychiatric disorders. Modeling phenotypes of neurodevelopmental and neurodegenerative diseases is partially restricted to developmental cell stages. Current analysis of patient-derived iPS cells and neural cells offers a range of potential methods, comprising genome, morphology, physiology, and transcriptome. Created with Biorender.com

network analyses detected abnormal protein levels and affected pathways in patients (Brennand et al. 2015; Li et al. 2015; Tobe et al. 2017). These robust and quantifiable techniques enable reproducible high-throughput approaches and can reveal biological pathways associated with psychiatric disorders in a human-based cellular model (Fig. 2).

2.3.2 Cellular Deficits in Neurodevelopment

Changes on the (epi-) genetic, transcriptional, and protein level are indicative of disturbed pathways in patients suffering from psychiatric disorders. The ideal workflow of stem cell-based models is the generation of authentic and reproducible cell types to mimic the aspects of the human central nervous system (CNS). Differentiating human iPS cells into NPCs and neuronal subtypes can recapitulate the brain development in vitro and further unveil disease-relevant cellular phenotypes. In addition to the choice of respective patients, it is critical to identify the appropriate cell type to study. The aim of numerous differentiation protocols is the efficient generation of homogenous cell systems, for example, toward glutamatergic, GABAergic, dopaminergic, and motor neurons, to enable the study of neurotransmitter-specific phenotypes (Wen et al. 2016). Moreover, it needs to be considered at which developmental cell stage the phenotype might be observable. Phenotypes of age-related diseases, such as AD, are more likely detectable in mature neurons, whereas neurodevelopmental phenotypes of psychiatric diseases might already be observed at precursor cell stages (Fig. 2). For neuropsychiatric disorders, several neurodevelopmental phenotypes have been observed during the differentiation process of patient-derived iPS cells into NPCs and neurons. The detected morphological phenotypes of human stem cell-based models for psychiatric diseases include smaller neuronal somas, reduced number of spines, and decreased neurite number and neurite lengths (Brennand et al. 2011; Cheung et al. 2011; Doers et al. 2014; Marchetto et al. 2010; Sheridan et al. 2011).

Moreover, some disease-specific phenotypes are already visible at the iPSC or NPC stage of in vitro cultivation. Altered proliferation, length of cell cycle, or deficits in the cytoskeleton organization of patient-derived iPS cells and NPCs were found in disease models of several psychiatric disorders, furthermore supported by gene expression analyses (Madison et al. 2015; Marchetto et al. 2017; Mariani et al. 2015; Yoon et al. 2014). Developmental deficits of iPSC-derived neurons in terms of differentiation efficiency and neuronal maturation found in vitro can give insights into the in vivo neurodevelopment of patients affected by psychiatric disorders (Madison et al. 2015; Robicsek et al. 2013). Considering the heterogeneous and complex genetic background of mental disorders, it is important that cellular phenotypes are robust and reproducible. For instance, a stem cell-based experiment should include three or more probands and controls with various lines from each individual. Human pluripotent stem cell-derived neural cells are in conclusion a relevant tool to model the complex pathology of psychiatric disorders.

2.3.3 Functionality of Neuronal Networks

Connecting genetic and transcriptomic results with developmental phenotypes observed during in vitro differentiation of patient-derived cells can give an important hint toward pathways involved in psychiatric diseases. An additional contribution to the patients' phenotypes concerns the functional metabolism and excitability of neurons and their interactions with glia cells. Metabolic abnormalities like oxidative stress or mitochondrial deficits were, for example, found in patients suffering from SCZ (Brennand et al. 2015; Paulsen BDa et al. 2012; Robicsek et al. 2013). Since synapses are the major communication points, disruption of synaptic connections might induce mental disorders in humans. Impaired synaptic activity in patients was identified using stem cell-based models, including defects in neurotransmitter release (Hook et al. 2014; Wen et al. 2014; Yu et al. 2014). Hyperexcitability and hyperactivity of patient-derived neurons were furthermore detected for BPD using patch-clamp recording and Ca^{2+}-imaging (Mertens et al. 2015b).

Furthermore, synaptic connections within the human brain are vulnerable to disruption and might trigger neuropsychiatric disorders. Decreased neuronal connectivity provoked by reduced synaptic connections and synaptic density in SCZ has been found in several postmortem studies and was further shown in iPS cell-derived neurons (Brennand et al. 2011). Even though NPCs and neurons derived from SCZ patients showed no apparent defects of protein expression and synaptic functionality or density, neuronal connectivity was disrupted. Interestingly, Brennand et al. took advantage of an assay that chases transneuronal spread of rabies which is usually transferred via synaptic contacts and hereby demonstrated reduced connectivity in patient-derived neurons compared to control neurons (Brennand et al. 2011). The multielectrode arrays (MEA) system is becoming an increasingly relevant tool to analyze neuronal activity and connective networks of iPS-derived neurons in vitro, demonstrating their capacity to self-organize into spontaneous and synchronous active networks (Xu et al. 2017). Moreover, MEA systems might be a useful tool to identify neurons that exhibit pacemaker properties and trigger synchronous activity of the neuronal network in a human cellular system (Illes et al. 2014).

Interactions in the human brain do not only occur between neurons. Glia cells represent the predominant brain population; therefore neuroglia interactions might have an important contribution to psychiatric diseases. Glia-associated phenotypes in psychiatric patients were found for different glial cells including astrocytes, oligodendrocytes, and microglia, indicating a higher complexity of the disease-causing mechanisms in the human brain (Cotter et al. 2001; Yamamuro et al. 2015). Abnormalities of all three glia cells are implicated, for example, in the disease mechanism of SCZ (Bernstein et al. 2015). (Non-) autologous co-culturing of patient-derived neurons and glia cells represent a valid system to investigate the influence and interaction between the neural subtypes. Differentiation of patient-derived iPS cells into *astrocytes* further indicated a crucial disease mechanism of glia cells in psychiatric disorders (Russo et al. 2017; Williams et al. 2014). Even though several protocols have been established to generate iPS cell-derived astrocytes, their implementation in disease modeling is still ongoing (Krencik et al. 2011; Shaltouki et al. 2013; Tcw et al. 2017). Early cell-cell interactions are required for astrocyte maturation, limiting the in vitro differentiations. Moreover, culture conditions in co-culture systems should be specially tailored to the physiological characteristics of astrocytes (Tcw et al. 2017). Since *microglia* are essential for physiological brain functionality, associations with psychiatric phenotypes were found (Derecki et al. 2012). Several approaches have been made to generate microglia out of human iPS cells (Abud et al. 2017; Douvaras et al. 2017; Muffat et al. 2016; Pandya et al. 2017) or to transdifferentiate human monocytes (Etemad et al. 2012; Ohgidani et al. 2017). Differentiation of human iPS cells into microglia-like cells is a major challenge due to their unique developmental origin. However, glial dysfunctions represent a hallmark of neuropsychiatric disorders, and including glia into new cell models might give new insights into their contribution to patients' phenotypes.

2.3.4 Neuronal Network Interactions in Three-Dimensional (3D) Cell Systems

Co-culture systems of brain cells are usually performed in an adherent and two-dimensional (2D) culture, lacking the 3D structure and cell-cell interactions those found in vivo. Moreover, the characteristics of neuropsychiatric disorders usually relate to reduced connectivity and circuit dysfunction. The shift from single-cell measurements to multidimensional analyses by use of brain organoids generated from human iPS cells revolutionized disease modeling of brain disorders (Lancaster et al. 2013). Brain organoids grown in suspension recapitulate specific features of the developing brain, including incorporation of various neural subtypes and spatial organization. Adding brain region-specific patterning factors to the organoid differentiation can form a more homogenous cell population than unpatterned organoids. Recently, 3D models of developing hippocampus (Sakaguchi et al. 2015), midbrain (Jo et al. 2016; Qian et al. 2016), forebrain (Birey et al. 2017), hypothalamus (Qian et al. 2016), and cerebellum (Muguruma et al. 2015) have been generated. On the contrary, unpatterned whole-brain organoids generate tissues with a more heterogeneous cell composition, as shown by immunohistochemistry and single-cell RNA sequencing (Camp et al. 2015; Lancaster et al. 2013; Quadrato et al. 2017). Moreover, neuronal cell migration within 3D cell systems has been studied through the fusion of region-specific brain organoids, suggesting high potential for robust neuronal network studies in human stem cell-based organoids (Bagley et al. 2017). Since the field is still young, available data for modeling psychiatric disorders using brain organoids is limited (Mariani et al. 2015; Wang et al. 2017). So far, 3D brain organoids have the greatest potential to model cell-cell-interactions with broad cellular diversity. Moreover, incorporation of glia cells into the organoids might promote the neuronal and circuit maturation that are impaired in neuropsychiatric disorders.

2.4 *Examples for Stem Cell-Based Disease Modeling of Neuropsychiatric Disorders*

Subsequently, exemplary studies on ASD (Table 1) and AD (Table 2) will be chosen, and major results of disease modeling experiments performed with the previously mentioned methods will be described.

Table 1 Current stem cell-based disease models of ASD and their major findings

Autism spectrum disorder	Phenotype	Major result	hPSCs	NPCs	Neurons	3D models
	Genome/ epigenome	Maintenance of modifications upon differentiation	Sheridan et al. (2011), Sugathan et al. (2014), Urbach et al. (2010), Mariani et al. (2015), Marchetto et al. (2010), Cheung et al. (2011) and Doers et al. (2011)		Marchetto et al. (2010)	
	Transcriptome	Differential expression of neurodevelopmental genes	Marchetto et al. (2017)	Sugathan et al. (2014), Paşca et al. (2011), Wang et al. (2015), Marchetto et al. (2017) and Wang et al. (2017)	Paşca et al. (2011), Wang et al. (2015), Marchetto et al. (2017), Wang et al. (2017), Shcheglovitov et al. (2013) and Baldino-Russo et al. (2017)	Mariani et al. (2015) and Wang et al. (2017)
	Proliferation/ differentiation	Altered proliferation at NPC stage	Mariani et al. (2015)	Marchetto et al. (2017) and Mariani et al. (2015)	Pasca et al. (2011), Mariani et al. (2015), Shcheglovitov et al. (2013) and Baldino-Russo et al. (2017)	Mariani et al. (2015)
	Neurite growth/ synaptogenesis	Reduced synaptogenesis and neurite outgrowth			Sheridan et al. (2011), Marchetto et al. (2017), Mariani et al. (2015), Marchetto et al. (2010), Cheung et al. (2011), Doers et al. (2011), Shcheglovitov et al. (2013) and Baldino-Russo et al. (2017)	Mariani et al. (2015)
	Electrophysiology	Reduced connectivity and network activity			Marchetto et al. (2010), (2017), Mariani et al. (2015),	Mariani et al. (2015)

(continued)

Table 1 (continued)

					Shcheglovitov et al. (2013) and Baldino-Russo et al. (2017)	
	Biochemical	Rescue of cellular phenotype		Paşca et al. (2011) and Marchetto et al. (2017)	Pasca et al. (2011), Marchetto et al. (2010), (2017), Shcheglovitov et al. (2013) and Baldino-Russo et al. (2017)	Mariani et al. (2015)
	Glia cells	Astrocytes contribute to neuronal phenotype	Sheridan et al. (2011), Baldino-Russo et al. (2017) and Williams et al. (2014)			

Table 2 Current stem cell-based disease models of AD and their major findings

Alzheimer's disease	Phenotype	Major result	hPSCs	NPCs	Neurons	3D models
	Genome/ epigenome					
	Transcriptome	Differential expression correlates with dysregulation in AD brains	Sproul et al. (2014)	Sproul et al. (2014)	Kondo et al. (2013)	
	Proliferation/ differentiation					
	Neurite growth/ synaptogenesis					
	Electrophysiology	Altered excitatory synapse activity			Honda et al. (2016)	
	Biochemical	Increased Aβ42/40 ratio and tau phosphorylation		Liu et al. (2012), Sproul et al. (2014)	Kondo et al. (2013), Israel et al. (2012), Yagi et al. (2011), Muratore et al. (2014), Choi et al. (2014), Raja et al. (2016), Shi et al. (2014), Liu et al. (2012), Koch et al. (2012), Sproul et al. (2014) and Ochalek et al. (2017)	Choi et al. (2014) and Raja et al. (2016)
	Glia cells	Aβ peptides in astrocytes, increased oxidative stress in astrocytes	Kondo et al. (2013) and Oksanen et al. (2017)			

2.4.1 Autism Spectrum Disorder

ASD is a behaviorally defined condition that is characterized by a broad spectrum of mild to severe disabilities in communication, social interaction, and stereotyped repetitive behavior. Only a minority of ASD cases is monogenic. However, GWAS of large patient cohorts identified numerous rare genetic variants associated with ASD (Autism Spectrum Disorders Working Group of The Psychiatric Genomics Consortium 2017). The primary pathological cause of ASD is thought to be altered neurodevelopment, leading to reduced neuronal circuitry, abnormal synaptogenesis, and dysregulation of the excitatory/inhibitory (E/I) balance (Persico et al. 2006). Nonetheless, ASD patients are phenotypically and etiologically heterogeneous, making it challenging to unveil the cellular mechanisms. Human iPS cell-derived neurons have been generated from monogenic ASD patients carrying mutations in MeCP2 (Rett syndrome), CACNA1C (Timothy syndrome), and SHANK3 (Phelan-McDermid syndrome), showing synaptic defects, altered calcium signaling, and electrophysiological aberrations (Marchetto et al. 2010; Paşca et al. 2011; Shcheglovitov et al. 2013). However, uncovering the phenotypes of complex ASD forms remains challenging.

A recent study of Marchetto et al. (2017) showed that early enlarged brain volume of sporadic ASD patients correlated with increased proliferation of iPS cell-derived NPCs. Based on genomic analyses, the altered proliferation of ASD NPCs resulted from changes in the canonical Wnt signaling pathway. Upon neuronal differentiation, they found a significant reduction of inhibitory neurons and impaired synaptogenesis. Neuronal network analysis of ASD neurons showed connectivity defects that might result from the observed imbalance of excitatory/inhibitory neurons and was further supported by RNA sequencing. In this study, patient-specific NPCs and neurons were utilized to examine the underlying molecular mechanisms of early brain enlargement in ASD patients (Marchetto et al. 2017).

Furthermore, early cortical development in patients with idiopathic ASD was modeled combining RNA sequencing and generation of iPS cell-derived telencephalic organoids. The transcriptome analysis of ASD patients revealed altered transcriptional regulation of cell proliferation, neuronal differentiation, and synaptic transmission. Subsequent analysis confirmed a significant decrease in cell cycle length of ASD-derived iPS cells and increased neuronal and synaptic differentiation in patient-specific organoids. Neuronal cell fate was altered in ASD-derived organoids, resulting in an imbalance between glutamatergic and GABAergic neurons. These differences were caused by an overexpression of the transcription factor FOXG1; knockdown of FOXG1 rescued the imbalance of GABA/glutamate neurons. Additionally, FOXG1 expression correlated with autism symptom severity of the patients, proposing a potential biomarker for sporadic ASD. This study emphasizes the importance of patient-specific disease modeling of idiopathic ASD and identifying molecular and cellular alterations using stem cell-based organoids (Mariani et al. 2015).

2.4.2 Alzheimer's Disease

In contrary, a psychiatric disorder that appears at advanced age is AD, affecting approximately 30 million people worldwide. AD is clinically characterized by progressive loss of memory, learning disability, and increasing difficulties in daily tasks with varying severity of neuropsychiatric symptoms. The pathological hallmarks of AD are extracellular plaques composed of insoluble amyloid-β (Aβ) peptides and intracellular neurofibrillary tangles formed by hyperphosphorylated tau protein associated with loss of neurons and synapses in the cerebral cortex and subcortical regions. Early-onset fAD is primarily caused by mutations in the amyloid precursor protein (*APP*), presenilin 1 (*PSEN1*), and presenilin 2 (*PSEN2*) genes and accounts for up to 5% of all AD cases. Modeling AD using human iPS cells was initiated with familial cases, caused by specific point mutations in the *APP*, *PSEN1*, or *PSEN2* gene. The patient-derived NPCs and neurons showed elevated secretion of longer Aβ peptides and altered expression and phosphorylation of tau compared to healthy control NPCs and neurons (Israel et al. 2012; Koch et al. 2012; Kondo et al. 2013; Liu et al. 2014; Muratore et al. 2014; Yagi et al. 2011; Sproul et al. 2014). Recently, elevated Aβ production and tau hyperphosphorylation were also shown for iPS cell-derived neurons derived from sporadic AD patients (Ochalek et al. 2017). Due to variability in differentiation efficiency, Israel et al. performed a fluorescence-activated cell sorting (FACS) step prior to differentiation and purification of NPCs. The accumulation of Aβ peptides was also found in adults suffering from Down syndrome caused by trisomy 21, due to increased expression of APP which is encoded on chromosome 21 (Shi et al. 2012). Overexpression of mutant *PSEN1* resulted in synaptic dysfunction of human ES cell-derived neurons, indicating additional disease-causing mechanisms besides protein aggregates (Honda et al. 2016). Kondo et al. were able to support the hypothesis that endoplasmic reticulum (ER) and oxidative stress, induced by Aβ aggregates, are associated with the pathogenesis of AD, in both familial and sporadic cases. Increased oxidative stress in astrocytes might also play an important role in the pathogenesis of AD, as indicated by previous studies (Kondo et al. 2013; Oksanen et al. 2017). However, a common challenge of modeling AD in vitro is the optimal timeframe to observe relevant phenotypes of age-related diseases (Fig. 2). Brain organoids generated from fAD patients that recapitulated AD pathology showed progressive increase in number and size of Aβ aggregates over time compared to control organoids (Choi et al. 2014; Raja et al. 2016). Therefore, organoids seem to be an important stem cell model to investigate age-related AD pathology in a time-dependent manner. Moreover, interference in the APP processing by inhibition of β- and γ-secretase decreased the number of Aβ aggregates, indicating that organoids might be a relevant model for future compound testing.

2.5 Untangling the Relationship Between Genotype and Phenotype

The increasing abundance of phenotypic data for neurological disorders of studies using human iPS cells makes it more and more difficult to correlate underlying mechanisms. A recent meta-analysis systematically searched 93 published articles, reporting neurological phenotypes of 31 neurological diseases using patient-specific cells, to unravel relationships between phenotypes and genotypes (Hollingsworth et al. 2017). With the iPS cell phenogenetic map project atlas (iPhemap), the accrued information is available online and will be continually updated with patient-derived models of neurodevelopmental and neurodegenerative diseases. Most of the included studies investigated neurodegenerative diseases with characterized somatic mutations (67%), followed by neurodevelopmental diseases (30%). Interestingly, differences were found in research practices and reporting of methods. Only a few studies used isogenic controls, and only 35% of the studies were conducted with three or more patient cell lines. Grouping of the observed phenotypes by a defined algorithm revealed increased or decreased cellular processes or products in 70% of the phenotypes included into the meta-analysis. Moreover, patient-derived oligodendrocytes and neurons were cell types with the largest number of reported phenotypes. The phenotype class "rescue/recovery from disease phenotypes after chemical treatment" was also mainly observed in neurons, indicating cell type-specific phenotypes in disease modeling. These findings were supported by the fact that neurodegenerative disease phenotypes were predominantly found at the neuronal stage, whereas neurodevelopmental phenotypes can be identified at early stages of differentiation (Fig. 2).

Network analysis of the phenotypes revealed novel gene ontology (GO) associations for several diseases and overlapping phenotypes between different diseases. Furthermore, the researchers included transcriptome data to determine the relationship between molecular and cellular phenotypes. Correlation between the molecular abnormalities and cellular phenotypes in iPS cells was found for neurodevelopmental disorders, implying an association between the molecular and cellular phenotypes. However, no correlation was found for neurodegenerative diseases. Significant molecular aberrations were identified at the neuronal level of both neurodevelopmental and neurodegenerative disorders, reflecting the cellular phenotypes observed by the included studies and implying high phenogenetic correlation. Taken together, iPhemap illustrates novel opportunities how phenogenetics can be used to improve future modeling of neurological disorders.

3 Challenges and Outlook

A large proportion of the worldwide population is affected by several neuropsychiatric disorders. Even though human iPS cells provide a broad platform for disease modeling of monogenic and idiopathic disorders, there are several technical challenges that limit the applicability of stem cell models. A major concern is the high variability between different patients or even between different clones of the same patient. Standardized protocols for reprogramming, cell culture, and differentiation are required for reproducible cell-based assays. A panel of control cell lines are necessary to determine disease-specific phenotypes. However, since human iPS cell culture and differentiation into disease-specific cell types is time-consuming and highly expensive, the sample size is often kept to a minimum.

Secondly, early cell-cell interactions are required for complete maturation of neurons and glia cells, missing in cell type-specific cultures and limiting disease modeling of whole-brain disorders like psychiatric conditions. Co-culture systems and brain organoids are expected to be a more suitable model to uncover connectivity phenotypes, especially in psychiatric disorders which are induced by altered interaction and communication between various cell types from different brain regions. Nonetheless, iPSC-derived brain organoids show high heterogeneity, variable structures, and cell composition, warranting optimization of current protocols.

Another concern in disease modeling using iPS cells that resemble fetal cell stage are questions relating to late-onset or neurodegenerative diseases, and that might be caused by aging and environmental factors. Variable possibilities have been proposed to overcome this challenge and accelerate the appearance of pathological phenotypes in cell culture, for instance, by exposure of the cells to environmental factors like oxidative stress or manipulation of intrinsic mechanisms (Miller et al. 2017; Vera et al. 2016). Finding new strategies to model the effects of aging and long-term environmental influences could reveal new pathological phenotypes of neurodegenerative disorders in the future.

Stem cell-based models represent a big opportunity for applications in translational medicine, providing a novel platform for discovering personalized therapies for patients and potential biomarkers to monitor the patients' disease progression. Developing technologies of automated cell culture on a large scale enables high-throughput compound testing. However, several aspects need to be considered prior to cell-based drug screening. For therapeutic implications, the patient-derived iPS cells should be well characterized genetically and phenotypically. Furthermore, choosing correct and homogeneous cell type and phenotype for measuring drug response is highly relevant. Nonetheless, previous studies successfully identified potential drugs for psychiatric disorders that were able to rescue the patients' phenotype (Brennand et al. 2011; Marchetto et al. 2010; Paşca et al. 2011; Shcheglovitov et al. 2013). Moreover, transplantation of iPS cells and derivates into diseased human tissue comes more and more into focus, though cell replacement therapy of neurological disorders is extremely challenging.

In conclusion, future stem cell-based disease modeling will generate a wealth of information about human neuropsychiatric disorders. With the recently developed techniques of cell reprogramming, differentiation protocols, genome editing, and high-throughput sequencing, previous studies represent only a fraction of the possibilities of patient-derived iPS cells in disease modeling and drug discovery. While there are still technical challenges that require improvement, the ability to investigate neuropsychiatric diseases in vitro at single-cell resolution and in 3D cell systems demonstrates that human stem cells are a powerful tool to promote preclinical studies in the scientific future.

Acknowledgments JK is an associated member of Research Training Grant GRK2162 of the Deutsche Forschungsgemeinschaft. Funding came from the German Federal Ministry of Education and Research (BMBF, 01GQ113, 01GM1520A, 01EK1609B and 01GM1905B to B.W.); the Bavarian Ministry of Science and the Arts in the framework of ForINTER and the Interdisciplinary Center for Clinical Research (IZKF), University Hospital Erlangen (E25).

References

Abud EM, Ramirez RN, Martinez ES, Healy LM, Nguyen CHH, Newman SA, Yeromin AV et al (2017) iPSC-derived human microglia-like cells to study neurological diseases. Neuron 94 (2):278–293.e9. https://doi.org/10.1016/j.neuron.2017.03.042

Autism Spectrum Disorders Working Group of The Psychiatric Genomics Consortium (2017) Meta-analysis of GWAS of over 16,000 individuals with autism spectrum disorder highlights a novel locus at 10q24.32 and a significant overlap with schizophrenia. Mol Autism 8(1):21. https://doi.org/10.1186/s13229-017-0137-9

Avior Y, Lezmi E, Yanuka D, Benvenisty N (2017) Modeling developmental and tumorigenic aspects of trilateral retinoblastoma via human embryonic stem cells. Stem Cell Rep 8 (5):1354–1365. https://doi.org/10.1016/j.stemcr.2017.03.005

Bagley JA, Reumann D, Bian S, Lévi-Strauss J, Knoblich JA (2017) Fused cerebral organoids model interactions between brain regions. Nat Methods 14(7):743–751. https://doi.org/10.1038/nmeth.4304

Bardy C, van den Hurk M, Kakaradov B, Erwin JA, Jaeger BN, Hernandez RV, Eames T et al (2016) Predicting the functional states of human iPSC-derived neurons with single-cell RNA-seq and electrophysiology. Mol Psychiatry 21(11):1573–1588. https://doi.org/10.1038/mp.2016.158

Bernstein H-G, Steiner J, Guest PC, Dobrowolny H, Bogerts B (2015) Glial cells as key players in schizophrenia pathology: recent insights and concepts of therapy. Schizophr Res 161(1):4–18. https://doi.org/10.1016/j.schres.2014.03.035

Birey F, Andersen J, Makinson CD, Islam S, Wu W, Nina H, Christina Fan H et al (2017) Assembly of functionally integrated human forebrain spheroids. Nature 545(7652):54–59. https://doi.org/10.1038/nature22330

Brennand K, Simone A, Jou J, Gelboin-Burkhart C, Tran N, Sangar S, Li Y et al (2011) Modelling schizophrenia using human induced pluripotent stem cells. Nature 473(7346):221–225. https://doi.org/10.1038/nature09915

Brennand K, Savas JN, Kim Y, Tran N, Simone A, Hashimoto-Torii K, Beaumont KG et al (2015) Phenotypic differences in hiPSC NPCs derived from patients with schizophrenia. Mol Psychiatry 20(3):361–368. https://doi.org/10.1038/mp.2014.22

Camp JG, Badsha F, Florio M, Kanton S, Gerber T, Wilsch-Bräuninger M, Lewitus E et al (2015) Human cerebral organoids recapitulate gene expression programs of fetal neocortex development. Proc Natl Acad Sci 112(51):201520760. https://doi.org/10.1073/pnas.1520760112

Chambers SM, Fasano CA, Papapetrou EP, Tomishima M, Sadelain M, Studer L (2009) Highly efficient neural conversion of human ES and iPS cells by dual inhibition of SMAD signaling. Nat Biotechnol 27(3):275–280. https://doi.org/10.1038/nbt.1529

Chen JS, Dagdas YS, Kleinstiver BP, Welch MM, Sousa AA, Harrington LB, Sternberg SH, Keith Joung J, Yildiz A, Doudna JA (2017) Enhanced proofreading governs CRISPR–Cas9 targeting accuracy. Nature 550:407. https://doi.org/10.1038/nature24268

Cheung AYL, Horvath LM, Grafodatskaya D, Pasceri P, Weksberg R, Hotta A, Carrel L, Ellis J (2011) Isolation of MECP2-null rett syndrome patient hiPS cells and isogenic controls through X-chromosome inactivation. Hum Mol Genet 20(11):2103–2115. https://doi.org/10.1093/hmg/ddr093

Choi SH, Kim YH, Hebisch M, Sliwinski C, Lee S, D'Avanzo C, Chen H et al (2014) A three-dimensional human neural cell culture model of Alzheimer's disease. Nature 515 (7526):274–278. https://doi.org/10.1038/nature13800

Colasante G, Lignani G, Rubio A, Medrihan L, Yekhlef L, Sessa A, Massimino L et al (2015) Rapid conversion of fibroblasts into functional forebrain GABAergic interneurons by direct genetic reprogramming. Cell Stem Cell 17(6):719–734. https://doi.org/10.1016/j.stem.2015.09.002

Cotter DR, Pariante CM, Everall IP, Cotter DR, Pariante CM, Everall IP (2001) Glial cell abnormalities in major psychiatric disorders: the evidence and implications. Brain Res Bull 55 (5):585–595. http://www.ncbi.nlm.nih.gov/pubmed/11576755

Davis RL, Weintraub H, Lassar AB, Davis RL, Weintraub H, Lassar AB (1987) Expression of a single transfected cDNA converts fibroblasts to myoblasts. Cell 51(6):987–1000. http://www.ncbi.nlm.nih.gov/pubmed/3690668

Derecki NC, Cronk JC, Lu Z, Xu E, Abbott SBG, Guyenet PG, Kipnis J (2012) Wild-type microglia arrest pathology in a mouse model of rett syndrome. Nature 484(7392):105–109. https://doi.org/10.1038/nature10907

Dever DP, Bak RO, Reinisch A, Camarena J, Washington G, Nicolas CE, Pavel-Dinu M et al (2016) CRISPR/Cas9 β-globin gene targeting in human haematopoietic stem cells. Nature 539 (7629):384–389. https://doi.org/10.1038/nature20134

Doers ME, Musser MT, Nichol R, Berndt ER, Baker M, Gomez TM, Zhang S-C, Abbeduto L, Bhattacharyya A (2014) iPSC-derived forebrain neurons from FXS individuals show defects in initial Neurite outgrowth. Stem Cells Dev 23(15):1777–1787. https://doi.org/10.1089/scd.2014.0030

Douvaras P, Sun B, Wang M, Kruglikov I, Lallos G, Zimmer M, Terrenoire C et al (2017) Directed differentiation of human pluripotent stem cells to microglia. Stem Cell Rep 8:1516–1524. https://doi.org/10.1016/j.stemcr.2017.04.023

Eiges R, Urbach A, Malcov M, Frumkin T, Schwartz T, Amit A, Yaron Y et al (2007) Developmental study of fragile X syndrome using human embryonic stem cells derived from Preimplantation genetically diagnosed embryos. Cell Stem Cell 1(5):568–577. https://doi.org/10.1016/j.stem.2007.09.001

Etemad S, Zamin RM, Ruitenberg MJ, Filgueira L (2012) A novel in vitro human microglia model: characterization of human monocyte-derived microglia. J Neurosci Methods 209(1):79–89. https://doi.org/10.1016/j.jneumeth.2012.05.025

Fogarty NME, McCarthy A, Snijders KE, Powell BE, Kubikova N, Blakeley P, Lea R et al (2017) Genome editing reveals a role for OCT4 in human embryogenesis. Nature 550(7674):67–73. https://doi.org/10.1038/nature24033

Gascón S, Masserdotti G, Russo GL, Götz M (2017) Direct neuronal reprogramming: achievements, hurdles, and new roads to success. Cell Stem Cell 21(1):18–34. https://doi.org/10.1016/j.stem.2017.06.011

Gratten J, Wray NR, Keller MC, Visscher PM (2014) Large-scale genomics unveils the genetic architecture of psychiatric disorders. Nat Neurosci 17(6):782–790. https://doi.org/10.1038/nn.3708

Hockemeyer D, Wang H, Kiani S, Lai CS, Gao Q, Cassady JP, Cost GJ et al (2017) Genetic engineering of human pluripotent cells using TALE nucleases. Nat Biotechnol 29:731. https://doi.org/10.1038/nbt.1927

Hollingsworth EW, Vaughn JE, Orack JC, Skinner C, Khouri J, Lizarraga SB, Hester ME, Watanabe F, Kosik KS, Imitola J (2017) iPhemap: an Atlas of phenotype to genotype relationships of human iPSC models of neurological diseases. EMBO Mol Med 9(12):1742–1762. https://doi.org/10.15252/emmm.201708191

Honda M, Minami I, Tooi N, Morone N, Nishioka H, Uemura K, Kinoshita A, Heuser JE, Nakatsuji N, Aiba K (2016) The modeling of Alzheimer's disease by the overexpression of mutant presenilin 1 in human embryonic stem cells. Biochem Biophys Res Commun 469 (3):587–592. https://doi.org/10.1016/j.bbrc.2015.12.025

Hook V, Brennand KJ, Kim Y, Toneff T, Funkelstein L, Lee KC, Ziegler M, Gage FH (2014) Human iPSC neurons display activity-dependent neurotransmitter secretion: aberrant catecholamine levels in Schizophrenia neurons. Stem Cell Rep 3(4):531–538. https://doi.org/10.1016/j.stemcr.2014.08.001

Hu W, Qiu B, Guan W, Wang Q, Wang M, Li W, Gao L et al (2015) Direct conversion of normal and Alzheimer's disease human fibroblasts into neuronal cells by small molecules. Cell Stem Cell 17(2):204–212. https://doi.org/10.1016/j.stem.2015.07.006

Illes S, Jakab M, Beyer F, Gelfert R, Couillard-Despres S, Schnitzler A, Ritter M, Aigner L (2014) Intrinsically active and pacemaker neurons in pluripotent stem cell-derived neuronal populations. Stem Cell Rep 2(3):323–336. https://doi.org/10.1016/j.stemcr.2014.01.006

Israel MA, Yuan SH, Bardy C, Reyna SM, Mu Y, Herrera C, Hefferan MP et al (2012) Probing sporadic and familial Alzheimer's disease using induced pluripotent stem cells. Nature 482 (7384):216–220. https://doi.org/10.1038/nature10821

Jinek M, Chylinski K, Fonfara I, Hauer M, Doudna JA, Charpentier E (2012) A programmable dual-RNA-guided DNA endonuclease in adaptive bacterial immunity. Science 337 (6096):816–821. https://doi.org/10.1126/science.1225829

Jo J, Xiao Y, Sun AX, Cukuroglu E, Tran H-D, Göke J, Tan ZY et al (2016) Midbrain-like organoids from human pluripotent stem cells contain functional dopaminergic and neuromelanin-producing neurons. Cell Stem Cell 19(2):248–257. https://doi.org/10.1016/j.stem.2016.07.005

Kim K, Doi A, Wen B, Ng K, Zhao R, Cahan P, Kim J et al (2010) Epigenetic memory in induced pluripotent stem cells. Nature 467(7313):285–290. https://doi.org/10.1038/nature09342

Koch P, Tamboli IY, Mertens J, Wunderlich P, Ladewig J, Stüber K, Esselmann H, Wiltfang J, Brüstle O, Walter J (2012) Presenilin-1 L166P mutant human pluripotent stem cell–derived neurons exhibit partial loss of γ-secretase activity in endogenous amyloid-β generation. Am J Pathol 180(6):2404–2416. https://doi.org/10.1016/j.ajpath.2012.02.012

Kondo T, Asai M, Tsukita K, Kutoku Y, Ohsawa Y, Sunada Y, Imamura K et al (2013) Modeling Alzheimer's disease with iPSCs reveals stress phenotypes associated with intracellular Aβ and differential drug responsiveness. Cell Stem Cell 12:487–496. https://doi.org/10.1016/j.stem.2013.01.009

Krencik R, Weick JP, Liu Y, Zhang Z-J, Zhang S-C (2011) Specification of transplantable astroglial subtypes from human pluripotent stem cells. Nat Biotechnol 29(6):528–534. https://doi.org/10.1038/nbt.1877

Kriks S, Shim J-W, Piao J, Ganat YM, Wakeman DR, Xie Z, Carrillo-Reid L et al (2011) Dopamine neurons derived from human ES cells efficiently engraft in animal models of Parkinson's disease. Nature 480(7378):547–551. https://doi.org/10.1038/nature10648

Lancaster MA, Renner M, Martin C-A, Wenzel D, Bicknell LS, Hurles ME, Homfray T, Penninger JM, Jackson AP, Knoblich JA (2013) Cerebral organoids model human brain development and microcephaly. Nature 501(7467):373–379. https://doi.org/10.1038/nature12517

Lee SH, Ripke S, Neale BM, Faraone SV, Purcell SM, Perlis RH, Mowry BJ et al (2013) Genetic relationship between five psychiatric disorders estimated from genome-wide SNPs. Nat Genet 45(9):984–994. https://doi.org/10.1038/ng.2711

Li Y, Wang H, Muffat J, Cheng AW, Orlando DA, Lovén J, Kwok S-m et al (2013) Global transcriptional and translational repression in human-embryonic-stem-cell-derived rett syndrome neurons. Cell Stem Cell 13(4):446–458. https://doi.org/10.1016/j.stem.2013.09.001

Li ZM, Shi M, Malty RH, Aoki H, Minic Z, Phanse S et al (2015) Identification of human neuronal protein complexes reveals biochemical activities and convergent mechanisms of action in autism spectrum disorders. Cell Syst 1(5):361–374. https://doi.org/10.1016/j.cels.2015.11.002

Liao J, Karnik R, Hongcang G, Ziller MJ, Clement K, Tsankov AM, Akopian V et al (2015) Targeted disruption of DNMT1, DNMT3A and DNMT3B in human embryonic stem cells. Nat Genet 47(5):469–478. https://doi.org/10.1038/ng.3258

Lin M, Zhao D, Hrabovsky A, Pedrosa E, Zheng D, Lachman HM (2014) Heat shock alters the expression of Schizophrenia and autism candidate genes in an induced pluripotent stem cell model of the human telencephalon. PloS One 9(4):e94968. https://doi.org/10.1371/journal.pone.0094968

Lister R, Pelizzola M, Kida YS, David Hawkins R, Nery JR, Hon G, Antosiewicz-Bourget J et al (2011) Hotspots of aberrant epigenomic reprogramming in human induced pluripotent stem cells. Nature 471(7336):68–73. https://doi.org/10.1038/nature09798

Liu Q, Waltz S, Woodruff G, Ouyang J, Israel MA, Herrera C, Sarsoza F et al (2014) Effect of potent γ-secretase modulator in human neurons derived from multiple presenilin 1-induced pluripotent stem cell mutant carriers. JAMA Neurol 71(12):1481–1489. https://doi.org/10.1001/jamaneurol.2014.2482

Madison JM, Zhou F, Nigam A, Hussain A, Barker DD, Nehme R, van der Ven K et al (2015) Characterization of bipolar disorder patient-specific induced pluripotent stem cells from a family reveals neurodevelopmental and mRNA expression abnormalities. Mol Psychiatry 20 (6):703–717. https://doi.org/10.1038/mp.2015.7

Mali P, Yang L, Esvelt KM, Aach J, Guell M, DiCarlo JE, Norville JE, Church GM (2013) RNA-guided human genome engineering via Cas9. Science 339(6121):823–826. https://doi.org/10.1126/science.1232033

Marchetto CC, Acab A, Yu D, Yeo GW, Mu Y, Chen G, Gage FH, Muotri AR (2010) A model for neural development and treatment of rett syndrome using human induced pluripotent stem cells. Cell 143(4):527–539. https://doi.org/10.1016/j.cell.2010.10.016

Marchetto MC, Belinson H, Tian Y, Freitas BC, Fu C, Vadodaria K, Beltrao-Braga P et al (2017) Altered proliferation and networks in neural cells derived from idiopathic autistic individuals. Mol Psychiatry 22(6):820–835. https://doi.org/10.1038/mp.2016.95

Mariani J, Coppola G, Zhang P, Abyzov A, Provini L, Tomasini L, Amenduni M et al (2015) FOXG1-dependent dysregulation of GABA/glutamate neuron differentiation in autism spectrum disorders. Cell 162(2):375–390. https://doi.org/10.1016/j.cell.2015.06.034

Maroof AM, Keros S, Tyson JA, Ying S-W, Ganat YM, Merkle FT, Liu B et al (2013) Directed differentiation and functional maturation of cortical interneurons from human embryonic stem cells. Cell Stem Cell 12(5):559–572. https://doi.org/10.1016/j.stem.2013.04.008

Mertens J, Paquola ACM, Manching K, Hatch E, Böhnke L, Ladjevardi S, McGrath S et al (2015a) Directly reprogrammed human neurons retain aging-associated Transcriptomic signatures and reveal age-related Nucleocytoplasmic defects. Cell Stem Cell 17(6):705–718. https://doi.org/10.1016/j.stem.2015.09.001

Mertens J, Wang Q-W, Kim Y, Yu DX, Pham S, Yang B, Zheng Y et al (2015b) Differential responses to lithium in hyperexcitable neurons from patients with bipolar disorder. Nature 527 (7576):95–99. https://doi.org/10.1038/nature15526

Mertens J, Marchetto MC, Bardy C, Gage FH (2016) Evaluating cell reprogramming, differentiation and conversion technologies in neuroscience. Nat Rev Neurosci 17(7):424–437. https://doi.org/10.1038/nrn.2016.46

Miller JD, Ganat YM, Kishinevsky S, Bowman RL, Liu B, Tu EY, Mandal P et al (2017) Human iPSC-based modeling of late-onset disease via progerin-induced aging. Cell Stem Cell 13:691–705. https://doi.org/10.1016/j.stem.2013.11.006

Muffat J, Li Y, Yuan B, Mitalipova M, Omer A, Corcoran S, Bakiasi G et al (2016) Efficient derivation of microglia-like cells from human pluripotent stem cells. Nat Med 22 (11):1358–1367. https://doi.org/10.1038/nm.4189

Muguruma K, Nishiyama A, Kawakami H, Hashimoto K, Sasai Y (2015) Self-organization of polarized cerebellar tissue in 3D culture of human pluripotent stem cells. Cell Rep 10 (4):537–550. https://doi.org/10.1016/j.celrep.2014.12.051

Muratore CR, Rice HC, Srikanth P, Callahan DG, Shin T, Benjamin LNP, Walsh DM, Selkoe DJ, Young-Pearse TL (2014) The familial Alzheimer's disease APPV717I mutation alters APP processing and Tau expression in iPSC-derived neurons. Hum Mol Genet 23(13):3523–3536. https://doi.org/10.1093/hmg/ddu064

Nicholas CR, Chen J, Tang Y, Southwell DG, Chalmers N, Vogt D, Arnold CM et al (2013) Functional maturation of hPSC-derived forebrain interneurons requires an extended timeline and mimics human neural development. Cell Stem Cell 12(5):573–586. https://doi.org/10.1016/j.stem.2013.04.005

Niclis JC, Pinar A, Haynes JM, Alsanie W, Jenny R, Dottori M, Cram DS (2013) Characterization of forebrain neurons derived from late-onset Huntington's disease human embryonic stem cell lines. Front Cell Neurosci 7:37. https://doi.org/10.3389/fncel.2013.00037

Nishino K, Toyoda M, Yamazaki-Inoue M, Fukawatase Y, Chikazawa E, Sakaguchi H, Akutsu H, Umezawa A (2011) DNA methylation dynamics in human induced pluripotent stem cells over time. PLoS Genet 7(5):e1002085. https://doi.org/10.1371/journal.pgen.1002085

Ochalek A, Mihalik B, Avci HX, Chandrasekaran A, Téglási A, Bock I, Giudice ML et al (2017) Neurons derived from sporadic Alzheimer's disease iPSCs reveal elevated TAU hyperphosphorylation, increased amyloid levels, and GSK3B activation. Alzheimer's Res Ther 9(1):90. https://doi.org/10.1186/s13195-017-0317-z

Ohgidani M, Kato TA, Setoyama D, Sagata N, Hashimoto R, Shigenobu K, Yoshida T et al (2017) Direct induction of ramified microglia-like cells from human monocytes: dynamic microglial dysfunction in Nasu-Hakola disease. Sci Rep 4:4957. https://doi.org/10.1038/srep04957

Oksanen M, Petersen AJ, Naumenko N, Puttonen K, Lehtonen Š, Olivé MG, Shakirzyanova A et al (2017) PSEN1 mutant iPSC-derived model reveals severe astrocyte pathology in Alzheimer's disease. Stem Cell Rep 9(6):1885–1897. https://doi.org/10.1016/j.stemcr.2017.10.016

Pandya H, Shen MJ, Ichikawa DM, Sedlock AB, Choi Y, Johnson KR, Kim G et al (2017) Differentiation of human and murine induced pluripotent stem cells to microglia-like cells. Nat Neurosci 20(5):753–759. https://doi.org/10.1038/nn.4534

Pang ZP, Yang N, Vierbuchen T, Ostermeier A, Fuentes DR, Yang TQ, Citri A et al (2011) Induction of human neuronal cells by defined transcription factors. Nature 476(7359):220–223. https://doi.org/10.1038/nature10202

Park I-H, Arora N, Huo H, Maherali N, Ahfeldt T, Shimamura A, Lensch MW, Cowan C, Hochedlinger K, Daley GQ (2008) Disease-specific induced pluripotent stem cells. Cell 134 (5):877–886. https://doi.org/10.1016/j.cell.2008.07.041

Paşca SP, Portmann T, Voineagu I, Yazawa M, Shcheglovitov A, Paşca AM, Cord B et al (2011) Using iPSC-derived neurons to uncover cellular phenotypes associated with timothy syndrome. Nat Med 17(12):1657–1662. https://doi.org/10.1038/nm.2576

Paulsen BDa S, De Moraes Maciel R, Galina A, Souza Da Silveira M, Dos Santos Souza C, Drummond H, Nascimento Pozzatto E et al (2012) Altered oxygen metabolism associated to neurogenesis of induced pluripotent stem cells derived from a schizophrenic patient. Cell Transplant 21(7):1547–1559. https://doi.org/10.3727/096368911X600957

Perrier AL, Tabar V, Barberi T, Rubio ME, Bruses J, Topf N, Harrison NL, Studer L (2004) Derivation of midbrain dopamine neurons from human embryonic stem cells. Proc Natl Acad Sci 101(34):12543–12548. https://doi.org/10.1073/pnas.0404700101

Persico AM, Bourgeron T, Persico AM, Bourgeron T, Persico AM, Bourgeron T (2006) Searching for ways out of the autism maze: genetic, epigenetic and environmental clues. Trends Neurosci 29(7):349–358. https://doi.org/10.1016/j.tins.2006.05.010

Pfisterer U, Kirkeby A, Torper O, Wood J, Nelander J, Dufour A, Björklund A, Lindvall O, Jakobsson J, Parmar M (2011) Direct conversion of human fibroblasts to dopaminergic neurons. Proc Natl Acad Sci U S A 108(25):10343–10348. https://doi.org/10.1073/pnas.1105135108

Purcell SM, Moran JL, Fromer M, Ruderfer D, Solovieff N, Roussos P, O'Dushlaine C et al (2014) A polygenic burden of rare disruptive mutations in schizophrenia. Nature 506(7487):185–190. https://doi.org/10.1038/nature12975

Qian X, Nguyen HN, Song MM, Hadiono C, Ogden SC, Hammack C, Yao B et al (2016) Brain-region-specific organoids using mini-bioreactors for modeling ZIKV exposure. Cell 165 (5):1238–1254. https://doi.org/10.1016/j.cell.2016.04.032

Qin H, Zhao A, Zhang C, Xiaobing F (2016) Epigenetic control of reprogramming and transdifferentiation by histone modifications. Stem Cell Rev 12(6):708–720. https://doi.org/10.1007/s12015-016-9682-4

Quadrato G, Nguyen T, Macosko EZ, Sherwood JL, Yang SM, Berger DR, Maria N et al (2017) Cell diversity and network dynamics in photosensitive human brain organoids. Nature 545 (7652):48–53. https://doi.org/10.1038/nature22047

Rackham OJL, Firas J, Fang H, Oates ME, Holmes ML, Knaupp AS, Harukazu FANTOM Consortium et al (2016) A predictive computational framework for direct reprogramming between human cell types. Nat Genet 48(3):331–335. https://doi.org/10.1038/ng.3487

Raja WK, Mungenast AE, Lin Y-T, Ko T, Abdurrob F, Seo J, Tsai L-H (2016) Self-organizing 3D human neural tissue derived from induced pluripotent stem cells recapitulate Alzheimer's disease phenotypes. PloS One 11(9):e0161969. https://doi.org/10.1371/journal.pone.0161969

Ripke S, O'Dushlaine C, Chambert K, Moran JL, Kähler AK, Akterin S, Bergen SE et al (2013) Genome-wide association analysis identifies 13 new risk loci for schizophrenia. Nat Genet 45 (10):1150–1159. https://doi.org/10.1038/ng.2742

Ripke S, Neale BM, Corvin A, Walters JTR, Farh K-H, Holmans PA, Lee P et al (2014) Biological insights from 108 schizophrenia-associated genetic loci. Nature 511(7510):421–427. https://doi.org/10.1038/nature13595

Robicsek O, Karry R, Petit I, Salman-Kesner N, Müller F-J, Klein E, Aberdam D, Ben-Shachar D (2013) Abnormal neuronal differentiation and mitochondrial dysfunction in hair follicle-derived induced pluripotent stem cells of schizophrenia patients. Mol Psychiatry 18(10):1067–1076. https://doi.org/10.1038/mp.2013.67

Russo FB, Freitas BC, Pignatari GC, Fernandes IR, Sebat J, Muotri AR, Beltrão-Braga PCB (2017) Modeling the interplay between neurons and astrocytes in autism using human induced pluripotent stem cells. Biol Psychiatry 83:569. https://doi.org/10.1016/j.biopsych.2017.09.021

Sakaguchi H, Kadoshima T, Soen M, Narii N, Ishida Y, Ohgushi M, Takahashi J, Eiraku M, Sasai Y (2015) Generation of functional hippocampal neurons from self-organizing human embryonic stem cell-derived dorsomedial telencephalic tissue. Nat Commun 6:8896. https://doi.org/10.1038/ncomms9896

Shaltouki A, Peng J, Liu Q, Rao MS, Zeng X (2013) Efficient generation of astrocytes from human pluripotent stem cells in defined conditions. Stem Cells 31(5):941–952. https://doi.org/10.1002/stem.1334

Shcheglovitov A, Shcheglovitova O, Yazawa M, Portmann T, Shu R, Sebastiano V, Krawisz A et al (2013) SHANK3 and IGF1 restore synaptic deficits in neurons from 22q13 deletion syndrome patients. Nature 503(7475):267–271. https://doi.org/10.1038/nature12618

Sheridan SD, Theriault KM, Reis SA, Zhou F, Madison JM, Daheron L, Loring JF, Haggarty SJ (2011) Epigenetic characterization of the FMR1 gene and aberrant neurodevelopment in human induced pluripotent stem cell models of fragile X syndrome. PLoS One 6(10):e26203. https://doi.org/10.1371/journal.pone.0026203

Shi Y, Kirwan P, Smith J, MacLean G, Orkin SH, Livesey FJ (2012) A human stem cell model of early Alzheimer's disease pathology in down syndrome. Sci Trans Med 4(124):124ra29. https://doi.org/10.1126/scitranslmed.3003771

Slaymaker IM, Gao L, Zetsche B, Scott DA, Yan WX, Zhang F (2016) Rationally engineered Cas9 nucleases with improved specificity. Science 351(6268):84–88. https://doi.org/10.1126/science.aad5227

Soldner F, Laganière J, Cheng AW, Hockemeyer D, Gao Q, Alagappan R, Khurana V et al (2011) Generation of isogenic pluripotent stem cells differing exclusively at two early onset Parkinson point mutations. Cell 146(2):318–331. https://doi.org/10.1016/j.cell.2011.06.019

Soldner F, Stelzer Y, Shivalila CS, Abraham BJ, Latourelle JC, Barrasa MI, Goldmann J, Myers RH, Young RA, Jaenisch R (2016) Parkinson-associated risk variant in distal enhancer of α-synuclein modulates target gene expression. Nature 533(7601):95–99. https://doi.org/10.1038/nature17939

Soliman MA, Aboharb F, Zeltner N, Studer L (2017) Pluripotent stem cells in neuropsychiatric disorders. Mol Psychiatry 22(10):1241–1249. https://doi.org/10.1038/mp.2017.40

Son EY, Ichida JK, Wainger BJ, Toma JS, Rafuse VF, Woolf CJ, Eggan K (2011) Conversion of mouse and human fibroblasts into functional spinal motor neurons. Cell Stem Cell 9 (3):205–218. https://doi.org/10.1016/j.stem.2011.07.014

Sproul AA, Jacob S, Pre D, Kim SH, Nestor MW, Navarro-Sobrino M, Santa-Maria I et al (2014) Characterization and molecular profiling of PSEN1 familial Alzheimer's disease iPSC-derived neural progenitors. PloS One 9(1):e84547. https://doi.org/10.1371/journal.pone.0084547

Sugathan A, Biagioli M, Golzio C, Erdin S, Blumenthal I, Manavalan P, Ragavendran A et al (2014) CHD8 regulates neurodevelopmental pathways associated with autism spectrum disorder in neural progenitors. Proc Natl Acad Sci U S A 111(42):E4468–E4477. https://doi.org/10.1073/pnas.1405266111

Takahashi K, Tanabe K, Ohnuki M, Narita M, Ichisaka T, Tomoda K, Yamanaka S (2007) Induction of pluripotent stem cells from adult human fibroblasts by defined factors. Cell 131 (5):861–872. https://doi.org/10.1016/j.cell.2007.11.019

Tcw J, Wang M, Pimenova AA, Bowles KR, Hartley BJ, Lacin E, Machlovi SI et al (2017) An efficient platform for astrocyte differentiation from human induced pluripotent stem cells. Stem Cell Rep 9(2):600–614. https://doi.org/10.1016/j.stemcr.2017.06.018

Thomson JA, Itskovitz-Eldor J, Shapiro SS, Waknitz MA, Swiergiel JJ, Marshall VS, Jones JM (1998) Embryonic stem cell lines derived from human blastocysts. Science 282 (5391):1145–1147. http://www.ncbi.nlm.nih.gov/pubmed/9804556

Tobe BTD, Crain AM, Winquist AM, Calabrese B, Makihara H, Zhao W-N, Lalonde J et al (2017) Probing the lithium-response pathway in hiPSCs implicates the phosphoregulatory set-point for a cytoskeletal modulator in bipolar pathogenesis. Proc Natl Acad Sci U S A 114(22):E4462–E4471. https://doi.org/10.1073/pnas.1700111114

Urbach MS, Nissim B, Urbach A, Schuldiner M, Benvenisty N (2004) Modeling for Lesch-Nyhan disease by gene targeting in human embryonic stem cells. Stem Cells 22(4):635–641. https://doi.org/10.1634/stemcells.22-4-635

Urbach A, Benvenisty N, Urbach A, Benvenisty N (2009) Studying early lethality of 45,XO (Turner's syndrome) embryos using human embryonic stem cells. PLoS One 4(1):e4175. https://doi.org/10.1371/journal.pone.0004175

Urbach A, Bar-Nur O, Daley GQ, Benvenisty N (2010) Differential modeling of fragile X syndrome by human embryonic stem cells and induced-pluripotent stem cells. Cell Stem Cell 6:407–411. https://doi.org/10.1016/j.stem.2010.04.005

Vadodaria KC, Mertens J, Paquola A, Bardy C, Li X, Jappelli R, Fung L et al (2016) Generation of functional human serotonergic neurons from fibroblasts. Mol Psychiatry 21(1):49–61. https://doi.org/10.1038/mp.2015.161

Vera E, Bosco N, Studer L, Byrd W, Shay JW, Gu Y, Alt FW et al (2016) Generating late-onset human iPSC-based disease models by inducing neuronal age-related phenotypes through telomerase manipulation. Cell Rep 17(4):1184–1192. https://doi.org/10.1016/j.celrep.2016.09.062

Vierbuchen T, Ostermeier A, Pang ZP, Kokubu Y, Südhof TC, Wernig M (2010) Direct conversion of fibroblasts to functional neurons by defined factors. Nature 463(7284):1035–1041. https://doi.org/10.1038/nature08797

Wang P, Lin M, Pedrosa E, Hrabovsky A, Zheng Z, Guo W, Lachman HM, Zheng D (2015) CRISPR/Cas9-mediated heterozygous knockout of the autism gene CHD8 and characterization of its transcriptional networks in neurodevelopment. Mol Autism 6:55. https://doi.org/10.1186/s13229-015-0048-6

Wang P, Mokhtari R, Pedrosa E, Kirschenbaum M, Bayrak C, Zheng D, Lachman HM (2017) CRISPR/Cas9-mediated heterozygous knockout of the autism gene CHD8 and characterization

of its transcriptional networks in cerebral organoids derived from iPS cells. Mol Autism 8(1):11. https://doi.org/10.1186/s13229-017-0124-1
Wen Z, Nguyen HN, Guo Z, Lalli MA, Wang X, Su Y, Kim N-S et al (2014) Synaptic dysregulation in a human iPS cell model of mental disorders. Nature 515(7527):414–418. https://doi.org/10.1038/nature13716
Wen Z, Christian KM, Song H, Ming G-L (2016) Modeling psychiatric disorders with patient-derived iPSCs. Curr Opin Neurobiol 36:118–127. https://doi.org/10.1016/j.conb.2015.11.003
Whiteford HA, Ferrari AJ, Degenhardt L, Feigin V, Vos T (2017) The global burden of mental, neurological and substance use disorders: an analysis from the global burden of disease study 2010. PloS One 10(2):e0116820. https://doi.org/10.1371/journal.pone.0116820
WHO (2014) The global burden of disease: 2004 update. WHO, Geneva. http://www.who.int/healthinfo/global_burden_disease/2004_report_update/en/
Williams EC, Zhong X, Mohamed A, Li R, Liu Y, Dong Q, Ananiev GE et al (2014) Mutant astrocytes differentiated from rett syndrome patients-specific iPSCs have adverse effects on wild-type neurons. Hum Mol Genet 23(11):2968–2980. https://doi.org/10.1093/hmg/ddu008
Xu X, Radulescu C, Utami K, Pouladi M (2017) Obtaining multi-electrode array recordings from human induced pluripotent stem cell-derived neurons. Bio-Protocol 7(22). https://doi.org/10.21769/BioProtoc.2609
Yagi T, Ito D, Okada Y, Akamatsu W, Nihei Y, Yoshizaki T, Yamanaka S, Okano H, Suzuki N (2011) Modeling familial Alzheimer's disease with induced pluripotent stem cells. Hum Mol Genet 20(23):4530–4539. https://doi.org/10.1093/hmg/ddr394
Yamamuro K, Kimoto S, Rosen KM, Kishimoto T, Makinodan M (2015) Potential primary roles of glial cells in the mechanisms of psychiatric disorders. Front Cell Neurosci 9:154. https://doi.org/10.3389/fncel.2015.00154
Yoon K-J, Nguyen HN, Ursini G, Zhang F, Kim N-S, Wen Z, Makri G et al (2014) Modeling a genetic risk for schizophrenia in iPSCs and mice reveals neural stem cell deficits associated with adherens junctions and polarity. Cell Stem Cell 15(1):79–91. https://doi.org/10.1016/j.stem.2014.05.003
Yu DX, Paolo F, Giorgio D, Yao J, Marchetto MC, Brennand K, Wright R et al (2014) Stem cell reports modeling hippocampal neurogenesis using human pluripotent stem cells. Stem Cell Rep 2:295–310. https://doi.org/10.1016/j.stemcr.2014.01.009

Role of MicroRNAs in Anxiety and Anxiety-Related Disorders

Conor P. Murphy and Nicolas Singewald

Contents

Abstract MicroRNAs as critical regulators of gene expression important for functions including neuronal development, synapse formation, and synaptic plasticity have been linked with the regulation of neurobiological systems that underlie anxiety processing in the brain. In this chapter, we give an update on associative evidence linking regulation of microRNAs with anxiety- and trauma-related

The original version of this chapter was revised. A correction to this chapter is available at https://doi.org/10.1007/7854_2019_118.

C. P. Murphy
Department of Pharmacology and Toxicology, Institute of Pharmacy, Center for Molecular Biosciences Innsbruck, University of Innsbruck, Innsbruck, Austria

N. Singewald (✉)
Department of Pharmacology and Toxicology, Institute of Pharmacy, Center for Molecular Biosciences Innsbruck, University of Innsbruck, Innsbruck, Austria

Department of Pharmacology and Toxicology, Institute of Pharmacy, Center for Molecular Biosciences Innsbruck, CCB – Center for Chemistry and Biomedicine, University of Innsbruck, Innsbruck, Austria
e-mail: nicolas.singewald@uibk.ac.at

© The Author(s) 2019

Curr Topics Behav Neurosci (2019) 42: 185–220
DOI 10.1007/7854_2019_109
Published Online: 5 September 2019

disorders. Moving beyond correlative research, functional studies have emerged recently that explore causal relationships between microRNA expression and anxiety-like behavior. It has been demonstrated that experimental up- or downregulation of the candidate microRNAs in important nodes of the anxiety neurocircuitry can indeed modulate anxiety-related behavior in animal models. Improved methodologies for assessing microRNA-mediated modulation have aided such functional studies, revealing a number of anxiety-regulating microRNAs including miR-15a, miR-17-92, miR-34, miR-101, miR-124, miR-135, and miR-155. Important functional target genes of these identified microRNAs are associated with specific neurotransmitter/neuromodulator signaling, neurotrophin (e.g., BDNF) expression and other aspects of synaptic plasticity, as well as with stress-regulatory/hypothalamic-pituitary-axis function. Furthermore, microRNAs have been revealed that are regulated in distinct brain regions following various anxiety-attenuating strategies. These include pharmacological treatments such as antidepressants and other drugs, as well as non-pharmacological interventions such as fear extinction/exposure therapy or positive stimuli such as exposure to environmental enrichment. These are first indications for a role for microRNAs in the mechanism of action of anxiolytic treatments. As research continues, there is much hope that a deeper understanding of the microRNA-mediated mechanisms underlying anxiety-related disorders could open up possibilities for future novel biomarker and treatment strategies.

Keywords Anxiety disorders · Anxiolytic · Epigenetics · microRNA · RDoc

1 Introduction

Physiological anxiety is a beneficial emotion in response to real or potential threats, but patients with pathological anxiety display a broad range of exaggerated and/or enduring symptoms, such as increased arousal/vigilance and excessive fear- and anxiety-related responses (for detailed discussion into differences and characteristics between fear and anxiety, see Perusini and Fanselow 2015) including autonomic changes such as heart rate increases and sweating, often in the absence of any real threat or danger (Dell'osso et al. 2010). Many of these symptoms are shared by different anxiety disorders, which are classified in the tenth edition of the International Classification of Diseases (ICD-10) (WHO 1993; Craske et al. 2017) and the fifth edition of the *Diagnostic and Statistical Manual of Mental Disorders* (DSM-5) (American Psychiatric Association 2013). They include specific phobias, panic disorder (with or without agoraphobia), social phobia, and generalized anxiety. Anxiety- and trauma-related disorders, such as post-traumatic stress disorder (PTSD), are grouped together in the ICD-10 (WHO 1993; Craske et al. 2017). The current diagnostic systems are based on subjective measures of client self-report, clinical observation, and clinical judgment, while the recently proposed Research Domain Criteria (RDoC) system defines psychopathologies as phenomena of multilevel neurobiological characteristics underlying dimensions of behavior from normal to abnormal (Anderzhanova et al. 2017). The RDoC system holds promise for better characterization of complex psychiatric disorders that are commonly comorbid, as in the case of anxiety often comorbid with depression and alcohol/drug dependence, and aims to improve translatability of studies from animals to humans by supporting the endophenotype-based comparison on a neurobiological basis across behavioral dimension (Anderzhanova et al. 2017). Anxiety-related disorders are the most common mental disorders, affecting up to 14% of Europeans each year (Wittchen et al. 2011). In developed countries, the lifetime prevalence of these disorders is up to 28% (Kessler et al. 2012). Anxiety disorders impact the quality of life and place a high emotional and financial burden on families and government bodies (Olesen et al. 2012; Wittchen et al. 2011). Furthermore, the high comorbidity that exists between anxiety and other psychiatric disorders, such as depression and bipolar disorder, makes treating these disorders even more complex (Pavlova et al. 2017; McMillan et al. 2017).

Both psychotherapy and pharmacotherapy are currently used to treat anxiety-related disorders and have been shown to be more effective than placebo. However, a considerable percentage of patients only exhibit partial long-term benefits following therapy, failing to achieve complete remission (Bandelow et al. 2007, 2012; Farach et al. 2012). Pharmacological treatments include selective serotonin reuptake inhibitors (SSRIs) and other antidepressants (Baldwin et al. 2014). Benzodiazepines are not recommended as first-line treatments due to their potential side effects such as increased risk of drug dependence and should be used only transiently. Disadvantages of the current first-line drug treatments include a delayed period before the onset of therapeutic benefits, suicidal tendencies, and risk of sexual dysfunction, often leading to discontinuation of treatment (Maron and Nutt 2015; Farb and Ratner 2014; Ravindran and Stein 2010). Hence, we are in need of novel interventions targeting causal factors and resulting in long-lasting benefits with lower incidence of side effects. Despite this clear need for improved treatments, the development of novel drugs for anxiety-related disorders has been rather unsuccessful (Singewald et al. 2015; Griebel and Holmes 2013). One of the most important factors involved in this is the still limited knowledge that exists about underlying mechanisms and pathways mediating and modulating anxiety-related behavior, although impressive advances have been made.

2 Neurobiology and Molecular Basis of Anxiety

Key to understanding anxiety processing in the brain is the elucidation of the neuronal correlates that are involved. Regulation of fear and anxiety behavior depends on neuronal activity and communication between relevant brain areas, and there is evidence that cellular plasticity is vital for appropriate inter-regional communications (Hill and Martinowich 2016). It has been found that there is a high degree of overlap between humans and animals in the neural circuitry underlying anxiety-related behaviors (Calhoon and Tye 2015). The use of modern methodologies such as optogenetics, among others, in rodents has led to the improved and more detailed characterization of the intricate interplay between the different brain loci forming circuits to mediate and modulate anxiety-like behavior (Tovote et al. 2015; Orsini and Maren 2012; Dias et al. 2013; Calhoon and Tye 2015). At the center of this circuitry is the extended amygdala, important for the emotional interpretation of environmental information as it arises, determining whether downstream fear (see below) or reward pathways (e.g., via the nucleus accumbens, NA) will be activated. The extended amygdala includes different amygdala subregions, including the basolateral amygdala (BLA) and the central amygdala (CeA; composed of the centro-lateral (CeL) and centro-medial (CeM) amygdala), as well as the bed nucleus of the stria terminalis (BNST), which coordinates inputs from the amygdala and hippocampus. Together with other regions, such as the hypothalamus (HYP), they also modulate the neuroendocrine stress response. Other brain regions implicated in the modulation of anxiety-related behaviors include the medial prefrontal cortex (mPFC); the highly interconnected hippocampus, which encodes contextual aspects important for anxiety, and important modulatory monoaminergic cell body areas; the locus coeruleus (LC); the raphe nucleus (RN); and the ventral tegmental area (VTA) (Fig. 1, top panel) (for detailed review, see Tovote et al. 2015). The signal processing between these brain areas and neuronal populations is driven by a finely tuned interplay of different neurotransmitters and neuromodulators, including neurotransmitters such as gamma-aminobutyric acid (GABA); glutamate; monoamines such as serotonin, norepinephrine, and dopamine; and a number of neuropeptides, including neuropeptide Y (NPY), neuropeptide S (NPS), endocannabinoids, cholecystokinin (CCK), and calcitonin gene-related peptide (CGRP), which have been associated with anxiety regulation and considered as drug targets (Bowers et al. 2012; Wang and Pereira 2016; Millan 2003). Many anxiety disorders are considered stress-related, and thus regulation of the corticotropin-releasing factor (CRH) system and of the hypothalamic-pituitary-adrenal (HPA) axis has been found to be important players, particularly in stress-induced anxiety (Shin and Liberzon 2010; Bartlett et al. 2017). Aberrant anxiety reactions are thought to arise from altered processing and deficiencies in such circuitries, including imbalances in neurotransmitter regulation and downstream signaling pathways (Calhoon and Tye 2015; Millan 2003). Mechanisms that can orchestrate and restore this delicate balance are sought in the quest for more rational therapeutic targets in anxiety treatment. MicroRNAs are such candidates (see below), having been shown to interact with such regulatory neurotransmitters and signaling cascades (see below and Martinetz 2016, for review).

The mechanisms leading to deficiencies in the networks processing anxiety are not understood in detail. Family and twin studies have indicated that anxiety disorders are multifactorial in cause, involving an interaction between genetic and environmental factors (Hettema et al. 2001, 2005; Klengel and Binder 2015; Sharma et al. 2016). Heritability in anxiety disorders is rather moderate but can amount to 30–48%, for example, in generalized anxiety/panic disorder (Hettema et al. 2001; Kendler et al. 1999; Shimada-Sugimoto et al. 2015). Evidence suggests that multiple genes with small effects are involved, with single gene variation insufficient to produce full anxiety pathology. Genome-wide association studies

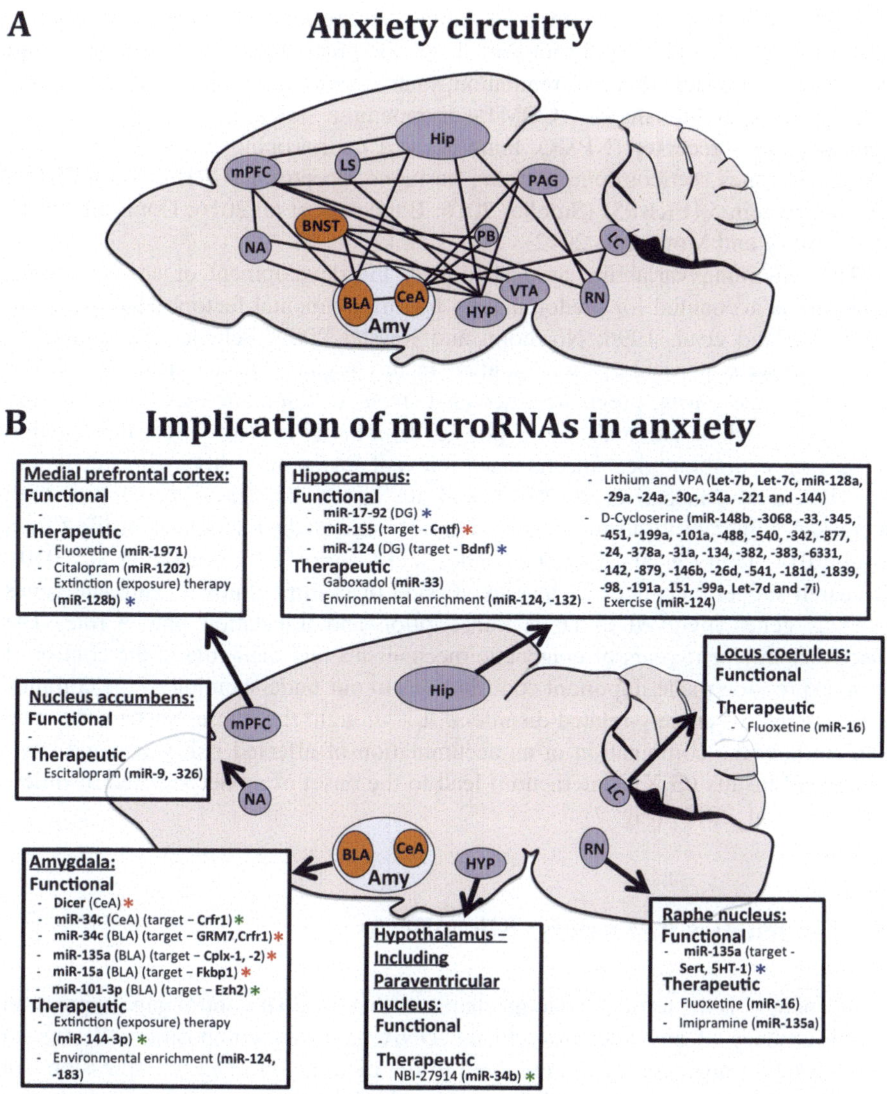

Fig. 1 Brain areas investigated for a (functional) role of microRNAs in anxiety. (**a**) Schematic of proposed anxiety circuitry based on references (Calhoon and Tye 2015; Tovote et al. 2015). (**b**) Schematic representation summarizes the most recent studies assessing microRNAs in the brain for regional, functional, and therapeutic implications in anxiety-related behaviors. Studies that assessed the implications of candidate microRNA expression using functional approaches are indicated with an asterisk (∗), while red (knockdown), green (overexpression/enhancement), and blue (both) indicate the direction of regulation. Bed nucleus of the striatum terminalis (BNST), medial prefrontal cortex (mPFC), hippocampus (Hip), amygdala (Amy; amygdala subregions: basolateral amygdala (BLA) and central amygdala (CeA)), nucleus accumbens (NA), raphe nucleus (RN), hypothalamus (HYP), locus coeruleus (LC), periaqueductal gray (PAG), parabrachial nucleus (PB), ventral tegmental area (VTA), and lateral septum (LS)

(GWAS) assessing the presence of single nucleotide polymorphisms (SNPs) have revealed a number of chromosomal risk loci, including genes associated with neurotransmitter and HPA axis regulation, such as serotonin receptor 1A (5HT-1A), catechol-O-methyltransferase (COMT), monoamine oxidase A (MAOA) genes, neuropeptide S receptor (NPSR), brain-derived neurotrophic factor (BDNF), and HPA axis genes corticotropin-releasing hormone receptor 1 (CRHR1) and FK506 binding protein 5 (FKBP5) (Smoller 2016; Bandelow et al. 2016; Domschke et al. 2012; Autry and Monteggia 2012).

The remaining variability associated with the development of anxiety-related disorders is accounted for predominantly by environmental factors (Hettema et al. 2005; Vieland et al. 1996; Norrholm and Ressler 2009; Schiele and Domschke 2017). Stress is particularly well studied in this regard (Dirven et al. 2017). For example, stress during pregnancy, neonatal stress, or traumatic experiences during childhood, such as parental loss or physical/emotional neglect, affect the function of various neurobiological mechanisms relevant for anxiety processing, including the HPA axis (Maccari et al. 2014), and are factors that have been linked with increased vulnerability to the development of anxiety disorders later in life (Stein et al. 2014; Drury et al. 2016; Fernandes and Osorio 2015; Klauke et al. 2010). Research suggests changes in gene expression of multiple known candidate genes such as genes involved in DNA transcription and translation play a role. The discovery in recent years of epigenetic mechanisms and their role in the control of gene expression made important contributions to our understanding of the complex development of anxiety-related disorders. It is thought that these mechanisms may explain how the combination of an accumulation of affected risk genes and environmental insults (G X E interaction) lead to the onset of anxiety-related disorders (Sharma et al. 2016) (Fig. 2).

3 Epigenetics in Anxiety: MicroRNAs

"Epigenetics" is the term given to mechanisms that together control gene expression in the absence of any alterations to the DNA base pair composition (Bjornsson et al. 2004). Epigenetic marks involved in the modulation of gene expression can be influenced by environmental factors, can be inherited to influence subsequent generations, but, importantly, can also be reversed (Jaenisch and Bird 2003), suggesting a dynamic modulation mediated by these marks. These mechanisms exert multiple levels of control over gene expression, ranging from DNA and histone modifications to posttranscriptional regulation via noncoding RNAs, including microRNAs (Dias et al. 2015; Cholewa-Waclaw et al. 2016), and they have emerged as critical regulators in nearly all neurobiological processes (Sevignani et al. 2006; Ong and Corces 2011). Environmental insults, such as stress, diet, or toxin exposure, can affect epigenetic mechanisms, which could explain their profound and long-lasting effects on multiple neurobiological systems and networks. Epigenetically mediated mechanisms could also shed light on individual variations

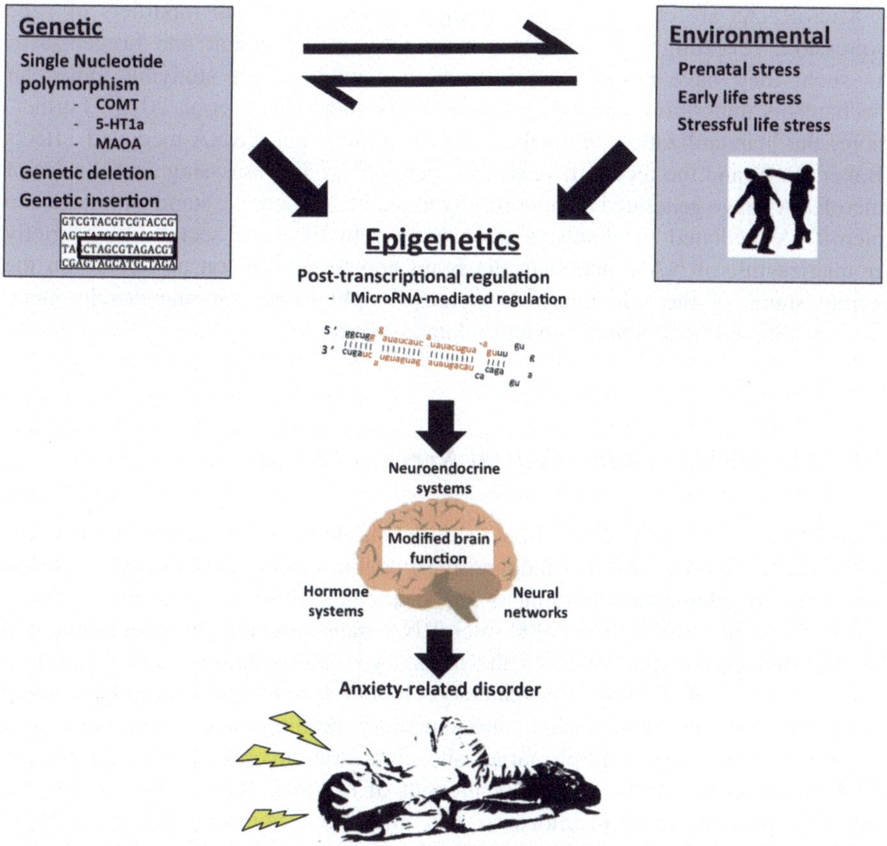

Fig. 2 Genetic/environmental/epigenetic influences in anxiety-related disorders. Schematic representation of gene × environment × epigenetics in the development of anxiety-related disorders. Genetic and environmental factors can lead to long-lasting changes in epigenetic mechanisms, resulting in changes in brain structure and also functional brain processing. These changes can subsequently lead to the development of anxiety-related disorders

in the development and treatment of some of the most common central nervous system (CNS) disorders (Yehuda and Bierer 2009; Landgrave-Gomez et al. 2015), including anxiety disorders (Nieto et al. 2016; O'Connor et al. 2016; Bartlett et al. 2017). Environmental influences associated with the development of anxiety disorders have been linked with distinct changes in epigenetic mechanisms, such as histone acetylation, DNA methylation, and noncoding RNA expression, which regulate the long-lasting expression of specific genes (Schiele and Domschke 2017). Genes targeted are, for example, associated with neurotransmission/neuromodulation, including GABAergic, glutamatergic, dopaminergic and serotonergic signaling, or genes associated with HPA axis functioning, many of them known to be aberrantly regulated in anxiety (Fass et al. 2014; Hunter 2012).

MicroRNAs are one of the best-studied posttranscriptional modifiers of gene expression, revealing their critical role in CNS development and homeostasis. As such, they have emerged as interesting candidates for studying molecular mechanisms underlying anxiety and anxiety disorders (Scott et al. 2015). Furthermore, the standardization of methods of examining microRNA-mediated effects (Baker 2010) and the recent increase in functional studies assessing causal roles of microRNAs have generated an opportunity to reassess where we stand with regard to microRNA-mediated mechanisms and anxiety. In the next sections, we briefly summarize microRNA function in the brain and then focus on an update on the current status of the role microRNAs are thought to play in the development, progression, and treatment of anxiety-related disorders.

3.1 *MicroRNAs in the Central Nervous System*

MicroRNAs are short (19–24-nucleotide) single-stranded ncRNAs that are genomically encoded and that mediate the posttranscriptional regulation of target genes via complementary binding to the messenger RNA seed sequence (Bartel 2009). There have been over 5,000 microRNA transcripts annotated in humans so far, together comprising 1–2% of the human genome (Londin et al. 2015). It is well documented that microRNAs can target hundreds of genes often within distinct functional networks via partial complementary binding and, furthermore, any given gene may have multiple target sites for different microRNAs (Lim et al. 2005; Wang et al. 2011). Originally thought of as "junk RNA", such as leftover remnants of past viral infections, it has now been revealed that microRNAs may control the expression of up to 60% of the protein-coding genes in the human genome in a highly complex network-like fashion (Krol et al. 2010b; Bartel 2004). The first microRNA to be discovered was lin-4, which was found in *Caenorhabditis elegans* using a genetic screening tool for gene variations associated with deficits in the temporal control of postembryonic development (Lee et al. 1993). In 2001 the true potential of these small noncoding RNAs was revealed when their role as regulators of gene expression was demonstrated (Lau et al. 2001). Since this initial breakthrough discovery, the emergence of microRNAs and their downstream effects on target genes has revealed them as "fine tuners" of gene expression, and this has subsequently become one of the best-studied mechanisms of posttranscriptional regulation.

3.2 *MicroRNA Induction, Biogenesis, Turnover, and Function*

Induction, biogenesis, turnover, and general function of microRNAs have been the subject of specific reviews; for a detailed discussion on this topic, please see Ruegger

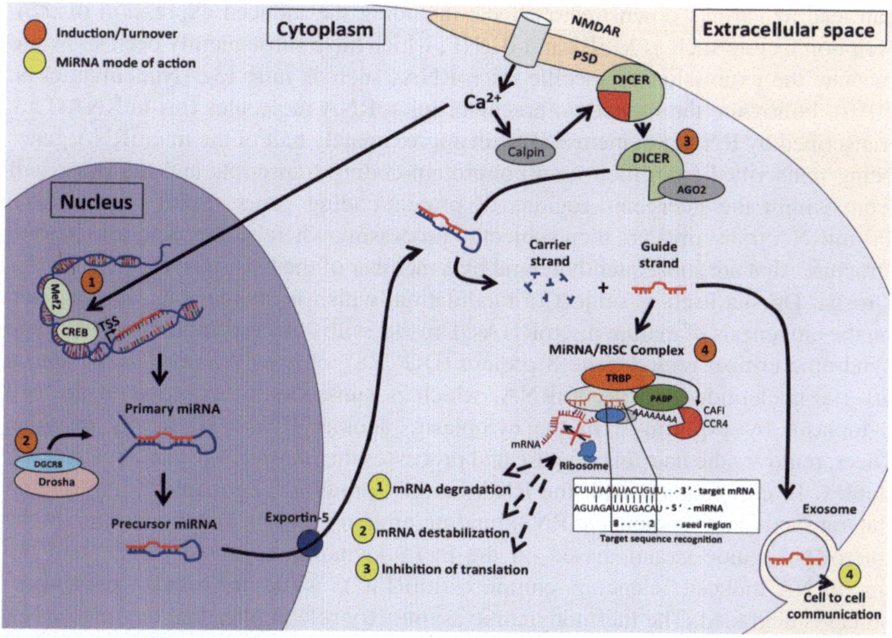

Fig. 3 MicroRNA induction, biogenesis, turnover, and function. MicroRNA expression is induced in a number of ways including via neuronal activation whereby (1: orange) Ca^{2+} influx influences the expression of transcriptions factors such as MEF2 and CREB, which subsequently have been shown to mediate the expression of a wide range of microRNAs. Following transcriptional induction, RNA polymerase II transcribes microRNAs as precursor microRNAs (pri-miRNA) from specific microRNA genes or from within the introns of protein-coding genes. These pri-miRNAs are then processed by a member of the RNase III enzyme family, Drosha, and along with other cofactors, including DiGeorge syndrome critical region gene 8 protein (DGCR8), cleave the pri-miRNA into a 70–100-nucleotide-long pre-miRNA. (2: orange) microRNA turnover is in part controlled by the regulation of Drosha where decreased expression reduces the ability of a cell to transcribe new microRNAs. Exportin-5 then transports the pre-miRNA to the cell cytoplasm where Dicer processes it into a microRNA duplex. (3: orange) Ca^{2+} influx is also associated with increased Calpain expression, which releases Dicer from the postsynaptic densities (PSD). Along with Argonaute 2 (Ago2) and transactivation-responsive RNA-binding protein (TRBP), Dicer incorporates the microRNA "guide" strand, based on the thermodynamic asymmetry rule, into the microRNA-induced silencing complex (miRISC). The co-processed "passenger" strand is degraded at this point. The microRNA and miRISC complex then induce (1 and 2: yellow) mRNA degradation/destabilization and/or (3: yellow) translational repression via target gene binding through a "seed sequence" located between nucleotides 2 and 8 on the microRNA guide strand (enclosed panel). PABP, poly(A)-binding protein. CCR4, C-C motif chemokine receptor 4

and Grosshans (2012), Vidigal and Ventura (2015), Krol et al. (2010b), Bartel (2004, 2009), and Sim et al. (2014). Here, we will give a brief account of these processes (*summarized in* Fig. 3).

In the nervous system, the induction of microRNA expression can occur via a number of different mechanisms of which the most studied is via neuronal activity (Sim et al. 2014). Generally speaking, Ca^{2+} influx following neuronal activation can lead to various downstream effects including the induced expression of transcription factors such as MEF2 and BDNF, which have subsequently been shown to mediate the expression of specific microRNAs such as miR-132 (Nudelman et al. 2010). Following the induction, precursor microRNA molecules (pri-miRNAs) are transcribed by RNA polymerase II, with approximately half of the microRNA genes being transcribed from their own nonprotein-coding transcripts and the other half from within the intergenic regions of protein-coding genes (Saini et al. 2007). Pri-miRNA transcripts are then subject to processing whereby they fold into hairpin structures that are subsequently bound by a member of the RNase III enzyme family, Drosha. Drosha itself is subject to modulation, which is another rate-limiting step on the biogenesis of mature microRNAs. Drosha, with cofactors including DiGeorge syndrome critical region gene 8 protein (DGCR8), cleaves the pri-miRNA into a 70–100-nucleotide-long pre-miRNA, which is subsequently exported to the cell cytoplasm by exportin-5. In the cytoplasm, another RNase III family enzyme, Dicer, removes the hairpin structure and processes the pre-miRNA into a microRNA duplex. In combination with the RNA-binding proteins, Argonaute 2 (Ago2) and the transactivation-responsive RNA-binding protein (TRBP), Dicer, incorporate the microRNA guide strand, based on the thermodynamic asymmetry rule, into the microRNA-induced silencing complex (miRISC), while the other "passenger" strand is degraded. The thermodynamic asymmetry rule implies that the microRNA strand with the least stable 5′ terminus will be chosen as the "guide" strand and incorporated into the miRISC complex, while the other "passenger" strand is degraded. Any given microRNA has a 5p and a 3p, which is dependent on what side of the hairpin each microRNA strand is situated. In principle, both the 5p and the 3p strand can act as the guide strand. Once incorporated into the miRISC complex, microRNAs then induce translational repression and/or mRNA degradation via binding with target genes through a "seed sequence" located between nucleotides 2 and 8 on the microRNA guide strand (Fig. 3, enclosed panel). A microRNA whose seed sequence has perfect complementarity with its target gene will have the greatest effect on the expression levels of this target.

While some microRNAs are expressed ubiquitously, such as let-7b, miR-17, and miR21, there are a number of microRNAs that are enriched in the brain, such as miR-128b (Lin et al. 2011), as well as microRNAs that are expressed in a temporal-, regional-, and cell-type-specific manner (Taguchi 2013), as shown for miR-7, miR-34a, and miR-132, among others, in the brain (Olsen et al. 2009). MicroRNA-mediated control over mechanisms such as gene expression regulation in the nervous system makes them essential mediators of important processes in the brain, including neuronal plasticity and learning and memory (Ruegger and Grosshans 2012). MicroRNA expression/turnover within neurons can be relatively fast. In the mouse retina, for example, light-sensitive microRNAs such as miR-204 and miR-211 and the microRNA-cluster miR-183/96/182 undergo rapid transcriptional repression, e.g., 30 min following adaption of the mouse to the dark chamber (Krol et al. 2010a). This microRNA regulation has subsequently been found to play a functional role, preventing retinal degeneration (Lumayag et al. 2013). A number of studies have further characterized microRNA turnover in the brain and have shown that microRNA expression is dynamic and under the control of a number of different mechanisms, including those involved in biogenesis, maturation, and degradation (Marzi et al. 2016) (Fig. 3). For example, one approach blocking the biogenesis of all new microRNAs, via ablation of Dicer, led to the discovery of varying half-lives of investigated microRNAs, ranging from 5 to 12 days (Gantier et al. 2011). Thus, microRNAs in general can be induced within minutes (e.g., by neuronal activity in neurons (Sim et al. 2014)) and can remain within cells in a stable condition for at least several days and maybe much longer (Gantier et al. 2011).

3.3 MicroRNAs as Modulators of Neuronal Plasticity and Learning and Memory

Adaptive responses to environmental stimuli are mediated via neuronal activity at a cellular level and involve a number of molecular changes altering synaptic dynamics, including local synaptic expression of plasticity-associated proteins. This expression is controlled by a number of synaptically located microRNAs, induced by neuronal activity, which ensures fine-tuned control over synaptic plasticity (Lugli et al. 2008; Hu et al. 2014; Schratt 2009; Gu et al. 2015) and learning and memory mechanisms (Wang et al. 2012). For example, miR-132 has emerged as an interesting candidate. High-frequency stimulation-induced long-term potentiation in the dentate gyrus reflecting enhanced synaptic plasticity is associated with increased expression levels of miR-132 (Wibrand et al. 2010). It was suggested that miR-132, depending on brain region and/or behavioral-specific signals, can act as a "proteome switch" and mediate different mechanisms of neuronal plasticity through its target genes, including the N-methyl-D-aspartate receptor subunit 2A (Bredy et al. 2011). The critical role played by microRNAs in these processes, which include learning and memory and related disorders, has been the topic of detailed reviews (please see Saab and Mansuy 2014; Hu and Li 2017; Aksoy-Aksel et al. 2014).

In relation to anxiety disorders, it should be mentioned that disturbances in learning and memory mechanisms are common in these disorders (Hemstedt et al. 2017), which are increasingly targeted by specific therapeutic interventions (e.g., Singewald et al. 2015). There is mounting evidence linking microRNAs to these processes (see below).

3.4 MicroRNAs in Anxiety and Anxiety-Related Disorders: Associative Studies

There is increasing evidence of microRNA dysregulation in disease. By far the best-studied field so far is cancer (for more detailed reviews, please see Kong et al. 2012; Nana-Sinkam and Croce 2013; Garzon et al. 2009). Concerning the brain, dysregulation of microRNAs has also been demonstrated in brain tumors, neurodegenerative diseases such as Alzheimer's and Parkinson's disease, and, more recently, psychiatric disorders (Spadaro and Bredy 2012; Wang et al. 2012; You et al. 2016; Cao et al. 2016; O'Connor et al. 2016). In particular, the investigation of microRNAs in psychiatric disorders has gained momentum due to their potential application as molecular biomarkers and/or their use in microRNA-specific therapies for psychiatric illness (Kichukova et al. 2015; Scott et al. 2015).

Because of the inaccessibility of brain tissues in living patients, human studies investigating the role of microRNAs in psychiatric patients rely on postmortem brain tissues obtained from patients with higher mortality rates due to higher risk-taking and/or higher suicide rates accompanying disorders such as bipolar disorder, schizophrenia, and major depressive disorder (Rao et al. 2013). Little is known regarding brain microRNA changes in anxiety disorders on their own, but there are studies in comorbid patients suffering from anxiety disorders and depression showing altered microRNA regulation (see below).

The majority of human studies examining anxiety-related disorders are limited to the assessment of differential microRNA expression in the periphery – mostly in blood and its constituents – and the correlation of these findings with the presence of disease (Table 1). For example, in generalized anxiety disorder (GAD) patients, circulating miR-663 and miR-4505 levels correlated with anxiety symptoms (Chen et al. 2016), and miR-29c expression increased in response to social anxiety in a social stress task, corresponding both to the stress experience and to alterations in ventromedial prefrontal cortex functional connectivity (Vaisvaser et al. 2016). Other insights came from postmortem samples of patients who had suffered from depression comorbid with anxiety, which revealed a downregulation of miR-135a in the raphe nucleus (Issler et al. 2014). Introducing further complexity, single nucleotide polymorphisms (SNPs) in the precursors or mature microRNAs could interfere with balanced processing in anxiety (for review see Malan-Muller et al. 2013). For example, the search for SNPs associated with specific microRNA candidates revealed a number of microRNA variations, including miR-22, miR-138-2, miR-148a, miR-339, and miR-488, which are associated with anxiety-related disorders such as panic disorder (Muinos-Gimeno et al. 2011). Some microRNA candidates with SNPs, such as miR-22, were also identified in a different cohort of panic disorder patients (Kim et al. 2015). As another example, polymorphisms in the Dicer1 gene have been linked with PTSD (Wingo et al. 2015). These studies demonstrated aberrant microRNA expression and possibly regulation in anxiety and provided important indication of a role of microRNAs in anxiety-related disorders. For more detailed discussion of these findings, please see recent reviews (Snijders et al. 2018; Martinetz 2016; Scott et al. 2015; Issler and Chen 2015; Malan-Muller et al. 2013; O'Connor et al. 2012, 2016; Hommers et al. 2015). These correlative studies are very important but have the disadvantage that it is hard to conclude whether microRNA expression and its subsequent posttranscriptional regulation causes the anxiety disorder or is a consequence of it. This is where the use of relevant animal models of anxiety is invaluable. Indeed, there are preclinical studies emerging that attempt to provide causal evidence for a functional role of microRNAs in anxiety. In the next sections, we focus primarily on these studies.

Table 1 Altered microRNA levels in human anxiety and anxiety-related disorders

MicroRNA(s)	Disorder	Tissue	Trial design	Results	Ref
miR-570, miR-219, miR-637, miR-668, miR-519a, miR-518f, miR-615, miR-125a, and miR-181c	PTSD	PBMC	30 PTSD vs 42 HC	miR-570, miR-219, miR-637, miR-668, miR-519a, miR-518f, and miR-615 expression is increased, while miR-125a and miR-181c expression is decreased in PTSD patients	Zhou et al. (2014)
miR-144 and miR-16	Anticipatory anxiety	Blood plasma	Ten healthy medical students	miR-144 and miR-16 expression in blood plasma increased in the run-up to a stressful exam, peaking on the day of the exam before returning to baseline expression following	Katsuura et al. (2012)
miR-144	Depression comorbid with anxiety	Blood plasma	169 patients vs 52 HC	miR-144 expression is lower in patients with depression and anxiety. Following group therapy, the expression levels of miR-144 increased	Wang et al. (2015)
miR-3130	PTSD/ depression	Blood	34 PTSD/ depression vs 20 healthy controls	miR-3130 is decreased in patients who suffer from PTSD/ depression	Wingo et al. (2015)
miR-29c	Social stress	PBMC	49 HC	Social stress was associated with increased expression of miR-29c in PBMCs, which furthermore correlated with increased connectivity of the vmPFC with the anterior insula and decreased connectivity with the left dorsolateral PFC	Vaisvaser et al. (2016)
miR-4484, miR-4674, miR-501, miR-663,	Generalized anxiety disorder	PBMC	76 patients vs 39 HC	miR-4484, miR-4674, miR-501, miR-663,	Chen et al. (2016)

(continued)

Table 1 (continued)

MicroRNA(s)	Disorder	Tissue	Trial design	Results	Ref
miR-4505, miR-1301, and miR-432				and miR-4505 expression was elevated in PBMCs of patients with generalized anxiety, while the expression levels of miR-1301 and miR-432 were decreased	
miR-15a	PTSD/ depression	Whole blood	Male individuals	miR-15a expression was increased in the peripheral blood following DEX administration or childhood trauma exposure	Volk et al. (2016)
miR-193a	PTSD (Male combat veterans)	PBMC	16 PTSD vs 17 HC	Expression levels of miR-193a are decreased in the PBMCs of combat veterans suffering from PTSD	Bam et al. (2016a)
Differential regulation of 190 microRNAs	PTSD (male combat veterans)	PBMC	24 PTSD vs 24 HC	In male combat veterans suffering from PTSD, 183 microRNAs exhibit decreased expression, while 7 microRNAs display enhanced expression in PBMCs	Bam et al. (2016b)
miR-19a, miR-101, miR-20b, miR-20a, miR-486, miR-128, miR-15b, and miR-125b	PTSD (Male combat veterans)	Whole blood	15 PTSD vs 9 HC	The expression of miR-19a, miR-101, miR-20b, and miR-20a are increased while miR-486, miR-128, miR-15b, and miR-125b exhibit decreased expression in the blood of combat veterans who suffer from PTSD	Martin et al. (2017)

3.5 MicroRNAs in Anxiety and Anxiety-Related Disorders: Studies Elucidating Functional/Causal Evidence

As mentioned above, microRNAs are abundant in the brain, where they appear to be differentially distributed, including the anxiety circuitry (Fig. 1), and even differently within neurons and distinct neuronal compartments. Functional implications of microRNAs in anxiety have been revealed using animal tests and models showing that down- or up-regulation of the candidate microRNAs in important nodes of the anxiety circuitry can indeed modulate anxiety-related behavior. The results of these studies are discussed below and are summarized in Table 2. In Fig. 1 (bottom panel), an overview is given of where in the brain such functional relationships have been investigated so far.

The first to explore the role of microRNAs in anxiety with a more functional approach were Haramati and colleagues, who locally depleted the microRNA-processing enzyme, Dicer, in the central amygdala of adult mice. As predicted, this depletion led to a drastically reduced ability for central amygdalar neurons to generate new mature microRNAs, which was found to be associated with increased anxiety-like behavior in the light/dark and open-field tests (Haramati et al. 2011). Since stress, as mentioned, is a major contributor to the development and triggering of anxiety-related behavior, stress-associated microRNA regulation has been particularly studied as a means of bridging stress with downstream pathology (Issler and Chen 2017). For example, in investigating mice following acute and chronic stress, miR-34c was revealed as a prominent stress-induced microRNA in the central amygdala. The authors suggest this microRNA plays a role in stress resilience and the development of anxiety. It was revealed that its upregulation reduced anxiety-like behavior via downregulating the expression of its target gene, corticotropin-releasing hormone receptor 1 (CRHR1), a key player in stress-induced anxiety. Furthermore, miR-34c expression was decreased in the central amygdala of Dicer-ablated neurons, which was associated with increased anxiety-like behavior (Haramati et al. 2011). Together these data highlight miR-34c expression in the central amygdala as a prominent microRNA involved in resilience to anxiety following stress. Attempting to further characterize the functional role of this microRNA, overexpression of miR-34c in the central amygdala produced an anxiolytic effect, demonstrating that manipulation of a single microRNA in one important node of the anxiety circuitry can drive a behavioral response (Haramati et al. 2011). Somewhat in contrast to these findings, Andolina and colleagues demonstrated that knockout (rather than overexpression) of miR-34 in the BLA is associated with resilience against stress-induced anxiety and with the facilitation of fear extinction (Andolina et al. 2016). Discrepancies in miR-34-mediated effects on anxiety behavior may be attributed to a number of reasons, including the fact that two different amygdalar subnuclei were investigated. Haramati et al. performed a lenti-virally mediated overexpression of miR-34c in the central amygdala, whereas Andolina et al. knocked out miR-34 in the BLA. This could suggest distinct regionally specific roles played by specific candidate microRNAs such as miR-34

Table 2 MicroRNAs functionally involved in anxiety-related behaviors

MicroRNA(s)	Species	Test	Region	Effect	Gene target(s)	Refs
Functional (anxiety paradigms)						
Dicer	Mouse	*Anxiety-like behavior*: Light-dark test, Open field test	Central amygdala	*CeA Dicer depletion* led to *increased anxiety-like behavior*	–	Haramati et al. (2011)
miR-15a	Mouse	*Stress-induced anxiety*: Elevated plus maze	Amygdala (basolateral amygdala)	Mice exposed to *chronic stress* were associated with increased *amygdalar miR-15a* expression. miR-15a *knockdown* in the *basolateral amygdala* led to *increased anxiety-like behavior* in mice, suggesting that miR-15a expression is important for *stress adaption*	Fkbp1	Volk et al. (2016)
miR-17-92	Mouse	*Anxiety- and depressive-like behavior*: Elevated plus maze, open field test, forced swim test, tail suspension test	Hippocampus (dentate gyrus)	*miR-17-92 depletion* in adult *neural progenitor* cells was associated with *decreased neurogenesis* in the *dentate gyrus* whereas *overexpression* led to *increased neurogenesis*. Furthermore, miR-17-92 *knockout* mice exhibit *increased anxiety-* and *depressive*-like behaviors while *overexpression* mice display *anxiolytic* and *antidepressive*-like behavior	–	Jin et al. (2016)
mir-34c	Mouse	*Stress-induced anxiety*: Light-dark test, elevated plus maze	Central amygdala	*Overexpression* of *miR-34c* led to the *protection* against stress-induced *anxiety*	Crfr1	Haramati et al. (2011)
miR-34	Mouse	*Stress-induced anxiety*: Light-dark test, elevated plus maze, open field test, fear extinction	Amygdala (basolateral amygdala)	*miR-34 knockout* was associated with *increased resilience* to *stress-induced anxiety*. Furthermore miR-34 knockout *facilitated fear extinction*	Grm7, 5HT-2c, Crfr1	Andolina et al. (2016)
miR-124a	Rats	*Neonatal isolation-induced anxiety*: Open field test, elevated plus maze	Hippocampus (dentate gyrus)	*miR-124a expression* is *increased* in the *dentate gyrus* following *neonatal isolation*, which is associated with *increased anxiety-like behavior*. Furthermore, *overexpression* of miR-124a in the *dentate gyrus* further *exacerbated anxiety-like behavior*	Bdnf	Bahi (2016)

miR-124a	Rats	*Neonatal isolation-induced anxiety*: Open field test, elevated plus maze	Hippocampus (dentate gyrus)	Virally mediated *decreased miR-124a expression* in the *dentate gyrus* was associated with *decreased anxiety-like behavior*	Bdnf	Bahi (2017)
miR-101-3p	Rats	*Trait-anxiety*: Open field test, elevated plus maze	Amygdala (basolateral amygdala)	Rats selectively bred for *high novelty responding* (low trait anxiety) exhibit lower expression levels of miR-101a-3p in the *amygdala, compared* to rats bred for *low novelty responding* (high trait anxiety). *Virally enhancing miR-101-3p* expression in *high novelty responding rats* led to *enhanced anxiety-like behavior* in open-field test and elevated-plus maze	Ezh2	Cohen et al. (2017)
miR-135a	Mouse	*Anxiety- and depressive-like behavior*: Social interaction following social defeat, elevated plus maze, light-dark test	Raphe nucleus	*Overexpression* of *miR-135a* in the *raphe nucleus* led to reduction in *anxiety* and *depressive-like behavior* following *social defeat*. Whereas *knock-down* of *miR-135a* led to *increased anxiety-like behavior*	Sert, 5HT-1a	Issler et al. (2014)
miR-135a	Mouse	*Anxiety-like behavior*: Elevated plus maze	Amygdala (basolateral amygdala)	*Knockdown* of *miR-135a* in the *basolateral amygdala* was associated with *increased anxiety-like behavior* and furthermore was shown to mediate *glutaminergic neurotransmission*	Cplx-1 and -2	Mannironi et al. (2017)
miR-155	Mouse	*Anxiety- and depressive-like behavior*: Elevated plus maze, open field test, forced swim test, tail suspension test	Hippocampus	*miR-155 depletion* was associated with *reduced anxiety-* and *depressive-like* behavior	Cntf	Fonken et al. (2016)

in the central amygdala vs the BLA. Further underlining the importance of this microRNA, members of the miR-34 precursor family have emerged as critical regulators in a number of amygdalar-dependent learning paradigms, including fear learning. For example, it has been demonstrated that miR-34a expression is increased in the amygdala 30 min following auditory fear conditioning, while knockdown of this microRNA in the basolateral amygdala impairs fear learning (Dias et al. 2014). These data suggest that the miR-34 precursor family plays a critical role in amygdalar-dependent memory formation, with dissociated effects on fear and extinction learning, whereas knockdown of miR-34 in the basolateral amygdala can facilitate fear extinction and at the same time impairs the formation of fear memory.

As discussed above, the RDoC initiative is fostering trans-diagnostic research investigating endophenotypes, to evaluate neurobiological mechanisms that may cut across traditional diagnoses. In line with the high comorbidity in anxiety disorders, involving multiple anxiety disorder categories and other comorbid conditions such as mood disorders, a number of studies assessing the functional role of microRNAs in animal models have investigated both anxiety- and depressive-like behaviors. Fonken and colleagues demonstrated that a genetic knockout of miR-155 in the hippocampus was associated with reduced anxiety- and depressive-like behavior (Fonken et al. 2016). Using a similar methodology, knockout of miR-17-92 in adult neural progenitor cells led to decreased neurogenesis in the dentate gyrus of the hippocampus and increased anxiety- and depressive-like behavior, while overexpression produced the opposite effects on neurogenesis and behavior (Jin et al. 2016). This is in line with reports linking enhanced trait anxiety with reduced hippocampal neurogenesis (Sah et al. 2012). The functional relationship between anxiety-related behavior and microRNA-mediated modulation has so far been demonstrated in three different brain areas in the anxiety circuitry: the amygdala, the hippocampus, and the raphe nucleus (Fig. 1, bottom panel). For example, overexpression of miR-135a in the raphe nucleus led to a reduction in anxiety- and depressive-like behavior elicited by social defeat, while miR-135a knockdown was associated with increased anxiety-like behavior (Issler et al. 2014). The authors found these effects to be mediated via interaction of miR-135a with the serotonergic system, targeting specifically the serotonin transporter (SERT) and serotonin receptor 1a (5HT-1a) genes in the dorsal raphe nucleus. As stated in the previous sections, microRNAs can target numerous genes and have varying effects in different cell types and tissues/brain areas due to distinct expression patterns and induction pathways. In agreement with this, a recent study demonstrated that knockdown of miR-135a in the basolateral amygdala also increased anxiety-like behavior, but in this brain area, it was found that miR-135a regulates anxiety via interaction with glutamatergic neurotransmission, targeting in particular complexin-2 (Mannironi et al. 2017).

Investigating microRNAs in trait anxiety in rats, Cohen and colleagues found that rats selectively bred for high novelty responding – which is associated with low trait anxiety – exhibit lower expression of miR-101a-3p in the amygdala compared with rats bred for low novelty responding that show high trait anxiety (Cohen et al. 2017).

Supporting the relation of miR-101-3p regulation to anxiety-like behavior, overexpression of miR-101-3p in the basolateral amygdala was indeed associated with enhanced anxiety-like behavior. In this relation it should be mentioned that recent results suggest that the gut microbiome is necessary for appropriate regulation of microRNA expression in brain regions implicated in anxiety-like behaviors including the amygdala and prefrontal cortex, adding an additional level of complexity (Hoban et al. 2017). Also in hippocampal areas, functional microRNA regulation was investigated. Using a model of neonatal isolation-induced anxiety, Bahi revealed that expression of the brain-enriched miR-124a was increased in the dentate gyrus of neonatally isolated Wistar rats (Bahi 2016). Overexpression of miR-124a in the dentate gyrus further exacerbated the anxiety-like behavior, demonstrating that the intensity of the anxiety response correlates with the expression level of miR-124a in the dentate gyrus. Supporting these findings, the author more recently demonstrated that virally mediated reduction of miR-124a expression in the dentate gyrus is associated with decreased anxiety-like behavior, which inversely correlates with the expression of its target gene, brain-derived neurotrophic factor (BDNF), in this brain area (Bahi 2017). BDNF is an activity-dependent neurotrophic factor that has been extensively implicated in fear and anxiety regulation (Andero et al. 2014; Andero and Ressler 2012). The increase in functional evidence linking microRNA regulation with anxiety supports a crucial role played by these "fine tuners" of gene expression and highlights them as possible novel therapeutic targets for the treatment of anxiety disorders.

3.6 MicroRNAs in the Treatment of Anxiety-Related Disorders

MicroRNAs have been investigated as therapeutic targets for numerous disorders, including in particular the treatment of cancer, with the first cancer-targeted microRNA, MRX34, entering clinical trials in patients with advanced hepatocellular carcinoma (Ceman and Saugstad 2011; Kong et al. 2012; Ling et al. 2013). Miravirsen, which targets miR-122, was recently demonstrated in a phase II trial to have no long-term safety issues and sustained its virological response, which is associated with reduced occurrence of liver failure, in the treatment of chronic hepatitis C (van der Ree et al. 2014). More recently, microRNAs have also been studied as potential targets in psychiatric disorders (Tardito et al. 2013; O'Connor et al. 2012, 2016). One idea here is that microRNA-mediated regulation of relevant biological processes is linked with the therapeutic effect of anxiety disorder treatments. As studies functionally assessing microRNAs in the therapeutic action of drugs are scarce, here we have also included correlative studies that specifically focused on drug-induced changes in microRNA expression associated with a therapeutic decrease in anxiety-like behavior.

3.7 MicroRNA Regulation in Established and Experimental Pharmacological Interventions Used in Anxiety Treatment

A number of studies have assessed the possible contribution of specific microRNAs to the therapeutic action of currently used interventions for anxiety disorders (summarized Fig. 1 (bottom panel)). Indeed, medications used in anxiety and affective disorders have been shown to modulate microRNA levels in the brain, although microRNAs were not their original target. One of the first of such studies demonstrated that long-term administration of different mood stabilizers led to altered expression of microRNAs in the hippocampus of rats when compared with saline-treated controls (Zhou et al. 2009). Following chronic oral treatment with valproate (VPA) or lithium, a number of microRNAs, including miR-29a, Let-7b, Let-7c, miR-128a, miR-24a, miR-30c, miR-34a, and miR-221, exhibited decreased hippocampal expression, while miR-144 expression was increased. Interestingly, VPA has shown promise in the treatment of some forms of anxiety, such as social anxiety disorder (Kinrys et al. 2003), and has previously also been shown to be useful as an adjunct facilitating the effect of fear extinction, the central mechanism of exposure-based cognitive behavioral therapy (Kuriyama et al. 2011; Whittle et al. 2013). Further investigating functional downstream target genes of miR-34a in primary hippocampal cultures, the authors revealed that administration of a miR-34a inhibitor, as well as lithium or VPA, increased the expression of miR-34a's target gene, metabotropic glutamate receptor 7 (GRM7). Furthermore, administration of an miR-34a precursor led to decreased expression levels of GRM7, suggesting a functional relationship between miR-34a and GRM7 expression levels and pointing toward one novel possibly common neurobiological effect of lithium and VPA with putative therapeutic significance.

Since this breakthrough study, the number of microRNAs linked with relevant downstream effects of therapeutic drugs commonly used in the treatment of anxiety and affective disorders has steadily increased. Antidepressants, such as selective serotonin reuptake inhibitors (SSRI) used as first-line treatments in different anxiety disorders (Baldwin et al. 2014), have been particularly investigated in this regard. Baudry and colleagues were the first to investigate SSRI effects following different modes of administration and revealed increased expression of miR-16 in the raphe nucleus and decreased expression in the locus coeruleus of mice (Baudry et al. 2010). Interestingly, the expression patterns of miR-16 in the raphe nucleus and locus coeruleus were inversely correlated with the expression levels of one of its target genes, the serotonin transporter (SERT). These data indicate that the therapeutic effect of fluoxetine could in part be mediated via differential miR-16 expression levels in monoaminergic neurons targeting SERT in the raphe nucleus and locus coeruleus (Baudry et al. 2010). Using a mouse model of PTSD, oral treatment with fluoxetine for 31 days reduced anxiety-related behavior in these mice, and this was associated with long-term effects on microRNA expression (Schmidt et al. 2013): the expression of miR-1971 was even after 74 days decreased in the prefrontal cortex in comparison with control groups, suggesting that the long-lasting beneficial effects on anxiety-related behaviors in fluoxetine-treated mice may be mediated in part by reduced miR-1971 expression. Taken together, these studies implicate the regulation of specific microRNA candidates (e.g., miR-16, miR-1971) in distinct brain regions in the therapeutic action of fluoxetine (see also Fig. 1).

Other SSRIs have also been investigated for a possible relation between their behavioral effects and microRNA regulation. It has been shown that maternal separation or chronic unpredictable stress-induced increase in depression-/anxiety-related behavior was associated with decreased levels of miR-326 expression in the nucleus accumbens and striatum (Zhang et al. 2015). Following chronic escitalopram treatment, the levels of miR-326 in both brain areas increased, thus returning to levels comparable with those of non-stressed or saline-treated rats. In investigating effects following tricyclic antidepressant treatment, it was shown that mice treated chronically with imipramine for 18–21 days display lower social avoidance, which was associated with increased expression of miR-135a in the raphe nucleus. As mentioned above, miR-135a targets the serotonergic system directly via binding to the serotonin transporter and 5HT-1a receptor gene (Issler et al. 2014). Demonstrating the functional involvement of miR-135a expression, its overexpression in serotonergic neurons was associated with decreased anxiety- and depression-like behavior following social defeat, while its knockdown led to increased anxiety- and depression-like behaviors (Issler et al. 2014).

Information concerning the effects of antidepressants on microRNA expression in the human brain is scarce. Interestingly, in patient's peripheral blood, the baseline expression of miR-1202 was predictive of treatment success, as patients who responded to treatment exhibited lower blood expression of miR-1202, which subsequently increased following an 8-week treatment with citalopram (Lopez et al. 2014). This brain-enriched microRNA was found to target the GRM4 gene, which shows decreased expression following treatment with antidepressants (Cruceanu et al. 2016). This observation suggests that miR-1202 contributes to antidepressant action via modulation of glutamatergic neurotransmission. Using human postmortem tissues from patients, Lopez and colleagues were able to demonstrate that major depressive disorder is associated with decreased expression of miR-1202 in the prefrontal cortex and in blood (Lopez et al. 2014). To our knowledge, this is the only study that links the dysregulation of a specific microRNA in the brain of psychiatric patients with the expression in blood and that shows dynamic regulation with treatment, thus demonstrating its possible usefulness as a biomarker of treatment success. Importantly, this study demonstrates long-term changes in specific microRNA expression in the brain associated with an affective disorder. In a recent follow-up study, the authors further explored this relationship between peripheral miR-1202 expression levels and brain activity in patients, following an 8-week treatment with desvenlafaxine (Lopez et al. 2017). Using functional magnetic resonance imaging, the authors found that lower peripheral levels of miR-1202 in desvenlafaxine-treated subjects were associated with increased neural connectivity between the posterior cingulate and the parietal, occipital, and prefrontal cortices.

Therapeutic drugs targeting primarily other neurotransmitter and hormone systems besides the serotonergic system have also been examined for the involvement of microRNA candidates in their mode of action. In one study, treatment with the antidepressants agomelatin, which targets melatonin receptors, and tianeptine, which targets glutamatergic, adenosine, and opioid systems, normalized stress-induced changes in microRNA levels in the dentate gyrus, including the expression of miR-181b, miR-9, and miR-411, among others (Patrício et al. 2015). Treatment with the CRHR1 antagonist NBI-27914 was shown to decrease anxiety-like behavior of mice in the elevated plus maze and open-field test. It was found that miR-34b targets the CRHR1 gene and it was thus demonstrated that administration of an agomir for miR-34b into the paraventricular nucleus of the hypothalamus mimicked the behavioral effect of the CRHR1 antagonist and resulted in reduced anxiety-like behavior (Zhu et al. 2017). Also by targeting GABAergic neurotransmission, microRNA expression changes. Administration of the GABA agonist gaboxadol was associated with reduced anxiety-like behavior and increased miR-33 expression in the dorsal hippocampus (Jovasevic et al. 2015). The pharmacological enhancement of therapy-relevant cognitive mechanisms (see also next chapter) via "cognitive enhancers" including the partial NMDA agonist D-cycloserine (DCS) is a more recent avenue for the treatment of anxiety-related disorders (for recent review see Singewald et al. 2015). In exploring a possible involvement of microRNAs in this effect, the expression of 32 microRNAs in the left dorsal hippocampus was found to be regulated after treatment with DCS. This interesting first finding needs to be followed up in more detail also in relation to a possible facilitation of extinction learning (Malan-Muller et al. 2017).

3.8 Non-pharmacological Interventions

Exposure-based cognitive behavioral therapy is a frequently used psychotherapeutic intervention for anxiety disorders. It involves the process of fear extinction, which results in the gradual reduction of fear- and anxiety-related responses to stimuli perceived as threatening to the patient (Abramowitz 2013; Milad and Quirk 2012; Craske and Stein 2016). Meta-analyses of cognitive behavioral therapy used in treating anxiety disorders show that it is efficacious for a number of anxiety-related disorders but has its limitations, including the problem that it produces insufficient long-lasting effects, and thus symptoms can return (Hofmann and Smits 2008; Tolin 2010; James et al. 2015; Gould et al. 2012). Importantly, anxiety patients are often associated with fear extinction deficits (Milad et al. 2009, 2013; Jovanovic et al. 2012; Michael et al. 2007; Duits et al. 2015; Wessa and Flor 2007). Thus, understanding the underlying mechanisms of deficient fear extinction (Graham and Milad 2011; Hill and Martinowich 2016; Baker et al. 2016; Maren and Holmes 2016) and how to overcome this deficiency may hold potential for the development of novel treatment strategies (for review see Holmes and Singewald 2013, Bukalo et al. 2014). Exploring the role of microRNAs in the rescue of deficient fear extinction, our group recently revealed an increased expression of miR-144-3p in the basolateral amygdala 2 h following fear extinction training. This increase was not seen in the very first phase of extinction training involving mainly fear retrieval, indicating that the effect was specific for extinction learning (Murphy et al. 2017). MiR-144-3p was found to target a number of genes implicated in the activation of plasticity-associated signaling cascades, including MAPK/ERK, PI3K/AKT, and NOTCH, which have been implicated in extinction mechanisms (Herry et al. 2006; Yang and Lu 2005; Dias et al. 2014). Confirming the functional significance of this, viral overexpression of miR-144-3p in the basolateral amygdala was able to rescue impaired fear extinction in extinction-impaired 129S1 mice and to prevent the return of fear in extinction-competent 129S6 mice (Murphy et al. 2017). Regulation of miR-144 in the context of anxiety has also been observed in humans. A recent study in healthy students found that whole blood miR-144 expression increased during the period preceding a stressful event (an effect involving, it is thought, anticipatory anxiety) before returning to baseline levels in the days following (Katsuura et al. 2012). In another study, miR-144-5p expression levels in the blood plasma of patients suffering from depression/anxiety were found to be lower in comparison with healthy controls (Wang et al. 2015). Compellingly, following 8 weeks of psychotherapy, miR-144-5p expression levels increased and were normalized to expression levels seen in healthy controls, suggesting that miR-144-5p expression levels could possibly be useful as a biomarker of treatment success.

While fear extinction is associated with the formation of a new fear-inhibitory memory, it is also important to note that during this process the original fear memory is brought into a labile state where it is subject to updating mechanisms (Flavell et al. 2013). It is during this period that the reconsolidation of the original fear memory can be disrupted or modulated, allowing for the transition to the predominant expression of a newly formed fear-inhibitory memory (Hemstedt et al. 2017). Exploring the role of microRNAs in this process, Lin and colleagues assessed the expression of microRNAs in the medial prefrontal cortex (Lin et al. 2011). Following fear extinction training, the expression of miR-128b was shown to be increased in the infralimbic subdivision of the medial prefrontal cortex (ILPFC). Interestingly, the authors found that miR-128b targets a number of plasticity-associated genes known to be important for fear memory retrieval (Lin et al. 2011). The fear-associated genes targeted by miR-128b include the regulator of calmodulin signaling (RCS), which has been demonstrated to inhibit calmodulin and to negatively regulate protein phosphatase calcineurin activity overall, leading to the disruption of fear memory consolidation/reconsolidation. Thus, increased miR-128b expression during fear extinction was suggested to facilitate the transition from the expression of fear to the expression of this newly formed fear-inhibitory memory. Indeed, overexpression of miR-128b in the ILPFC was able to enhance the effects of fear extinction, thus indicating that increased miR-128b expression could enhance exposure-based cognitive behavioral therapy via enhancing the expression of fear-inhibitory memories.

Beneficial environmental influences have been shown to contribute to successful outcomes of anxiety treatments (Kheirbek et al. 2012). In preclinical models, environmental enrichment, for example, has been used to model positive environmental stimulation, producing anxiolytic effects (Hendriksen et al. 2014; Huttenrauch et al. 2016). Interestingly, Ragu Varman and colleagues demonstrated increased expression of miR-183 and miR-124 in the amygdala following 30 days of environmental enrichment (Ragu Varman et al. 2013). Expression of miR-124 was also regulated in the dentate gyrus in response to environmental enrichment, and this was associated with increased neurogenesis (Brenes et al. 2016). Another non-pharmacological means that can improve anxiety symptomology is physical exercise. Subjecting male mice to voluntary wheel running reduced anxiety-like behavior and suppressed reinstatement of fear measured in their juvenile offspring, demonstrating transgenerational transmission of changes in anxiety-like behavior. The sperm of the mice following voluntary wheel running was found to have decreased expression of miR-19b and miR-455, while miR-133a was increased (Short et al. 2017). Furthermore, long-term exercise in mice, which was associated with increased resilience to stress, led to the decreased expression of miR-124 in the hippocampus and the increased expression of glucocorticoid receptors Nr3c1 and Nr3c1-1f (Pan-Vazquez et al. 2015). Interestingly, a recent genome-wide association study of positive emotion in humans identified a genetic variant proposed to mediate positive emotionality via the nucleus accumbens and miR-181 (Wingo et al. 2017). We have learned from these findings, taken together, that the beneficial effects of established as well as experimental therapeutic interventions for the treatment of anxiety-related disorders are associated with the regulation of microRNAs. As our understanding of microRNA-mediated therapeutic effects increases, novel therapeutic interventions targeting such mechanisms can be developed to improve treatments in anxiety-related disorders.

4 Conclusion and Further Perspectives

Epigenetic mechanisms, in particular microRNA-mediated modulation of gene expression and its dysregulation in disease states, have become a topic of sustained interest in the search for novel understanding and/or development of therapies for some of the most common diseases. In this chapter we have summarized the evidence linking microRNA regulation with anxiety-related behaviors and anxiety disorders. It has been found that specific microRNAs may play various roles in the development, progression, and treatment of anxiety. First studies have provided functional/causal evidence for such a role. Identified microRNAs include miR-15a, miR-17-92, the miR-34 family of microRNAs, miR-101, miR-124, miR-135, and miR-155, which influence anxiety behavior in a brain area-dependent manner. As an interesting example, miR-124 expression was shown to be upregulated in stress-induced anxiety, and its experimental targeted downregulation in the dentate gyrus led to a reduction in anxiety-like behavior, mediated in part via a miR-124a/BNDF interaction. Such functional knowledge aids the interpretation of associative studies and the identification of false positives. In this context, it was reported that running-induced anxiolysis is associated with hippocampal downregulation of miR-124.

In addition, we have summarized studies that examined how pharmacological and non-pharmacological interventions affect the expression of microRNAs. It was found that SSRIs commonly used to treat many anxiety disorders are associated with altered expression of a number of microRNAs in the brain, such as miR-16 and miR-1971. Non-pharmacological interventions, such as fear extinction, modeling exposure-based CBT, are associated with distinct changes in miR-144 and miR-128b in the BLA and infralimbic cortex of the mPFC, respectively. This new understanding has been aided by recent methodological advancements in assessing microRNAs functionally. In general, the field has moved beyond correlative studies whereby the presence of disease is associated with the altered expression of microRNA(s) and toward a more functional/causal understanding of microRNAs in distinct brain regions and of their roles in anxiety. Hence, the current data suggest that microRNAs may (in part) mediate therapeutic effects of established drugs and psychotherapeutic approaches used currently to treat these disorders.

This relation suggests that microRNAs could possibly be specifically targeted in the treatment of anxiety, although the therapeutic use of microRNAs for psychiatric diseases is in its very early stages. The initial problem encountered in such an approach – the fact that microRNAs have multiple targets – may in fact be an advantage, given that deregulation in anxiety disorders involves networks rather than single effector molecules. Thus, by targeting a single "key" microRNA, it is possible to modulate entire relevant gene networks. Target specificity may be determined by the observed highly distinct regional, cellular, and temporal expression of some of the candidate microRNAs in relevant nodes of the affected network. Along these lines, microRNA-based interventions targeting candidates such as miR-128b and miR-144 have the potential to augment the efficacy of exposure-based CBT through targeting plasticity-associated genes in extinction learning relevant brain areas such as the infralimbic cortex and basolateral amygdala, respectively. However, there is much work to be done before any microRNA-based therapy for anxiety can be used in a clinical setting. Answers to questions regarding oral bioavailability, the ability of microRNA-based therapeutics to cross the blood-brain barrier, and the presence of off-target effects currently elude us and are required to be solved before a push for microRNA-based interventions can be attempted. Therapeutic methodologies targeting microRNAs, such as virally mediated gene therapy or synthetic locked nucleic acid (LNA) agomirs/antagomirs, are associated with increased risks and thus require careful safety studies that consider off-target effects and other potential problems (Scott et al. 2015). In an effort to circumvent the limitations of virally mediated interventions, targeting mRNAs via locked nucleic acid probes, which are chemically modified to increase thermal stability and hybridization affinity, is a novel and possibly less invasive method of microRNA administration.

It is important to note that while to date numerous microRNAs have been implicated in anxiety, there is often little overlap within animal studies or between animal and human studies. This is in part due to the various experimental setups, the different behaviors assessed, and the different time points investigated. An important step forward would be to elucidate emerging microRNA candidates that play a role across various experimental setups and behaviors. Furthermore, it is unlikely that a single microRNA candidate alone will be critical in the development, progression, or treatment of anxiety; rather, a specific microRNA "signature" is more likely to be of importance in the various stages (Scott et al. 2015). In support of this, studies have revealed microRNA signatures for cancer, and more recently such signatures have emerged in preclinical models of stress-related disorders (Balakathiresan et al. 2014; Jiang et al. 2015).

In summary, we have provided an update on the most recent research functionally linking microRNAs with anxiety-related behaviors. In particular, it is through these more functional studies that we will be able to move forward in search of novel therapeutic interventions for psychiatric disorders such as anxiety disorders. A more complete regional, temporal, and subcellular picture of the role of microRNAs in anxiety will have the potential to open up novel ways of using this knowledge in the development of microRNA-based therapeutics for the improved treatment of patients suffering from anxiety-related disorders (Aten et al. 2019; Hommers et al. 2018; Sun et al. 2019; Snijders et al. 2018).

Acknowledgments Research of the authors' work related to this book chapter is supported by the Austrian Science Fund (FWF) (Grant No, SFB F4410 (to NS)), Signal Processing in Neurons (SPIN) (FWF Grant No, DK W-1206 (to NS)), and the European Molecular Biology Organization (ASTF Grant No. 459-2015 (to CPM)). The authors report no financial interests or conflicts of interest.

References

Abramowitz JS (2013) The practice of exposure therapy: relevance of cognitive-behavioral theory and extinction theory. Behav Ther 44:548–558

Aksoy-Aksel A, Zampa F, Schratt G (2014) MicroRNAs and synaptic plasticity--a mutual relationship. Philos Trans R Soc Lond Ser B Biol Sci 369:20130515

American Psychiatric Association (2013) Diagnostic and statistical manual of mental disorders, 5th edn. American Psychiatric Publishing, Washington

Andero R, Ressler KJ (2012) Fear extinction and BDNF: translating animal models of PTSD to the clinic. Genes Brain Behav 11:503–512

Andero R, Dias BG, Ressler KJ (2014) A role for Tac2, NkB, and Nk3 receptor in normal and dysregulated fear memory consolidation. Neuron 83:444–454

Anderzhanova E, Kirmeier T, Wotjak CT (2017) Animal models in psychiatric research: the RDoC system as a new framework for endophenotype-oriented translational neuroscience. Neurobiol Stress 7:47–56

Andolina D, DI Segni M, Bisicchia E, D'alessandro F, Cestari V, Ventura A, Concepcion C, Puglisi-Allegra S, Ventura R (2016) Effects of lack of microRNA-34 on the neural circuitry underlying the stress response and anxiety. Neuropharmacology 107:305–316

Aten S, Page CE, Kalidindi A, Wheaton K, Niraula A, Godbout JP, Hoyt KR, Obrietan K (2019) miR-132/212 is induced by stress and its dysregulation triggers anxiety-related behavior. Neuropharmacology 144:256–270

Autry AE, Monteggia LM (2012) Brain-derived Neurotrophic factor and neuropsychiatric disorders. Pharmacol Rev 64:238–258

Bahi A (2016) Sustained lentiviral-mediated overexpression of microRNA124a in the dentate gyrus exacerbates anxiety- and autism-like behaviors associated with neonatal isolation in rats. Behav Brain Res 311:298–308

Bahi A (2017) Hippocampal BDNF overexpression or microR124a silencing reduces anxiety- and autism-like behaviors in rats. Behav Brain Res 326:281–290

Baker M (2010) MicroRNA profiling: separating signal from noise. Nat Method 7:687–692

Baker KD, Bisby MA, Richardson R (2016) Impaired fear extinction in adolescent rodents: behavioural and neural analyses. Neurosci Biobehav Rev 70:59–73

Balakathiresan NS, Chandran R, Bhomia M, Jia M, Li H, Maheshwari RK (2014) Serum and amygdala microRNA signatures of posttraumatic stress: fear correlation and biomarker potential. J Psychiatr Res 57:65–73

Baldwin DS, Anderson IM, Nutt DJ, Allgulander C, Bandelow B, den Boer JA, Christmas DM, Davies S, Fineberg N, Lidbetter N, Malizia A, McCrone P, Nabarro D, O'neill C, Scott J, van der Wee N, Wittchen HU (2014) Evidence-based pharmacological treatment of anxiety disorders, post-traumatic stress disorder and obsessive-compulsive disorder: a revision of the 2005 guidelines from the British Association for Psychopharmacology. J Psychopharmacol 28:403–439

Bam M, Yang X, Zhou J, Ginsberg JP, Leyden Q, Nagarkatti PS, Nagarkatti M (2016a) Evidence for epigenetic regulation of pro-inflammatory cytokines, interleukin-12 and interferon gamma, in peripheral blood mononuclear cells from PTSD patients. J Neuroimmune Pharmacol 11:168–181

Bam M, Yang X, Zumbrun EE, Zhong Y, Zhou J, Ginsberg JP, Leyden Q, Zhang J, Nagarkatti PS, Nagarkatti M (2016b) Dysregulated immune system networks in war veterans with PTSD is an outcome of altered miRNA expression and DNA methylation. Sci Rep 6:31209

Bandelow B, Seidler-Brandler U, Becker A, Wedekind D, Ruther E (2007) Meta-analysis of randomized controlled comparisons of psychopharmacological and psychological treatments for anxiety disorders. World J Biol Psychiatry 8:175–187

Bandelow B, Sher L, Bunevicius R, Hollander E, Kasper S, Zohar J, Moller HJ (2012) Guidelines for the pharmacological treatment of anxiety disorders, obsessive-compulsive disorder and posttraumatic stress disorder in primary care. Int J Psychiatry Clin Pract 16:77–84

Bandelow B, Baldwin D, Abelli M, Altamura C, Dell'osso B, Domschke K, Fineberg NA, Grunblatt E, Jarema M, Maron E, Nutt D, Pini S, Vaghi MM, Wichniak A, Zai G, Riederer P (2016) Biological markers for anxiety disorders, OCD and PTSD - a consensus statement. Part I: neuroimaging and genetics. World J Biol Psychiatry 17:321–365

Bartel DP (2004) MicroRNAs: genomics, biogenesis, mechanism, and function. Cell 116:281–297

Bartel DP (2009) MicroRNAs: target recognition and regulatory functions. Cell 136:215–233

Bartlett AA, Singh R, Hunter RG (2017) Anxiety and epigenetics. Adv Exp Med Biol 978:145–166

Baudry A, Mouillet-Richard S, Schneider B, Launay JM, Kellermann O (2010) miR-16 targets the serotonin transporter: a new facet for adaptive responses to antidepressants. Science 329:1537–1541

Bjornsson HT, Fallin MD, Feinberg AP (2004) An integrated epigenetic and genetic approach to common human disease. Trends Genet 20:350–358

Bowers ME, Choi DC, Ressler KJ (2012) Neuropeptide regulation of fear and anxiety: implications of cholecystokinin, endogenous opioids, and neuropeptide Y. Physiol Behav 107:699–710

Bredy TW, Lin Q, Wei W, Baker-Andresen D, Mattick JS (2011) MicroRNA regulation of neural plasticity and memory. Neurobiol Learn Mem 96:89–94

Brenes JC, Lackinger M, Hoglinger GU, Schratt G, Schwarting RK, Wohr M (2016) Differential effects of social and physical environmental enrichment on brain plasticity, cognition, and ultrasonic communication in rats. J Comp Neurol 524:1586–1607

Bukalo O, Pinard CR, Holmes A (2014) Mechanisms to medicines: elucidating neural and molecular substrates of fear extinction to identify novel treatments for anxiety disorders. Br J Pharmacol 171:4690–4718
Calhoon GG, Tye KM (2015) Resolving the neural circuits of anxiety. Nat Neurosci 18:1394–1404
Cao DD, Li L, Chan WY (2016) MicroRNAs: key regulators in the central nervous system and their implication in neurological diseases. Int J Mol Sci 17:E842
Ceman S, Saugstad J (2011) MicroRNAs: meta-controllers of gene expression in synaptic activity emerge as genetic and diagnostic markers of human disease. Pharmacol Ther 130:26–37
Chen S-D, Sun X-Y, Niu W, Kong L-M, He M-J, Fan H-M, Li W-S, Zhong A-F, Zhang L-Y, Lu J (2016) Correlation between the level of microRNA expression in peripheral blood mononuclear cells and symptomatology in patients with generalized anxiety disorder. Compr Psychiatry 69:216–224
Cholewa-Waclaw J, Bird A, von Schimmelmann M, Schaefer A, Yu H, Song H, Madabhushi R, Tsai LH (2016) The role of epigenetic mechanisms in the regulation of gene expression in the nervous system. J Neurosci 36:11427–11434
Cohen JL, Jackson NL, Ballestas ME, Webb WM, Lubin FD, Clinton SM (2017) Amygdalar expression of the microRNA miR-101a and its target Ezh2 contribute to rodent anxiety-like behaviour. Eur J Neurosci 46:2241
Craske MG, Stein MB (2016) Anxiety. Lancet 388:3048
Craske MG, Stein MB, Eley TC, Milad MR, Holmes A, Rapee RM, Wittchen HU (2017) Anxiety disorders. Nat Rev Dis Primers 3:17024
Cruceanu C, Lopez JP, Tsai WT, Turecki G (2016) Dysregulation of the glutamatergic receptors after antidepressant treatment in human neural progenitor cells. Mol Psychiatry 22:1228–1229
Dell'osso B, Buoli M, Baldwin DS, Altamura AC (2010) Serotonin norepinephrine reuptake inhibitors (SNRIs) in anxiety disorders: a comprehensive review of their clinical efficacy. Hum Psychopharmacol 25:17–29
Dias BG, Banerjee SB, Goodman JV, Ressler KJ (2013) Towards new approaches to disorders of fear and anxiety. Curr Opin Neurobiol 23:346–352
Dias BG, Goodman JV, Ahluwalia R, Easton AE, Andero R, Ressler KJ (2014) Amygdala-dependent fear memory consolidation via miR-34a and Notch signaling. Neuron 83:906–918
Dias BG, Maddox SA, Klengel T, Ressler KJ (2015) Epigenetic mechanisms underlying learning and the inheritance of learned behaviors. Trends Neurosci 38:96–107
Dirven BCJ, Homberg JR, Kozicz T, Henckens M (2017) Epigenetic programming of the neuroendocrine stress response by adult life stress. J Mol Endocrinol 59:R11–r31
Domschke K, Tidow N, Kuithan H, Schwarte K, Klauke B, Ambree O, Reif A, Schmidt H, Arolt V, Kersting A, Zwanzger P, Deckert J (2012) Monoamine oxidase A gene DNA hypomethylation – a risk factor for panic disorder? Int J Neuropsychopharmacol 15:1217–1228
Drury SS, Sanchez MM, Gonzalez A (2016) When mothering goes awry: challenges and opportunities for utilizing evidence across rodent, nonhuman primate and human studies to better define the biological consequences of negative early caregiving. Horm Behav 77:182–192
Duits P, Cath DC, Lissek S, Hox JJ, Hamm AO, Engelhard IM, van den Hout MA, Baas JM (2015) Updated meta-analysis of classical fear conditioning in the anxiety disorders. Depress Anxiety 32:239–253
Farach FJ, Pruitt LD, Jun JJ, Jerud AB, Zoellner LA, Roy-Byrne PP (2012) Pharmacological treatment of anxiety disorders: current treatments and future directions. J Anxiety Disord 26:833–843
Farb DH, Ratner MH (2014) Targeting the modulation of neural circuitry for the treatment of anxiety disorders. Pharmacol Rev 66:1002–1032
Fass DM, Schroeder FA, Perlis RH, Haggarty SJ (2014) Epigenetic mechanisms in mood disorders: targeting neuroplasticity. Neuroscience 264:112–130
Fernandes V, Osorio FL (2015) Are there associations between early emotional trauma and anxiety disorders? Evidence from a systematic literature review and meta-analysis. Eur Psychiatry 30:756–764

Flavell CR, Lambert EA, Winters BD, Bredy TW (2013) Mechanisms governing the reactivation-dependent destabilization of memories and their role in extinction. Front Behav Neurosci 7:214

Fonken LK, Gaudet AD, Gaier KR, Nelson RJ, Popovich PG (2016) MicroRNA-155 deletion reduces anxiety- and depressive-like behaviors in mice. Psychoneuroendocrinology 63:362–369

Gantier MP, McCoy CE, Rusinova I, Saulep D, Wang D, Xu D, Irving AT, Behlke MA, Hertzog PJ, Mackay F, Williams BR (2011) Analysis of microRNA turnover in mammalian cells following Dicer1 ablation. Nucleic Acids Res 39:5692–5703

Garzon R, Calin GA, Croce CM (2009) MicroRNAs in cancer. Annu Rev Med 60:167–179

Gould RL, Coulson MC, Howard RJ (2012) Efficacy of cognitive behavioral therapy for anxiety disorders in older people: a meta-analysis and meta-regression of randomized controlled trials. J Am Geriatr Soc 60:218–229

Graham BM, Milad MR (2011) The study of fear extinction: implications for anxiety disorders. Am J Psychiatry 168:1255–1265

Griebel G, Holmes A (2013) 50 years of hurdles and hope in anxiolytic drug discovery. Nat Rev Drug Discov 12:667–687

Gu QH, Yu D, Hu Z, Liu X, Yang Y, Luo Y, Zhu J, Li Z (2015) miR-26a and miR-384-5p are required for LTP maintenance and spine enlargement. Nat Commun 6:6789

Haramati S, Navon I, Issler O, Ezra-Nevo G, Gil S, Zwang R, Hornstein E, Chen A (2011) MicroRNA as repressors of stress-induced anxiety: the case of amygdalar miR-34. J Neurosci 31:14191–14203

Hemstedt TJ, Lattal KM, Wood MA (2017) Reconsolidation and extinction: using epigenetic signatures to challenge conventional wisdom. Neurobiol Learn Mem 142:55

Hendriksen H, Olivier B, Oosting RS (2014) From non-pharmacological treatments for post-traumatic stress disorder to novel therapeutic targets. Eur J Pharmacol 732:139–158

Herry C, Trifilieff P, Micheau J, Luthi A, Mons N (2006) Extinction of auditory fear conditioning requires MAPK/ERK activation in the basolateral amygdala. Eur J Neurosci 24:261–269

Hettema JM, Neale MC, Kendler KS (2001) A review and meta-analysis of the genetic epidemiology of anxiety disorders. Am J Psychiatry 158:1568–1578

Hettema JM, Prescott CA, Myers JM, Neale MC, Kendler KS (2005) The structure of genetic and environmental risk factors for anxiety disorders in men and women. Arch Gen Psychiatry 62:182–189

Hill JL, Martinowich K (2016) Activity-dependent signaling: influence on plasticity in circuits controlling fear-related behavior. Curr Opin Neurobiol 36:59–65

Hoban AE, Stilling RM, Moloney GM, Moloney RD, Shanahan F, Dinan TG, Cryan JF, Clarke G (2017) Microbial regulation of microRNA expression in the amygdala and prefrontal cortex. Microbiome 5:102

Hofmann SG, Smits JA (2008) Cognitive-behavioral therapy for adult anxiety disorders: a meta-analysis of randomized placebo-controlled trials. J Clin Psychiatry 69:621–632

Holmes A, Singewald N (2013) Individual differences in recovery from traumatic fear. Trends Neurosci 36:23–31

Hommers LG, Domschke K, Deckert J (2015) Heterogeneity and individuality: microRNAs in mental disorders. J Neural Transm 122:79–97

Hommers LG, Richter J, Yang Y, Raab A, Baumann C, Lang K, Schiele MA, Weber H, Wittmann A, Wolf C, Alpers GW, Arolt V, Domschke K, Fehm L, Fydrich T, Gerlach A, Gloster AT, Hamm AO, Helbig-Lang S, Kircher T, Lang T, Pané-Farré CA, Pauli P, Pfleiderer B, Reif A, Romanos M, Straube B, Ströhle A, Wittchen HU, Frantz S, Ertl G, Lohse MJ, Lueken U, Deckert J (2018) A functional genetic variation of SLC6A2 repressor hsa-miR-579-3p upregulates sympathetic noradrenergic processes of fear and anxiety. Transl Psychiatry 8(1):226

Hu Z, Li Z (2017) miRNAs in synapse development and synaptic plasticity. Curr Opin Neurobiol 45:24–31

Hu Z, Yu D, Gu QH, Yang Y, Tu K, Zhu J, Li Z (2014) miR-191 and miR-135 are required for long-lasting spine remodelling associated with synaptic long-term depression. Nat Commun 5:3263

Hunter RG (2012) Epigenetic effects of stress and corticosteroids in the brain. Front Cell Neurosci 6:18

Huttenrauch M, Salinas G, Wirths O (2016) Effects of long-term environmental enrichment on anxiety, memory, hippocampal plasticity and overall brain gene expression in C57BL6 mice. Front Mol Neurosci 9:62

Issler O, Chen A (2015) Determining the role of microRNAs in psychiatric disorders. Nat Rev Neurosci 16:201–212

Issler O, Chen A (2017) The role of microRNAs in stress-induced psychopathologies A2. In: Fink G (ed) Stress: neuroendocrinology and neurobiology. Academic Press, San Diego

Issler O, Haramati S, Paul ED, Maeno H, Navon I, Zwang R, Gil S, Mayberg HS, Dunlop BW, Menke A, Awatramani R, Binder EB, Deneris ES, Lowry CA, Chen A (2014) MicroRNA 135 is essential for chronic stress resiliency, antidepressant efficacy, and intact serotonergic activity. Neuron 83:344–360

Jaenisch R, Bird A (2003) Epigenetic regulation of gene expression: how the genome integrates intrinsic and environmental signals. Nat Genet 33(Suppl):245–254

James AC, James G, Cowdrey FA, Soler A, Choke A (2015) Cognitive behavioural therapy for anxiety disorders in children and adolescents. Cochrane Database Syst Rev 6:Cd004690

Jiang X, Du L, Wang L, Li J, Liu Y, Zheng G, Qu A, Zhang X, Pan H, Yang Y, Wang C (2015) Serum microRNA expression signatures identified from genome-wide microRNA profiling serve as novel noninvasive biomarkers for diagnosis and recurrence of bladder cancer. Int J Cancer 136:854–862

Jin J, Kim SN, Liu X, Zhang H, Zhang C, Seo JS, Kim Y, Sun T (2016) miR-17-92 cluster regulates adult hippocampal neurogenesis, anxiety, and depression. Cell Rep 16:1653–1663

Jovanovic T, Kazama A, Bachevalier J, Davis M (2012) Impaired safety signal learning may be a biomarker of PTSD. Neuropharmacology 62:695–704

Jovasevic V, Corcoran KA, Leaderbrand K, Yamawaki N, Guedea AL, Chen HJ, Shepherd GM, Radulovic J (2015) GABAergic mechanisms regulated by miR-33 encode state-dependent fear. Nat Neurosci 18:1265–1271

Katsuura S, Kuwano Y, Yamagishi N, Kurokawa K, Kajita K, Akaike Y, Nishida K, Masuda K, Tanahashi T, Rokutan K (2012) MicroRNAs miR-144/144* and miR-16 in peripheral blood are potential biomarkers for naturalistic stress in healthy Japanese medical students. Neurosci Lett 516:79–84

Kendler KS, Karkowski LM, Prescott CA (1999) Fears and phobias: reliability and heritability. Psychol Med 29:539–553

Kessler RC, Petukhova M, Sampson NA, Zaslavsky AM, Wittchen HU (2012) Twelve-month and lifetime prevalence and lifetime morbid risk of anxiety and mood disorders in the United States. Int J Methods Psychiatr Res 21:169–184

Kheirbek MA, Klemenhagen KC, Sahay A, Hen R (2012) Neurogenesis and generalization: a new approach to stratify and treat anxiety disorders. Nat Neurosci 15:1613–1620

Kichukova TM, Popov NT, Ivanov HY, Vachev TI (2015) Circulating microRNAs as a novel class of potential diagnostic biomarkers in neuropsychiatric disorders. Folia Med 57:159–172

Kim B, Kim MK, Kim SW, Kim KM, Kim HS, An HJ, Kim JO, Choi TK, Kim NK, Lee SH (2015) Association of human microRNAs miR-22 and miR-491 polymorphisms with panic disorder with or without agoraphobia in a Korean population. J Affect Disord 188:118–126

Kinrys G, Pollack MH, Simon NM, Worthington JJ, Nardi AE, Versiani M (2003) Valproic acid for the treatment of social anxiety disorder. Int Clin Psychopharmacol 18:169–172

Klauke B, Deckert J, Reif A, Pauli P, Domschke K (2010) Life events in panic disorder-an update on "candidate stressors". Depress Anxiety 27:716–730

Klengel T, Binder EB (2015) Epigenetics of stress-related psychiatric disorders and gene × environment interactions. Neuron 86:1343–1357

Kong YW, Ferland-McCollough D, Jackson TJ, Bushell M (2012) microRNAs in cancer management. Lancet Oncol 13:e249–e258

Krol J, Busskamp V, Markiewicz I, Stadler MB, Ribi S, Richter J, Duebel J, Bicker S, Fehling HJ, Schubeler D, Oertner TG, Schratt G, Bibel M, Roska B, Filipowicz W (2010a) Characterizing light-regulated retinal microRNAs reveals rapid turnover as a common property of neuronal microRNAs. Cell 141:618–631

Krol J, Loedige I, Filipowicz W (2010b) The widespread regulation of microRNA biogenesis, function and decay. Nat Rev Genet 11:597–610

Kuriyama K, Honma M, Soshi T, Fujii T, Kim Y (2011) Effect of d-cycloserine and valproic acid on the extinction of reinstated fear-conditioned responses and habituation of fear conditioning in healthy humans: a randomized controlled trial. Psychopharmacology 218:589–597

Landgrave-Gomez J, Mercado-Gomez O, Guevara-Guzman R (2015) Epigenetic mechanisms in neurological and neurodegenerative diseases. Front Cell Neurosci 9:58

Lau NC, Lim LP, Weinstein EG, Bartel DP (2001) An abundant class of tiny RNAs with probable regulatory roles in Caenorhabditis elegans. Science 294:858–862

Lee RC, Feinbaum RL, Ambros V (1993) The C. elegans heterochronic gene lin-4 encodes small RNAs with antisense complementarity to lin-14. Cell 75:843–854

Lim LP, Lau NC, Garrett-Engele P, Grimson A, Schelter JM, Castle J, Bartel DP, Linsley PS, Johnson JM (2005) Microarray analysis shows that some microRNAs downregulate large numbers of target mRNAs. Nature 433:769–773

Lin Q, Wei W, Coelho CM, Li X, Baker-Andresen D, Dudley K, Ratnu VS, Boskovic Z, Kobor MS, Sun YE, Bredy TW (2011) The brain-specific microRNA miR-128b regulates the formation of fear-extinction memory. Nat Neurosci 14:1115–1117

Ling H, Fabbri M, Calin GA (2013) MicroRNAs and other non-coding RNAs as targets for anticancer drug development. Nat Rev Drug Discov 12:847–865

Londin E, Loher P, Telonis AG, Quann K, Clark P, Jing Y, Hatzimichael E, Kirino Y, Honda S, Lally M, Ramratnam B, Comstock CES, Knudsen KE, Gomella L, Spaeth GL, Hark L, Katz LJ, Witkiewicz A, Rostami A, Jimenez SA, Hollingsworth MA, Yeh JJ, Shaw CA, McKenzie SE, Bray P, Nelson PT, Zupo S, van Roosbroeck K, Keating MJ, Calin GA, Yeo C, Jimbo M, Cozzitorto J, Brody JR, Delgrosso K, Mattick JS, Fortina P, Rigoutsos I (2015) Analysis of 13 cell types reveals evidence for the expression of numerous novel primate- and tissue-specific microRNAs. Proc Natl Acad Sci U S A 112:E1106–E1115

Lopez JP, Lim R, Cruceanu C, Crapper L, Fasano C, Labonte B, Maussion G, Yang JP, Yerko V, Vigneault E, El Mestikawy S, Mechawar N, Pavlidis P, Turecki G (2014) miR-1202 is a primate-specific and brain-enriched microRNA involved in major depression and antidepressant treatment. Nat Med 20:764–768

Lopez JP, Pereira F, Richard-Devantoy S, Berlim M, Chachamovich E, Fiori LM, Niola P, Turecki G, Jollant F (2017) Co-variation of peripheral levels of miR-1202 and brain activity and connectivity during antidepressant treatment. Neuropsychopharmacology 42:2043

Lugli G, Torvik VI, Larson J, Smalheiser NR (2008) Expression of microRNAs and their precursors in synaptic fractions of adult mouse forebrain. J Neurochem 106:650–661

Lumayag S, Haldin CE, Corbett NJ, Wahlin KJ, Cowan C, Turturro S, Larsen PE, Kovacs B, Witmer PD, Valle D, Zack DJ, Nicholson DA, Xu S (2013) Inactivation of the microRNA-183/96/182 cluster results in syndromic retinal degeneration. Proc Natl Acad Sci U S A 110:E507–E516

Maccari S, Krugers HJ, Morley-Fletcher S, Szyf M, Brunton PJ (2014) The consequences of early-life adversity: neurobiological, behavioural and epigenetic adaptations. J Neuroendocrinol 26:707–723

Malan-Muller S, Hemmings SM, Seedat S (2013) Big effects of small RNAs: a review of microRNAs in anxiety. Mol Neurobiol 47:726–739

Malan-Muller S, Fairbairn L, Hart S, Daniels WMU, Jalali Sefid Dashti M, Kidd M, Seedat S, Gamieldien J, Hemmings SMJ (2017) The role of microRNAs in the therapeutic action of D-cycloserine in a post-traumatic stress disorder animal model: an exploratory study. Psychiatr Genet 27:139

Mannironi C, Biundo A, Rajendran S, de Vito F, Saba L, Caioli S, Zona C, Ciotti T, Caristi S, Perlas E, del Vecchio G, Bozzoni I, Rinaldi A, Mele A, Presutti C (2017) miR-135a regulates synaptic transmission and anxiety-like behavior in amygdala. Mol Neurobiol 55:3301–3315

Maren S, Holmes A (2016) Stress and fear extinction. Neuropsychopharmacology 41:58–79

Maron E, Nutt D (2015) Biological predictors of pharmacological therapy in anxiety disorders. Dialogues Clin Neurosci 17:305–317
Martin CG, Kim H, Yun S, Livingston W, Fetta J, Mysliwiec V, Baxter T, Gill JM (2017) Circulating miRNA associated with posttraumatic stress disorder in a cohort of military combat veterans. Psychiatry Res 251:261–265
Martinetz S (2016) MicroRNA's impact on neurotransmitter and neuropeptide systems: small but mighty mediators of anxiety. Pflugers Arch 468:1061–1069
Marzi MJ, Ghini F, Cerruti B, de Pretis S, Bonetti P, Giacomelli C, Gorski MM, Kress T, Pelizzola M, Muller H, Amati B, Nicassio F (2016) Degradation dynamics of microRNAs revealed by a novel pulse-chase approach. Genome Res 26:554–565
McMillan KA, Asmundson GJG, Sareen J (2017) Comorbid PTSD and social anxiety disorder: associations with quality of life and suicide attempts. J Nerv Ment Dis 205:732–737
Michael T, Blechert J, Vriends N, Margraf J, Wilhelm FH (2007) Fear conditioning in panic disorder: enhanced resistance to extinction. J Abnorm Psychol 116:612–617
Milad MR, Quirk GJ (2012) Fear extinction as a model for translational neuroscience: ten years of progress. Annu Rev Psychol 63:129–151
Milad MR, Pitman RK, Ellis CB, Gold AL, Shin LM, Lasko NB, Zeidan MA, Handwerger K, Orr SP, Rauch SL (2009) Neurobiological basis of failure to recall extinction memory in posttraumatic stress disorder. Biol Psychiatry 66:1075–1082
Milad MR, Furtak SC, Greenberg JL, Keshaviah A, Im JJ, Falkenstein MJ, Jenike M, Rauch SL, Wilhelm S (2013) Deficits in conditioned fear extinction in obsessive-compulsive disorder and neurobiological changes in the fear circuit. JAMA Psychiat 70:608–618; quiz 554
Millan MJ (2003) The neurobiology and control of anxious states. Prog Neurobiol 70:83–244
Muinos-Gimeno M, Espinosa-Parrilla Y, Guidi M, Kagerbauer B, Sipila T, Maron E, Pettai K, Kananen L, Navines R, Martin-Santos R, Gratacos M, Metspalu A, Hovatta I, Estivill X (2011) Human microRNAs miR-22, miR-138-2, miR-148a, and miR-488 are associated with panic disorder and regulate several anxiety candidate genes and related pathways. Biol Psychiatry 69:526–533
Murphy CP, Li X, Maurer V, Oberhauser M, Gstir R, Wearick-Silva LE, Viola TW, Schafferer S, Grassi-Oliveira R, Whittle N, Huttenhofer A, Bredy TW, Singewald N (2017) MicroRNA-mediated rescue of fear extinction memory by miR-144-3p in extinction-impaired mice. Biol Psychiatry 81:979–989
Nana-Sinkam SP, Croce CM (2013) Clinical applications for microRNAs in cancer. Clin Pharmacol Ther 93:98–104
Nieto SJ, Patriquin MA, Nielsen DA, Kosten TA (2016) Don't worry; be informed about the epigenetics of anxiety. Pharmacol Biochem Behav 146:60–72
Norrholm SD, Ressler KJ (2009) Genetics of anxiety and trauma-related disorders. Neuroscience 164:272–287
Nudelman AS, Dirocco DP, Lambert TJ, Garelick MG, Le J, Nathanson NM, Storm DR (2010) Neuronal activity rapidly induces transcription of the CREB-regulated microRNA-132, in vivo. Hippocampus 20:492–498
O'Connor RM, Dinan TG, Cryan JF (2012) Little things on which happiness depends: microRNAs as novel therapeutic targets for the treatment of anxiety and depression. Mol Psychiatry 17:359–376
O'Connor RM, Gururajan A, Dinan TG, Kenny PJ, Cryan JF (2016) All roads Lead to the miRNome: miRNAs have a central role in the molecular pathophysiology of psychiatric disorders. Trends Pharmacol Sci 37:1029–1044
Olesen J, Gustavsson A, Svensson M, Wittchen HU, Jonsson B, CDBE2010 Study Group; European Brain Council (2012) The economic cost of brain disorders in Europe. Eur J Neurol 19:155–162
Olsen L, Klausen M, Helboe L, Nielsen FC, Werge T (2009) MicroRNAs show mutually exclusive expression patterns in the brain of adult male rats. PLoS One 4:e7225

Ong CT, Corces VG (2011) Enhancer function: new insights into the regulation of tissue-specific gene expression. Nat Rev Genet 12:283–293

Orsini CA, Maren S (2012) Neural and cellular mechanisms of fear and extinction memory formation. Neurosci Biobehav Rev 36:1773–1802

Pan-Vazquez A, Rye N, Ameri M, McSparron B, Smallwood G, Bickerdyke J, Rathbone A, Dajas-Bailador F, Toledo-Rodriguez M (2015) Impact of voluntary exercise and housing conditions on hippocampal glucocorticoid receptor, miR-124 and anxiety. Mol Brain 8:40

Patrício P, Mateus-Pinheiro A, Irmler M, Alves ND, Machado-Santos AR, Morais M, Correia JS, Korostynski M, Piechota M, Stoffel R, Beckers J, Bessa JM, Almeida OFX, Sousa N, Pinto L (2015) Differential and converging molecular mechanisms of Antidepressants' action in the hippocampal dentate Gyrus. Neuropsychopharmacology 40:338–349

Pavlova B, Perlis RH, Mantere O, Sellgren CM, Isometsa E, Mitchell PB, Alda M, Uher R (2017) Prevalence of current anxiety disorders in people with bipolar disorder during euthymia: a meta-analysis. Psychol Med 47:1107–1115

Perusini JN, Fanselow MS (2015) Neurobehavioral perspectives on the distinction between fear and anxiety. Learn Mem 22:417–425

Ragu Varman D, Marimuthu G, Rajan KE (2013) Environmental enrichment upregulates micro-RNA-183 and alters acetylcholinesterase splice variants to reduce anxiety-like behavior in the little Indian field mouse (Mus booduga). J Neurosci Res 91:426–435

Rao P, Benito E, Fischer A (2013) MicroRNAs as biomarkers for CNS disease. Front Mol Neurosci 6:39

Ravindran LN, Stein MB (2010) The pharmacologic treatment of anxiety disorders: a review of progress. J Clin Psychiatry 71:839–854

Ruegger S, Grosshans H (2012) MicroRNA turnover: when, how, and why. Trends Biochem Sci 37:436–446

Saab BJ, Mansuy IM (2014) Neuroepigenetics of memory formation and impairment: the role of microRNAs. Neuropharmacology 80:61–69

Sah A, Schmuckermair C, Sartori SB, Gaburro S, Kandasamy M, Irschick R, Klimaschewski L, Landgraf R, Aigner L, Singewald N (2012) Anxiety- rather than depression-like behavior is associated with adult neurogenesis in a female mouse model of higher trait anxiety- and comorbid depression-like behavior. Transl Psychiatry 2:e171

Saini HK, Griffiths-Jones S, Enright AJ (2007) Genomic analysis of human microRNA transcripts. Proc Natl Acad Sci U S A 104:17719–17724

Schiele MA, Domschke K (2017) Epigenetics at the crossroads between genes, environment and resilience in anxiety disorders. Genes Brain Behav 17:e12423

Schmidt U, Herrmann L, Hagl K, Novak B, Huber C, Holsboer F, Wotjak CT, Buell DR (2013) Therapeutic action of fluoxetine is associated with a reduction in prefrontal cortical miR-1971 expression levels in a mouse model of posttraumatic stress disorder. Front Psych 4:66

Schratt G (2009) microRNAs at the synapse. Nat Rev Neurosci 10:842–849

Scott KA, Hoban AE, Clarke G, Moloney GM, Dinan TG, Cryan JF (2015) Thinking small: towards microRNA-based therapeutics for anxiety disorders. Expert Opin Investig Drugs 24:529–542

Sevignani C, Calin GA, Siracusa LD, Croce CM (2006) Mammalian microRNAs: a small world for fine-tuning gene expression. Mamm Genome 17:189–202

Sharma S, Powers A, Bradley B, Ressler KJ (2016) Gene x environment determinants of stress- and anxiety-related disorders. Annu Rev Psychol 67:239–261

Shimada-Sugimoto M, Otowa T, Hettema JM (2015) Genetics of anxiety disorders: genetic epidemiological and molecular studies in humans. Psychiatry Clin Neurosci 69:388–401

Shin LM, Liberzon I (2010) The neurocircuitry of fear, stress, and anxiety disorders. Neuropsychopharmacology 35:169–191

Short AK, Yeshurun S, Powell R, Perreau VM, Fox A, Kim JH, Pang TY, Hannan AJ (2017) Exercise alters mouse sperm small noncoding RNAs and induces a transgenerational modification of male offspring conditioned fear and anxiety. Transl Psychiatry 7:e1114

Sim SE, Bakes J, Kaang BK (2014) Neuronal activity-dependent regulation of MicroRNAs. Mol Cells 37:511–517

Singewald N, Schmuckermair C, Whittle N, Holmes A, Ressler KJ (2015) Pharmacology of cognitive enhancers for exposure-based therapy of fear, anxiety and trauma-related disorders. Pharmacol Ther 149:150–190

Smoller JW (2016) The genetics of stress-related disorders: PTSD, depression, and anxiety disorders. Neuropsychopharmacology 41:297–319

Snijders C, de Nijs L, Baker DG, Hauger RL, van den Hove D, Kenis G, Nievergelt CM, Boks MP, Vermetten E, Gage FH, Rutten BPF (2018) MicroRNAs in post-traumatic stress disorder. Curr Top Behav Neurosci 38:23–46

Spadaro PA, Bredy TW (2012) Emerging role of non-coding RNA in neural plasticity, cognitive function, and neuropsychiatric disorders. Front Genet 3:132

Stein A, Pearson RM, Goodman SH, Rapa E, Rahman A, McCallum M, Howard LM, Pariante CM (2014) Effects of perinatal mental disorders on the fetus and child. Lancet 384:1800–1819

Sun Y, Lu W, Du K, Wang JH (2019) microRNA and mRNA profiles in the amygdala are relevant to fear memory induced by physical or psychological stress. J Neurophysiol. https://doi.org/10.1152/jn.00215.2019

Taguchi YH (2013) MicroRNA-mediated regulation of target genes in several brain regions is correlated to both microRNA-targeting-specific promoter methylation and differential microRNA expression. BioData Min 6:11

Tardito D, Mallei A, Popoli M (2013) Lost in translation. New unexplored avenues for neuropsychopharmacology: epigenetics and microRNAs. Expert Opin Investig Drugs 22:217–233

Tolin DF (2010) Is cognitive-behavioral therapy more effective than other therapies? A meta-analytic review. Clin Psychol Rev 30:710–720

Tovote P, Fadok JP, Luthi A (2015) Neuronal circuits for fear and anxiety. Nat Rev Neurosci 16:317–331

Vaisvaser S, Modai S, Farberov L, Lin T, Sharon H, Gilam A, Volk N, Admon R, Edry L, Fruchter E, Wald I, Bar-Haim Y, Tarrasch R, Chen A, Shomron N, Hendler T (2016) Neuro-epigenetic indications of acute stress response in humans: the case of MicroRNA-29c. PLoS One 11:e0146236

van der Ree MH, van der Meer AJ, de Bruijne J, Maan R, van Vliet A, Welzel TM, Zeuzem S, Lawitz EJ, Rodriguez-Torres M, Kupcova V, Wiercinska-Drapalo A, Hodges MR, Janssen HLA, Reesink HW (2014) Long-term safety and efficacy of microRNA-targeted therapy in chronic hepatitis C patients. Antivir Res 111:53–59

Vidigal JA, Ventura A (2015) The biological functions of miRNAs: lessons from in vivo studies. Trends Cell Biol 25:137–147

Vieland VJ, Goodman DW, Chapman T, Fyer AJ (1996) New segregation analysis of panic disorder. Am J Med Genet 67:147–153

Volk N, Pape JC, Engel M, Zannas AS, Cattane N, Cattaneo A, Binder EB, Chen A (2016) Amygdalar MicroRNA-15a is essential for coping with chronic stress. Cell Rep 17:1882–1891

Wang Y, Li X, Hu H (2011) Transcriptional regulation of co-expressed microRNA target genes. Genomics 98:445–452

Wang F, Pereira A (2016) Neuromodulation, emotional feelings and affective disorders. Mens Sana Monogr 14:5–29

Wang W, Kwon EJ, Tsai LH (2012) MicroRNAs in learning, memory, and neurological diseases. Learn Mem 19:359–368

Wang X, Sundquist K, Hedelius A, Palmer K, Memon AA, Sundquist J (2015) Circulating microRNA-144-5p is associated with depressive disorders. Clin Epigenetics 7:69

Wessa M, Flor H (2007) Failure of extinction of fear responses in posttraumatic stress disorder: evidence from second-order conditioning. Am J Psychiatry 164:1684–1692

Whittle N, Schmuckermair C, Gunduz Cinar O, Hauschild M, Ferraguti F, Holmes A, Singewald N (2013) Deep brain stimulation, histone deacetylase inhibitors and glutamatergic drugs rescue resistance to fear extinction in a genetic mouse model. Neuropharmacology 64:414–423

WHO (1993) The ICD-10 classification of mental and behavioral disorders: diagnostic criteria for research. World Health Organization, Geneva

Wibrand K, Panja D, Tiron A, Ofte ML, Skaftnesmo KO, Lee CS, Pena JT, Tuschl T, Bramham CR (2010) Differential regulation of mature and precursor microRNA expression by NMDA and metabotropic glutamate receptor activation during LTP in the adult dentate gyrus in vivo. Eur J Neurosci 31:636–645

Wingo AP, Almli LM, Stevens JS, Klengel T, Uddin M, Li Y, Bustamante AC, Lori A, Koen N, Stein DJ, Smith AK, Aiello AE, Koenen KC, Wildman DE, Galea S, Bradley B, Binder EB, Jin P, Gibson G, Ressler KJ (2015) DICER1 and microRNA regulation in post-traumatic stress disorder with comorbid depression. Nat Commun 6:10106

Wingo AP, Almli LM, Stevens JS, Jovanovic T, Wingo TS, Tharp G, Li Y, Lori A, Briscione M, Jin P, Binder EB, Bradley B, Gibson G, Ressler KJ (2017) Genome-wide association study of positive emotion identifies a genetic variant and a role for microRNAs. Mol Psychiatry 22:774–783

Wittchen HU, Jacobi F, Rehm J, Gustavsson A, Svensson M, Jonsson B, Olesen J, Allgulander C, Alonso J, Faravelli C, Fratiglioni L, Jennum P, Lieb R, Maercker A, van Os J, Preisig M, Salvador-Carulla L, Simon R, Steinhausen HC (2011) The size and burden of mental disorders and other disorders of the brain in Europe 2010. Eur Neuropsychopharmacol 21:655–679

Yang YL, Lu KT (2005) Facilitation of conditioned fear extinction by d-cycloserine is mediated by mitogen-activated protein kinase and phosphatidylinositol 3-kinase cascades and requires de novo protein synthesis in basolateral nucleus of amygdala. Neuroscience 134:247–260

Yehuda R, Bierer LM (2009) The relevance of epigenetics to PTSD: implications for the DSM-V. J Trauma Stress 22:427–434

You HJ, Park JH, Pareja-Galeano H, Lucia A, Shin JI (2016) Targeting MicroRNAs involved in the BDNF signaling impairment in neurodegenerative diseases. NeuroMolecular Med 18:540

Zhang Y, Wang Y, Wang L, Bai M, Zhang X, Zhu X (2015) Dopamine receptor D2 and associated microRNAs are involved in stress susceptibility and resistance to escitalopram treatment. Int J Neuropsychopharmacol 18:pyv025

Zhou R, Yuan P, Wang Y, Hunsberger JG, Elkahloun A, Wei Y, Damschroder-Williams P, Du J, Chen G, Manji HK (2009) Evidence for selective microRNAs and their effectors as common long-term targets for the actions of mood stabilizers. Neuropsychopharmacology 34:1395–1405

Zhou J, Nagarkatti P, Zhong Y, Ginsberg JP, Singh NP, Zhang J, Nagarkatti M (2014) Dysregulation in microRNA expression is associated with alterations in immune functions in combat veterans with post-traumatic stress disorder. PLoS One 9:e94075

Zhu J, Chen Z, Tian J, Meng Z, Ju M, Wu G, Tian Z (2017) miR-34b attenuates trauma-induced anxiety-like behavior by targeting CRHR1. Int J Mol Med 40:90–100

Open Access This chapter is licensed under the terms of the Creative Commons Attribution 4.0 International License (http://creativecommons.org/licenses/by/4.0/), which permits use, sharing, adaptation, distribution and reproduction in any medium or format, as long as you give appropriate credit to the original author(s) and the source, provide a link to the Creative Commons licence and indicate if changes were made.

The images or other third party material in this chapter are included in the chapter's Creative Commons licence, unless indicated otherwise in a credit line to the material. If material is not included in the chapter's Creative Commons licence and your intended use is not permitted by statutory regulation or exceeds the permitted use, you will need to obtain permission directly from the copyright holder.

Transposable Elements

G. Guffanti, A. Bartlett, P. DeCrescenzo, F. Macciardi, and R. Hunter

Contents

Abstract Transposable elements (TEs) are low-complexity elements (e.g., LINEs, SINEs, SVAs, and HERVs) that make up to two-thirds of the human genome. There is mounting evidence that TEs play an essential role in molecular functions that influence genomic plasticity and gene expression regulation. With the advent of next-generation sequencing approaches, our understanding of the relationship between TEs and psychiatric disorders will greatly improve. In this chapter, the Authors comprehensively summarize the state-of the-art of TE research in animal models and humans supporting a framework in which TEs play a functional role in mechanisms affecting a variety of behaviors, including neurodevelopmental, neuro-psychiatric, and neurodegenerative disorders. Finally, the Authors discuss recent

G. Guffanti (✉) and P. DeCrescenzo
McLean Hospital - Harvard Medical School, Belmont, MA, USA
e-mail: gguffantimasetti@mclean.harvard.edu

A. Bartlett and R. Hunter
Department of Psychology, University of Massachusetts, Boston, Boston, MA, USA

F. Macciardi
Department of Psychiatry and Human Behavior, University of California, Irvine, Irvine, CA, USA

© Springer Nature Switzerland AG 2019

Curr Topics Behav Neurosci (2019) 42: 221–246
DOI 10.1007/7854_2019_112
Published Online: 12 October 2019

therapeutic applications raised from the increasing experimental evidence on TE functional mechanisms.

Keywords Epigenomic regulation · Exapation · Neuropsychiatric disorders · Therapeutic targeting · Transposable elements

1 Introduction

Transposable elements (TEs) refer to a class of DNA sequences that can move to different locations within the genome. Current estimates are that TEs comprehensively make up to between half (Kuhn et al. 2007; Lander et al. 2001) and two-thirds (de Koning et al. 2011) of the human genome. These elements consist of low-complexity sequences that are classified into two major classes based on their mechanism to copy or move throughout the genome: retrotransposons, also called class I, and DNA transposons, class II. Class I TEs function through a copy-paste mechanism requiring an RNA intermediate, endonuclease, and reverse transcriptase for transposition (Wang and Kunze 2015). In contrast, class II TEs transpose via a cut-and-paste mechanism requiring only endonuclease activity for excision and insertion (Wang and Kunze 2015). Class I TEs are further classified into two groups based on the presence or absence of long terminal repeats (LTRs). The majority of TEs are represented by non-LTR retrotransposons that include long interspersed nuclear elements (LINEs), short interspersed nuclear elements (SINEs), and SINE-VNTR-Alu (SVAs). In particular, there are three major classes of LINEs (up to 6 K bp), LINE-1, LINE-2, and LINE-3, of which only LINE-1 retained the ability to transpose in the contemporary genome (Brouha et al. 2003). SINEs are the second most abundant class of TEs after LINEs, are relatively shorter (300 bp), and are mostly represented by the Alus family. SVA elements are structurally complex, derived from different genomic repeats, an Alu-like sequence, and a SINE with a size ranging between 600 and 4 K bp. LTR transposons are represented by endogenous retroviruses (ERVs), characterized by long sequences (up to 10 K bp), although the majority of elements are truncated. Believed to be remnants of ancient germline viral infections, ERVs are especially active prenatally and may play a critical role in neurodevelopment. Finally, DNA transposons are apparently not mobile. Their sequence includes a protein called transposase, which is bound by terminal inverted repeats (TIRs) responsible for the excision and reintegration of the elements into the genome.

With the advent of modern molecular tools, such as RNA sequencing, to detect and classify TEs on genome-wide scale, it became possible to further study their distribution and functional role. It is now apparent that TEs may play a critical role in molecular processes that influence genomic plasticity and gene expression regulation. Although further evidence is needed to more fully establish the nature and mechanisms of their functions, emerging evidence supports TEs as a novel and still largely understudied category of noncoding RNAs with putative regulatory role. In this chapter, we will review the main principles of TE origins and biogenesis and then present our current understanding of their mechanisms of function. Then, we review evidence from preclinical and clinical studies supporting a role for TEs in neuropsychiatric, neurodevelopmental, and neurodegenerative disorders. Finally, we review what is known about TEs as drug targets and their potential relevance for pharmacological treatments.

2 Origins and Biogenesis

2.1 Evolution

TEs are present in most known organisms, including prokaryotes. However, in prokaryotic organisms, they comprise a relatively small fraction of a genome, mainly containing protein-coding genes. In many multicellular eukaryotes, the opposite is true, and TEs often represent the most substantial fraction of the genome by order of magnitude (Fedoroff 2012; Goodier and Kazazian 2008). TEs are now known to be a significant influence on genome structure and evolution in most eukaryotic species (Fedoroff 2012). While TEs are present in most taxa, it is only in eukaryotic organisms that they have come to comprise a majority of the genome. In prokaryotes, the selective pressures imposed by the need for high rates of replication ensure that genomes remain small. Further, prokaryotes lack the bioenergetic capacity to support large genomes. Lane and Martin have argued that the mitochondrial endosymbiosis at the root of eukaryotic evolution permits higher ratios of energy production per gene and thus more extensive, complex genomes and cells (Lane and Martin 2010). Protein production represents the most considerable part of cellular energy budgets (Harold 1986). Mitochondrial endosymbiosis permitted the expression of roughly five orders of magnitude more protein than is possible for prokaryotes. This energetic wealth also meant that eukaryotic genomes could support many more competitive noncoding elements than was possible in prokaryotes. This would, of course, have placed pressure on eukaryotes to develop mechanisms to control these potentially parasitic elements, and that control, in turn, may have permitted the evolution of extensive regulatory systems based on noncoding RNAs (ncRNAs) derived from TEs as well as other features of complex eukaryotic genomes such as molecular epigenetic modifications of histones (Bartlett and Hunter 2018; Daskalakis et al. 2018; Fedoroff 2012; Friedli and Trono 2015; Hunter et al. 2013, 2015; Wheeler 2013).

The mitochondrial endosymbiotic event is thought to have driven the entry of type II introns, arguably the earliest group of transposon-like elements, into eukaryotic genomes. These were thought to derive from the mitochondrial genome itself and likely accompanied the transfer of most of the mitochondrial coding genome to the nuclear genome. It is likely that many other TEs also entered eukaryotic genomes during this early phase in eukaryotic genome evolution, perhaps including the LINEs, whose origin is ancient but as yet poorly understood. Given the recent finding that LINE-1 is required for the transition from the two-cell stage to the blastocyst during embryonic development in mammals (Percharde et al. 2018), it seems likely that the LINEs may have entered eukaryotic genomes around the time that multicellularity emerged about 600 million years ago (Eickbush and Jamburuthugoda 2008). Other elements appear to have derived from the host genome itself. Most SINEs derive from ribosomal or transfer RNAs. The primate Alu SINEs, for example, derive from 7SL RNAs (Ullu and Tschudi 1984). Long terminal repeat TEs and endogenous retroviruses derive from ancient retroviral infections, some relatively recent and undomesticated. Murine leukemia virus (MLV) retroelements entered the murine genome roughly 1.5 million years ago (Boeke and Stoye 1997), and intracisternal A-particle (IAP) elements, which also entered murid genomes in the past few million years, still have members capable of generating virus particles (Ribet et al. 2008).

In contrast to protein coding genes, TEs are often not conserved across even closely related taxa. Species within the same genus may have different and unique classes of TEs in their genomes. Chimpanzees, for instance, have two families of TEs not present in the human genome (Polavarapu et al. 2006). TEs are likely to be significant drivers of speciation events due to their capacity to drive both large-scale changes in chromosomal organization and to their capacity to create new genes and new chromatin landscapes. Evidence suggests that bursts of transposon activity accompany many speciation events (Belyayev 2014; Fedoroff 2012; Lane and Martin 2010; Senerchia et al. 2015). These differences mean that both the total percentage of the genome occupied by TEs and the clades of TEs differ substantially across species, contributing to pronounced differences in genome size and function (Fedoroff 2012). Similarly rates of TE insertion and deletion vary considerably between species. Drosophila shows fairly high rates of TE insertion (as high as one per meiosis) for I elements (Bucheton et al. 1992), yet the fraction of the Drosophila genome occupied by TEs is roughly 4% (Fedoroff 2012) due to the effects of selection and deletion (Eickbush and Furano 2002). In contrast, humans and mice have much larger fractions of their genomes occupied by retrotransposons (Lander et al. 2001; Mouse Genome Sequencing Consortium et al. 2002) and a similar rate of insertion to the fruit fly, though there is a much lower rate of deletion (Eickbush and Furano 2002).

2.2 *Exaptation*

While it has long been argued that transposons are either genomic parasites or junk, the evidence is accumulating that they have acquired novel functional roles in a number of instances in many if not most multicellular genomes. This type of phenomenon is usually referred to as "exaptation," a terminology initially introduced to describe the acquisition of a new function of a trait during evolution and then used for DNA sequences such as TEs (Gould and Vrba 1982). The most persuasive argument that most of the genome is junk was articulated by Susumu Ohno's calculation that the genome could contain on the order of 20,000 selectable loci, based on observed rates of deleterious mutations (Ohno 1972). However, the finding that more than 80% of the genome is transcribed raised doubts as to the extent that Ohno's hypothesis was correct, at least beyond those parts of the genome that code for protein (ENCODE Project Consortium 2012). The underlying assumptions of Ohno's argument are founded in the biochemical constraints imposed by the genetic code, which requires a high level of sequence fidelity in order to avoid producing a codon which codes for an inappropriate amino acid and thus causes a defect in the structure of the translated protein. RNA genes, which do not have this constraint, would not necessarily suffer to the same degree from a single point mutation. Indeed, it appears that ncRNA genes are under selection pressure for structure, rather than sequence (Nitsche and Stadler 2017), and some estimates have put the number of selectable ncRNA genes as high as four million (Smith et al. 2013). As roughly 80% of confirmed ncRNA sequence is TE-derived, it follows that much if not most expressed TEs may be functional (Kelley and Rinn 2012). While this likely represents a high bound on the number of functional ncRNA elements in the genome, it nonetheless represents an essential argument for the function of TEs specifically and ncRNA in general (Bartlett and Hunter 2018; Daskalakis et al. 2018).

We have argued elsewhere that the transposome represents an "adaptive toolkit" for metazoan genomes, perhaps providing a replacement for the extensive lateral gene transfer present in prokaryotic organisms (Bartlett and Hunter 2018; Hunter et al. 2015; Lapp and Hunter 2016; Sonea and Mathieu 2000). Though at present the extent to which the transposome is functional is a major undecided question in biology, it is clear that TEs have been exapted to useful purpose in a number of cases. The V(D)J recombination system responsible for the generation of antibody diversity in the mammalian immune system is the result of the domestication of retrotransposons to useful purposes (Carmona and Schatz 2017), and the placenta also derives from exapted transposons (Dupressoir et al. 2012; Frank and Feschotte 2017; Lynch et al. 2011, 2015). Within the nervous system, LINE-1 retrotransposons have been shown to contribute to neuronal diversity during development and adult neurogenesis (Erwin et al. 2014) and have been shown to be regulated in part by environmental stress and maternal care (Bedrosian et al. 2018; Muotri et al. 2009). The mouse B2 SINE, which acts as a transcriptional regulator through a direct interaction with RNA polymerase II, is regulated in the brain by both stress and heat shock (Hunter et al. 2012; Kugel and Goodrich 2006; Ponicsan et al. 2010). Perhaps the most persuasive argument for the functional importance of retrotransposon RNA comes not from the brain but from studies of early development. LINE-1 appears to be required for successful transition from the 2–4 cell stages of embryonic development (Percharde et al. 2018), noted above. Whether or not this requirement for TE expression is not established for other developmental stages, it seems likely that the transposome will prove to play a role in more, rather than less, of our biology, the more it is explored.

3 Mechanisms of Function

Barbara McClintock first described DNA transposons as "controlling elements" regulating the "time and type of activity of individual genes" (Mcclintock 1956). While initially disregarded as "junk," the regulatory properties of TEs are recently of increased focus (Chuong et al. 2017). Since their initial observation, much evidence has accumulated suggesting exaptation of TEs both at the level of genomic structure and transcriptional regulation. Recent studies describe the role of TE in influencing genomic plasticity as well as their exaptation into transcriptional regulators such as new promoters, enhancers or insulators, and sources of new transcription factor binding sites (TFBS), respectively.

Genomic Plasticity Despite the genome's set up mechanisms to regulate TE activity, some TEs escape repression and introduce new insertions both in germline cells during embryonic development and in somatic tissues later in life. While the majority of LINE-1 referenced in the genome annotation is retrotranspositionally inactive, there are about 100 retrotransposition-competent LINEs in an average diploid human genome (Brouha et al. 2003). The rate of LINE-1 retrotransposition in humans is between 1/270 and 1/95 births (between 0.3 and 1%, respectively) (Ewing and Kazazian 2011). New TE insertions can affect genome integrity in several different ways (Cordaux and Batzer 2009; Guffanti et al. 2014). First, TE insertions in exonic regions can often lead to gene dysregulation and damage (Wang and Kunze 2015). Second, TEs insertion can create new exons or new genes by carrying along non-retrotransposon DNA sequences (Sela et al. 2010). Third, TEs can lead to a phenomenon called transduction, which refers to the duplication of the 3′ or 5′ flanking sequence along with the new copy of the original TE (Goodier et al. 2000; Pickeral et al. 2000; Tica et al. 2016). Fourth, TEs can induce LINE-1 nonallelic homologous and nonhomologous recombination events resulting in significant duplication and deletions of genomic DNA sequences (Campbell et al. 2014; Startek et al. 2015). Finally, TEs have been implicated in the generation of microsatellites from homopolymeric genomics tracts (Ahmed and Liang 2012). Overall, the role of TE in influencing genomic plasticity has been recognized as one of the major driving forces of the emergence of copy number variations (CNV) (Bose et al. 2014), whose effects have been widely associated to diseases of the nervous system (Kirov 2015).

Transcriptional Regulation While not yet fully elucidated, an emerging role for TEs in regulation of biological processes is increasingly evident. Beyond the genomic effects of these elements, a role for the RNA of class I elements is emerging (de Souza et al. 2013; Long et al. 2017). Several possible mechanisms for retrotransposon RNA in transcriptional regulation have been hypothesized. These include RNA recruitment of protein complexes and RNA inhibition by acting as a "decoy" (i.e., transcription factor "sink"), indirectly through the act of transcription itself and through nucleosome organization. Central to this type of regulatory mechanisms is the binding of transcription factor (TF) proteins to specific functional sequences, also referred to as "motifs," within TEs. The presence of functional motifs within TEs has been documented for over 20 years (Brosius 2003). In each case, class I TEs may play a critical role in rewiring the complex dynamic of transcriptional regulation.

Promoters Derived from TEs Most TEs contain *cis*-regulatory sequences that function as promoters. In addition to providing the elements recognized by host cell machinery to control TE transposition (Chuong et al. 2017), these sequences have the potential to be exapted as promoters for neighboring genes. For example, LTRs, which are HERVs promoter sequences flanking the coding elements essential for viral replication, have the ability to recruit RNA polymerase II (Pol II), which is the enzyme essential to initiate transcription (Mager and Stoye 2015). Fully autonomous LINEs capable of retrotransposition feature Pol II promoter sequences in the 5′ UTR as well as an antisense promoter. Similarly, SINEs which are thought to have evolved from Pol III-transcribed genes contain Pol III-specific recruitment sequences (Batzer and Deininger 2002). For example, active human Alu SINE elements contain both a Pol III promoter and Pol III-specific stop sequence (Batzer and Deininger 2002; Chu et al. 1995). These sequences may drive the expression on neighboring genes once exapted into alternative promoters (Richardson et al. 2015; Speek 2001; Swergold 1990). SVAs represent an evolutionarily young, primate-specific, composite retrotransposon featuring elements of Alu SINE, R-SINE, and VNTRs (Lapp and Hunter 2016; Shen et al. 1994). Beyond structural homology, hallmarks of SVAs include a polyadenylation signal at the 3′ end as well as recognition sequences for LINE-1 endonuclease suggesting that LINE-1 activity likely facilitates SVA abundance (Hancks et al. 2011). Interestingly, these elements appear relatively intact compared to LINE-1 elements which largely feature 5′ truncation (Wang et al. 2005).

TEs appear to have coevolved with host systems to respond to environmental stimuli including heat shock, infection, and stress (Hunter et al. 2012; Karijolich et al. 2015; Steller and Pirrotta 1986). Interestingly, single-cell RNA-sequencing has shown that B2 SINE expression increases following neuronal activation in vitro (Lacar et al. 2016). The heat shock response induces large changes in the transcription upregulating heat shock proteins while downregulating expression of housekeeping genes (DiDomenico et al. 1982; Findly and Pederson 1981; Gilmour and Lis 1985; O'Brien and Lis 1993). B2 SINE expression is increased as much as 40-fold as part of this response (Fornace and Mitchell 1986; Li et al. 1999; Liu et al. 1995). When B2 expression is inhibited, the heat shock response becomes dysregulated (i.e., housekeeping genes remain upregulated) (Allen et al. 2004). B2 SINE accumulation appears to be permissive for selectively downregulating Pol II-driven expression (Espinoza et al. 2004). This group also found that B2 RNA directly interacts with Pol II through the 3′ stem loop and inhibited transcription via single-stranded regions (Espinoza et al. 2007). Interestingly, Alus, which are similarly upregulated in response to heat shock, also appear to bind Pol II and inhibit transcription (Liu et al. 1995; Yakovchuk et al. 2009).

Despite the large fraction of the genome occupied, many of the TEs scattered throughout the genome are truncated and incapable of replication; yet the implications of the regulatory elements they still possess remain not fully realized (Feschotte et al. 2002; Lander et al. 2001).

Enhancers Derived from TEs TEs can be exapted into enhancers by donating sequences that can regulate the expression of neighbor genes. Enhancers are constituted by an array of transcription factor binding sites (TFBS) and are usually located far away from the promoter region of their target gene. Also defined as distal regulatory regions, enhancers regulate the expression of genes by interacting with transcription factors. The first evidence of TE exaptation into enhancers started to appear in the early 1990s and implicated mostly SINE (Hambor et al. 1993; Santangelo et al. 2007) and LINE (Yang et al. 1998). More recently, genome-wide investigation allowed to better characterize the TE-derived landscape of enhancers. Marino-Ramirez and colleagues found that TE sequences reside in 23% of regions of DNaseI-hypersensitive sites (DHS), which indicate regions of open chromatin involved in the regulation of transcription typical of promoters and enhancers. TE sequences in these regions displayed a high degree of correlation with nearby genes in CD^{4+} T cells (Mariño-Ramírez and Jordan 2006). Furthermore, genome-wide histone modification analysis in two cell types from the ENCODE project predicted the locations of thousands of TE-derived enhancers, with a minimum (~3%) overlap between the two cell types (Huda et al. 2011). These findings indicate that TE-derived sequences have the potential to be exapted into enhancers in a cell-type specific way and that most likely the enhancer-like function is mediated by epigenetic regulatory mechanisms. Another study by the same group of researchers showed that MIRs, one of the most ancient classes of TEs, were evolutionary co-opted into enhancers with a wide range of TFBS sequence motifs, including those associated with transcription factors C-JUN, NF-E2, and ZNF274 (Jjingo et al. 2014). A few, but consistent findings, revealed that HERVs might act as enhancers of protein-coding gene expression, for example, ERVs of the family -H have been shown to act as scaffold to recruit p300 and OCT4 to drive the expression of neighboring genes in the maintenance of stem cell identity (Lu et al. 2014); the ERV sequence at the promoter region of *PRODH* gene, previously associated with schizophrenia, has been shown to upregulate its expression when found in an under-methylated state (Suntsova et al. 2013).

Insulators Derived from TEs TEs can become insulators of a nearby transcriptional unit either by preventing the spread of heterochromatin from one locus to another or by acting as enhancer blockers. Perhaps the most persuasive example of this mechanism is provided by B2 SINE acting as transcriptional as well as epigenetic roadblocks at the murine growth hormone (GH) locus, silencing its expression (Hambor et al. 1993; Santangelo et al. 2007). The mouse growth hormone (GH) gene does not contain a canonical locus control region but is flanked by a B2 SINE element. Deletion of the B2 SINE element upstream of the GH transcription start site suggested that the B2 SINE was necessary for developmentally regulated GH expression. During mouse embryogenesis, levels of H3K9me3 proximal to GH locus in the pituitary decline while H3K9me2 levels increase. Specifically, H3K9 valence transitions at the B2 SINE element upstream of the GH gene. This B2 SINE features flanking Pol II and Pol III promoters that appear permissive for overlapping transcription of the element. Pol III transcription appears to facilitate euchromatic H3K14 acetylation, while Pol II transcription prevents H3K9me3 maintenance (Lunyak et al. 2007). This demonstrates a dynamic role for the B2 SINE element in both acting as a Pol II transcriptional roadblock and regulating local chromatin structure.

New Regulatory Regions Derived from TEs Retrotransposon RNA inhibition of transcription factor binding has not yet been determined. However, a number of emerging lncRNAs have been shown to act as "decoys" for protein regulators of transcription (Long et al. 2017). For example, the lncRNA Gas5 acts as a "riborepressor" through interaction of its glucocorticoid response element (GRE) with the DNA-binding domain of the glucocorticoid receptor (Kino et al. 2010). The GRE contained within the Gas5 RNA serves to sequester available receptors from genomic interaction. Given the abundance of transcription factor motifs found in class I TEs, it stands to reason that transcription factors bind similarly to retrotransposon RNA. As methods for preserving RNA-protein immunoprecipitation continue to improve, it is likely that "decoy" TE RNAs will be identified (Mili and Steitz 2004).

4 Preclinical Studies

Evidence for associations between TE activity and human brain disorders is fairly extensive (Guffanti et al. 2014; Reilly et al. 2013), but data from preclinical models of these diseases remains relatively sparse.

Stress contributes to many human neuropsychiatric diseases, and several studies have demonstrated stress-induced regulation of TEs in the brain (Hunter et al. 2013). Evidence is emerging that stress, maternal behavior, and exercise alter rates of transposition in the mouse hippocampus (Bedrosian et al. 2018; Muotri et al. 2009). Exposure to acute stress drives downregulation of IAP-LTR and B2 SINE in the rat hippocampus and upregulation of IAP-LTR and LINE-1 in the cerebellum. Further, in the hippocampus, global increases in the repressive histone mark H3K9me3 was observed across a large number of retrotransposon loci (Hunter et al. 2012). Recently, it was observed that stress increases the N6 methyladenine modification in mouse prefrontal cortex, roughly half of which seemed to be targeted to LINEs and associated with a decline in RNA expression from those elements (Yao et al. 2017). Cappucci and colleagues also observed strain-specific changes in LINE-1 expression in the mouse hippocampus after 5 days of repeated stress (Cappucci et al. 2018). Another study found that at least a subset of LINEs are also transcribed under normal conditions in the rat hippocampus, suggesting a role in nonstress neuronal physiology as well (Mukherjee et al. 2016). As noted in the section on exaptation, heat shock upregulates B2 SINE RNA expression in the mouse brain (Kugel and Goodrich 2006).

Cortical spreading depression (CSD) is associated with disorders such as migraine and epilepsy. In rodent models, CSD has been associated with changes in both DNA methylation and H3K9me3 of LINE1 in cortical genomes (Drongitis et al. 2016; Rana et al. 2012).

The possibility that retrotransposons play a role in psychotic disorders dates to at least the 1980s (Crow 1987), and the idea has received increasing empirical support with the advent of modern genomic approaches. While animal models of psychotic disorders have substantial limitations (Kesby et al. 2018; Mattei et al. 2015), there is nonetheless preclinical evidence supporting a role for retroelements in the pathogenesis of psychosis. One example is the finding that the insertion of a MusD retroelement into the *Slc6a5* gene induces a behavioral phenotype similar to that of other rodent schizophrenia models (Bogdanik et al. 2012). Overexpression of HERV-W env in U251 glioma cells alters expression of two genes associated with risk of schizophrenia, brain neurotrophic factor (*BNDF*) and the dopamine D3 receptor (*DRD3*) (Huang et al. 2011). Exogenous infections with cytomegalovirus and *T. gondii* can activate HERV-W transcription in cultured fibroblasts and neurons, respectively (Frank et al. 2006; Nelson et al. 1999). Interestingly, valproic acid treatment increases the expression of HERV elements in human cell lines, perhaps via its effects on histone acetylation (Diem et al. 2012).

TEs have also been proposed to play a role in neurodegenerative disorders. In Drosophila models of ALS and frontotemporal dementia, transposon activation via TDP-43 mediated suppression of small-RNA-mediated silencing (Krug et al. 2017; Li et al. 2012, 2013). These findings have found further support from work in both human cells and chicks (Li et al. 2015; Voigt et al. 2010). Deletion of the Uhrf1 gene in the developing mouse cerebellum results in severe neurodegeneration due to overexpression of the murid IAP endogenous retrovirus (Ramesh et al. 2016). Interestingly, mercury, which can induce neurodegeneration, causes increases LINE-1 expression in human neuroblastoma lines (Habibi et al. 2014). Perhaps most significantly, dysregulation of Alu expression has been demonstrated to mechanistically contribute to macular degeneration in both humans and model systems (Kerur et al. 2013; Tarallo et al. 2012). The latter studies demonstrate that disruption of small-RNA-mediated silencing of SINE RNA via downregulation of Dicer-1 expression results in degeneration via an NLRP3 inflammasome-mediated pathway.

These findings support the idea that TEs have a role to play in a variety of brain disorders. The observational data in patient populations is increasing in depth and strength as well. However, much further research is needed to build strong preclinical models of TE-driven pathology if we are to develop a clear mechanistic understanding of diseases induced by TEs.

5 TEs in Development and Disease

There is an emerging link between TEs and psychiatric disorders. TEs have been implicated in (1) neuropsychiatric disorders, such as schizophrenia (Weis et al. 2007), bipolar disorder (Weis et al. 2007; Yolken et al. 2000), major depressive disorder (MDD) (Weis et al. 2007), post-traumatic disorder (PTSD) (Rusiecki et al. 2012), and substance use disorders (Okudaira et al. 2014; Ponomarev et al. 2012); (2) neurodevelopmental disorders, such as attention deficit hyperactivity disorder (ADHD) (Balestrieri et al. 2014; D'Agati et al. 2016), autism (Balestrieri et al. 2012), and Rett syndrome (Muotri et al. 2010); and (3) neurodegenerative disorders, such as Alzheimer's disease (Huang et al. 2006), frontotemporal lobar degeneration (FTLD) (Li et al. 2012), ataxia telangiectasia (Coufal et al. 2011), and X-linked dystonia parkinsonism (Hancks et al. 2011; Makino et al. 2007). Here, we discuss the most significant findings.

5.1 Neuropsychiatric Disorders

Schizophrenia Multiple studies have shown differences in the expression of TEs both in neural tissue, blood, and cerebrospinal fluid (CSF) between people with schizophrenia and healthy controls. Yolken et al. (2000) first discovered a link between ERVs and schizophrenia (Yolken et al. 2000). Specifically, they found an increase in MSRV (multiple sclerosis retrovirus), homologs of ERVs initially identified in the body fluids of individuals with multiple sclerosis and in the postmortem frontal cortical tissue taken from people with schizophrenia, relative to healthy controls. This finding was followed by the detection of increased levels of human ERV-W by the same group of researchers, both in postmortem brain tissues and CSF (Karlsson et al. 2001). In an additional follow-up study, the researchers found a statistically significant increase in levels of the *gag* protein of ERV-W in the plasma of people with schizophrenia-spectrum disorders (schizophrenia, schizoaffective psychosis, and schizophreniform disorder) compared to healthy controls (Karlsson et al. 2004). Several other studies reported on the increased transcriptional activation of different types of ERVs in schizophrenia patients, like HERV-K10 in neural tissue (Frank et al. 2005), HERV-W in blood (Huang et al. 2006, 2010; Perron et al. 2012), and in postmortem anterior cingulate gyrus (ACG), and in many hippocampal structures (Weis et al. 2007). In a follow-up study, Huang et al. (2011) transfected human glioma cells with pCMV-*env*, a plasmid containing an HERV-*env* sequence. They found that increases in the copy number of HERV-W transcripts were associated with increased expression of BDNF, DRD3, NTRK2, and CREB in vitro (Huang et al. 2011). This finding represents the first evidence suggesting that retroviral RNA can impact the expression of genes involved with the development of schizophrenia. All of these studies were conducted focusing on one specific family/type of TEs, based on a study design characteristic of candidates selected a priori by the researchers.

With the advent of new technologies, and specifically next-generation whole genome-sequencing, it was finally possible to extend the investigation to the entire genome of schizophrenia patients in different tissues and in some cases even in single neurons. Bundo et al. (2014) found an increase in LINE-1 insertions in neurons from postmortem prefrontal cortex of schizophrenia patients and in induced pluripotent stem cell (iPSC)-derived neurons containing 22q11 deletions, a known genetic risk factor for the disease (Bundo et al. 2014). Many of these insertions were mapping to genes related to synapse formation and to genes whose mutations are associated with schizophrenia. Further, using animal models, they demonstrated that environmental triggers, such as perinatal environmental risk factors, can lead to increase LINE-1 retrotransposition. This finding corroborates the hypothesis that a prenatal genetic vulnerability might be exacerbated by an environmental risk exposure later in life, also known as the two-hit model of schizophrenia. LINE-1 insertions in schizophrenia-relevant genes were also detected in the blood DNA in a schizophrenia-affected family (Guffanti et al. 2016). This was the first evidence of Mendelian inheritance of LINE-1 insertions, very similar to any other structural genomic variant. Most of these insertions were human-specific LINE-1 (L1HS) insertions, mapping ubiquitously within and outside known genes and ncRNAs. Doyle et al. (2017) found an increase in intragenic LINE-1 insertions in the DNA derived from postmortem dorsolateral prefrontal cortical tissue taken from individuals with schizophrenia compared to healthy controls (Doyle et al. 2017). These LINE-1 insertions were primarily found within genes related to the postsynaptic membrane and to cell projections, suggesting that LINE-1 insertions in these genes may lead to a disruption of the development or function of these neuronal structures (Doyle et al. 2017).

It is worth noting that most of the patients with schizophrenia-spectrum disorders who take part in these studies have been on psychotropic medication at some point in their lives and that psychotropic medications have been shown to be among the potential triggers of TE expression. In conclusion, the evidence reviewed so far support the hypothesis that TEs such as LINE-1s and human ERV-Ws contribute to the development of schizophrenia. It is also possible that an exogenous infection, such as a virus or infection, serves as a trigger for both inflammation and endogenous retroviral activity in the pathogenesis of schizophrenia (Karlsson et al. 2004; Leboyer et al. 2013).

Bipolar Disorder The link between TEs and bipolar disorder (BP) was first noted by Yolken et al. (2000), who found an increase in the expression of human ERV-K in postmortem frontal cortical tissue taken from patients with bipolar disorder relative to healthy controls (Yolken et al. 2000). Frank et al. (2005) also found an increase in HERV-K10 expression in neural tissue from patients with BP (Frank et al. 2005). Weis et al. (2007) found that people with BP displayed increased expression of the *gag* protein of human ERV-W compared to controls in many hippocampal structures (Weis et al. 2007). Perron et al. (2012) found increased expression of the *env* protein of human ERV-W in people with BP relative to controls, coupled with a lower genomic copy number of the *env* protein (Perron et al. 2012). Interestingly, subjects with BP had increased *env* expression both in relation to healthy controls and to patients with schizophrenia. Given that environmental input can trigger *env* expression, the combination of its increased expression and reduced copy number in patients with BP could be related to an infectious or viral trigger, which could lead to increased activation of this TE.

MDD and PTSD Weis et al. (2007) found decreased expression of human ERV-W (*gag* protein) in the postmortem anterior cingulate gyrus, as well as in the CA2 interneurons, axons, and dendrites of the hippocampus of major depressive disorder (MDD) patients compared to healthy controls (Weis et al. 2007). Rusiecki et al. (2012) assessed the overall methylation of LINE-1s and Alus in US soldiers before and after deployment (Rusiecki et al. 2012). Cases were defined as soldiers who developed PTSD after deployment; controls were defined as soldiers who did not develop PTSD after deployment. Soldiers with any other psychiatric disorder were excluded from this study. There was no difference in serum LINE-1 methylation between cases and controls before deployment. After deployment, LINE-1 methylation in controls increased, while it stayed at baseline levels in cases. The authors speculate that the increase in LINE-1 methylation, and probable consequent decrease in LINE-1 expression, under extreme stress serves as a protective factor for the development of PTSD. Before deployment, cases had greater Alu methylation than controls. After deployment, cases had increased methylation relative to their pre-deployment levels and relative to control post-deployment levels. Although these results weren't statistically significant, they are worth noting as this represents a first-of-its-kind finding. This study indicates the possibility that serum Alu hypermethylation and LINE-1 hypomethylation might increase vulnerability to the effects of psychological stress, while Alu hypomethylation and LINE-1 hypermethylation might increase resilience in the face of it.

Substance Use Disorders Ponomarev et al. (2012) analyzed TEs in postmortem amygdala and superior frontal cortex taken from individuals with alcohol use disorder (AUD) and healthy controls (Ponomarev et al. 2012). They found an increase in the expression of long terminal repeats (LTRs), which represent the promoter sequences of human ERVs, in patients with AUD, coupled with a decrease in methylation. Overall, this evidence suggests that the increase in LTR expression might be related to reduced gene silencing. While it is possible that human ERV hypomethylation is a predisposing factor for the development of AUD, previous research suggests that such hypomethylation could also be a consequence of alcohol use. The authors of the study also noted that human ERV overexpression might increase the levels of the human ERV-encoded genes and the envelope glycoprotein gene (*env*), which have been associated with neurotoxic effects of alcohol, including cytotoxicity to oligodendrocytes and myelin degeneration.

Another study links TEs to neurotoxicity, in which LINE-1 retrotransposition is induced by several neurotoxic drugs. Okudaira et al. (2014) found that treating human neuroblast cell lines with methamphetamine caused LINE-1 retrotransposition

(Okudaira et al. 2014). There was no LINE-1 retrotransposition in non-neuronal cells in response to methamphetamine. Cocaine also induced LINE-1 retrotransposition in neuronal cell lines. In another experiment, D4T, a compound that inhibits reverse transcriptase, prevented LINE-1 retrotransposition when neuronal cell lines were exposed to methamphetamine and to cocaine. Additionally, the authors demonstrated that neither methamphetamine nor cocaine induced LINE-1 retrotransposition when CREB functional activity was silenced via specific siRNA (Okudaira et al. 2014). This finding suggests that CREB phosphorylation is also critical for LINE-1 retrotransposition in response to these drugs. Further experiments indicated that the open reading frame 1 (ORF1) of LINE-1s is recruited and translocated to chromatin when neuronal cell lines are exposed to methamphetamine and cocaine. The application of siRNAs that block CREB phosphorylation also prevented ORF1 translocation to chromatin. This study demonstrates that methamphetamine and cocaine can serve as physiological stressors that impact the activity of TEs through various mechanisms. Doyle et al. (2017) found de novo LINE-1 insertions in postmortem dorsomedial prefrontal cortex from individuals who had been addicted to cocaine (Doyle et al. 2017). Many of these LINE-1 insertions were associated with genes involved with cocaine addiction, leading to the hypothesis that these insertions may predispose people to develop cocaine addiction.

5.2 *Neurodevelopmental Disorders*

Autism and Rett Syndrome In 2010, Muotri et al. (2010) were the first to show that LINE-1 introduces new insertions in neuronal progenitor cells (NPCs) during neurodevelopment (Muotri et al. 2010). The authors induced pluripotent stems cells (iPSC) from fibroblasts of children with Rett syndrome (RTT) with a particular frameshift mutation in the MeCP2 gene and observed increased LINE-1 retrotranposition compared to NPCs derived from a healthy control. Additionally, increased LINE-1 ORF1 expression was found in postmortem brain tissue taken from individuals with RTT relative to healthy controls.

Balestrieri et al. (2012) found increased expression of human ERV-H and reduced human ERV-W in the blood of children with autism spectrum disorder (ASD) relative to healthy controls (Balestrieri et al. 2012). Within people with ASD, elevated human ERV-H expression was significantly associated with severely impaired communication and motor skills, as measured by a psychometric test of these constructs. Extending research on TEs from peripheral tissue to central nervous system tissue, Shpyleva et al. (2018) found an increase in LINE-1 expression in postmortem cerebellar tissue taken from children with autism relative to healthy controls (Shpyleva et al. 2018). This increased expression was associated with reduced DNA repressive mechanisms (reduced histone H3K9me3 in ORF1 and ORF2, both of which are overexpressed in cerebellar tissue taken from individuals with autism). In this study, increased LINE-1 ORF1 and ORF2 expression was linked to increased oxidative stress (as indicated by a reduction in glutathione/GSSH redox status), which is one potential trigger for LINE-1 retrotransposition (Shpyleva et al. 2018).

Attention Deficit Hyperactivity Disorder (ADHD) Relative to healthy controls, Balestrieri et al. (2014) found elevated expression of the envelope glycoprotein gene (*env*) of human ERV-H in blood samples from children with ADHD who had never taken any psychiatric medication, suggesting that there might be an inflammatory and/or neurotoxic component to ADHD that may be influenced by TE overexpression (Balestrieri et al. 2014). D'Agati et al. (2016) conducted a follow-up study in which an adolescent male with ADHD was treated with methylphenidate, a brain stimulant used to treat ADHD (D'Agati et al. 2016). Since methylphenidate increases availability of dopamine and norepinephrine in the brain, it is possible that this is the primary mechanism through which methylphenidate treats the symptoms of ADHD. However, it is also possible that methylphenidate impacts ADHD through inducing epigenetic changes that may or may not be related to neurotransmitter availability. After 6 months of treatment with methylphenidate, D'Agati and colleagues found a significant reduction in symptoms of ADHD and in human ERV-H env expression in this patient (D'Agati et al. 2016). Expression of human ERV-H env in this patient was compared to that in a small sample of healthy controls, and the patient's posttreatment level was below that of the healthy controls. This suggests that methylphenidate has the potential to strongly downregulate human ERV-H env expression. This downregulation could impact expression of genes typically associated with ADHD. It could potentially also lead to a reduction in inflammation and/or neurotoxicity, which could also be related to the reduction in symptoms.

5.3 *Neurodegenerative Disorders*

Alzheimer's Disease TEs transcriptional activity is usually under epigenetic control to monitor their potentially harmful reactivation. Several studies revealed that brain aging might lead to deteriorating TE surveillance. In Drosophila transgenic models of Alzheimer's disease (AD), this has been shown to lead to retrotransposon activation (Li et al. 2013; Maxwell et al. 2011; Wood et al. 2016). Despite the corroborating evidence from an established fly model relevant to AD, TE activation has not been extensively studied in humans up until very recently. A recent study examined TE profiles in 600 human cortical postmortem transcriptomes and found that TEs are substantially reactivated in AD patients' brains (Guo et al. 2018). The analysis of TE profiles in the postmortem dorsal lateral prefrontal cortex (DLPFC) of AD revealed that nine different types of TEs (regardless of their relative copy number or specific location within the genome), belonging to different classes and families, were significantly associated with the overall histologic count of protein Tau-derived neurofibrillary tangles, perhaps the most significant hallmarks of AD. This finding, corroborated by an experiment in a transgenic Drosophila model, leads to the hypothesis that chromatin relaxation induced by Tau protein might be responsible for derepression of silenced TEs. Although further evidence should be accumulated, this is the first evidence shedding some light on the potential causal link between AD Tau pathology and TE reactivation in human brains.

Frontotemporal Dementia (FTLD) As highlighted in a study focusing on neurodegenerative disorders, there is a putative causal link between TE overexpression and TAR DNA-binding protein 43 (TDP-43), hnRNP-like RNA-binding protein implicated in numerous cellular functions including alternative splicing and regulation of mRNA stability. The accumulation of TDP-43 is a hallmark of several neurodegenerative disorders, such as AD and frontotemporal lobar degeneration (FTLD). The study, a meta-analysis of cross-linking immunoprecipitation sequencing (CLIP-seq) datasets in mouse and human investigating the RNA targets of TDP-43 protein, revealed a significant enrichment in target sequences that derive from many classes of TEs in healthy controls compared to FTLD patients (Li et al. 2012). These findings support a possible mechanism of TE silencing by which TDP-43 would normally function to regulate TE expression.

Ataxia Telangiectasia Ataxia telangiectasia is a primary immunodeficiency disease whose patient displays progressive neurodegeneration and eventual death in the second or third decade of their life. AT is a rare inherited disorder and is inherited as autosomal recessive pattern. The causal gene has been identified, ATM, a serine/threonine kinase that is a sensor of cellular DNA damage. Mutations that deactivate the gene result in AT. In 2011, Coufal and colleagues demonstrated an increase in LINE-1 retrotransposition in cells that contained reduced levels of the mutant copies of the ATM genes (Coufal et al. 2011). The researcher further observed an increase in LINE-1 copy number in postmortem brain tissue derived from AT patients. These findings supported the unprecedented hypothesis that LINE-1 might play a role in modulating the biological pathways revolving around repair of DNA damage.

X-Linked Dystonia Parkinsonism One of the few examples of disease involving a TE of the SVA family comes from the rare disorder of X-linked dystonia parkinsonism, a neurodegenerative movement disorder characterized by adult-onset parkinsonism and dystonia. Thanks to their sequence fidelity, SVA elements can dynamically affect genomic stability by facilitating inversions, duplications, and deletions leading the allelic imbalance (Hancks et al. 2011). As such, SVA elements are highly regulated by epigenetic factors (e.g., DNA methylation) which can affect the expression of neighboring genes. In the case of X-linked dystonia parkinsonism, SVA insertion and hypermethylation of the disease causal *TAF1* gene led to its reduced expression in the caudate (Makino et al. 2007).

6 Therapeutic Targeting of Transposable Elements

As reviewed more extensively above, TEs may have the potential to affect the expression of other genes through a variety of mechanisms. The ability of TEs to influence gene expression and protein production makes them potential targets for drug development. Although the development of TE-associated drugs is still in its infancy, we review here, as proofs of principle, some examples derived from cancer and immune system disorders, in which this approach has already been successful. In contrast to single nucleotide or structural variation, reference TEs are shared by all patients, offering the advantage that they could be applied to entire patient populations.

The most experimentally interesting example is the therapeutic use of an antibody to target the expression of human ERV-W-Env in type 1 diabetes (T1D). Epidemiological studies revealed that pathogenicity is mediated by abnormally expressed human ERV-Env proteins in the blood and postmortem acinar pancreatic cells of T1D patients (Curtin et al. 2018). A monoclonal antibody (IgG4) has been developed to target and neutralize human ERV-W-Env in vitro and in vivo specifically. This pharmacological approach is currently under clinical development in T1D. If further stages of clinical efficacy will be successful, these results may lead to the development of innovative treatments for a disease associated with dysregulation of human ERVs expression.

Another example is the therapeutic use of inhibitors of reverse transcriptase (RT), the enzyme encoded as part of the open reading frame 2 of LINE-1 elements. RT has recently been proposed as a target for anticancer drugs, specifically prostate cancer. The mechanism involves its inhibition either via RNA interference-dependent signaling of active LINE-1 elements or using RT inhibitory drugs. As a result, the proliferation of cancer cells was significantly reduced, promoting their differentiation and antagonizing tumor progression in animal models (Houede et al. 2018; Sciamanna et al. 2016).

The effect of drugs that function as DNA methyltransferase (DNMT) inhibitors, generally used for epigenetic cancer therapy, on TEs is remarkable. These drugs are used to reactivate epigenetically silenced tumor suppressor genes. Unfortunately, not much is known about the sequence specificity of such a TE/tumor suppressor gene mechanism, and a possible emerging side effect could be the pharmacologically induced reactivation of epigenetically silenced TEs. A study revealed that the demethylation, and reactivation, of a LINE-1 antisense promoter, was able to drive the transcription of neighboring genes, even if the promoter of the gene itself was accurately silenced by the DNMT inhibitors (Weber et al. 2010). This example opens up a critical role of TEs in epigenetic therapy, whether in consideration of side effects or, hopefully in the future, of specific drug approaches designed on TE prerogative to provide alternative mechanisms to reactivate the transcription of particular genes.

These few examples prove that TEs may be experimentally pursued as potential therapeutic targets. We are only beginning to understand that there are many possible mechanisms for using TEs to manipulate gene expression and pathophysiological pathways. Once we can identify TEs involved in disease, their sequence could be used to design oligonucleotides that can inhibit or enhance the expression of a protein and modulate the symptoms of the disease, depending on the direction of the effect of the mechanism implicated.

7 Future Directions and Conclusions

The experimental evidence reviewed in previous sections of this chapter supports a role for TEs RNA-mediated epigenetic regulation and in the etiopathogenesis of neuropsychiatric diseases. TEs involved tend to include both long and short elements, with the most persuasive evidence pointing to insertional mutagenesis by LINE-1s and dysregulated transcriptional activity of HERVs. Much of the data available to date are still mostly anecdotal and merely descriptive. An ideal experiment would demonstrate the mechanism by which TEs mediate epigenetic effects and would show how these effects would be inhibited in the absence of the candidate TE. While such tests still present a lot of challenges, these are undoubtedly critical to unequivocally uncovering the functional mechanisms of TE regulatory function, and would allow the field to move forward. These data are especially relevant given the growing interest in the idea that the exposure to external factors, both biological and psychological stressors, can influence individuals' well-being and that TEs can mediate these processes. Few areas of biological science are more thought-provoking than the field of TEs. The underlying notion that TEs may be something other than silent "junk" DNA sequences runs counter to what was previously believed by scientists not long ago. Progress in this field now has the opportunity to advance quickly, almost as abundantly as controversies and debates. Preliminary evidence supports the increasing fascination with the many applications that arise from an increased understanding of TEs functional mechanisms.

References

Ahmed M, Liang P (2012) Transposable elements are a significant contributor to tandem repeats in the human genome. Comp Funct Genomics 2012:1–7. https://doi.org/10.1155/2012/947089

Allen TA, Von Kaenel S, Goodrich JA, Kugel JF (2004) The SINE-encoded mouse B2 RNA represses mRNA transcription in response to heat shock. Nat Struct Mol Biol 11(9):816–821. https://doi.org/10.1038/nsmb813

Balestrieri E, Arpino C, Matteucci C, Sorrentino R, Pica F, Alessandrelli R et al (2012) HERVs expression in autism spectrum disorders. PLoS One 7(11):e48831. https://doi.org/10.1371/journal.pone.0048831

Balestrieri E, Pitzianti M, Matteucci C, D'Agati E, Sorrentino R, Baratta A et al (2014) Human endogenous retroviruses and ADHD. World J Biol Psychiatry 15(6):499–504. https://doi.org/10.3109/15622975.2013.862345

Bartlett AA, Hunter RG (2018) Transposons, stress and the functions of the deep genome. Front Neuroendocrinol 49:170–174. https://doi.org/10.1016/j.yfrne.2018.03.002

Batzer MA, Deininger PL (2002) Alu repeats and human genomic diversity. Nat Rev Genet 3 (5):370–379. https://doi.org/10.1038/nrg798

Bedrosian TA, Quayle C, Novaresi N, Gage FH (2018) Early life experience drives structural variation of neural genomes in mice. Science 359(6382):1395–1399. https://doi.org/10.1126/science.aah3378

Belyayev A (2014) Bursts of transposable elements as an evolutionary driving force. J Evol Biol 27 (12):2573–2584. https://doi.org/10.1111/jeb.12513

Boeke JD, Stoye JP (1997) Retrotransposons, endogenous retroviruses, and the evolution of retroelements. In: Coffin JM, Hughes SH, Varmus HE (eds) Retroviruses. Cold Spring Harbor Laboratory Press, Cold Spring Harbor. http://www.ncbi.nlm.nih.gov/books/NBK19468/

Bogdanik LP, Chapman HD, Miers KE, Serreze DV, Burgess RW (2012) A MusD retrotransposon insertion in the mouse Slc6a5 gene causes alterations in neuromuscular junction maturation and behavioral phenotypes. PLoS One 7(1):e30217. https://doi.org/10.1371/journal.pone.0030217

Bose P, Hermetz KE, Conneely KN, Rudd MK (2014) Tandem repeats and G-rich sequences are enriched at human CNV breakpoints. PLoS One 9(7):e101607. https://doi.org/10.1371/journal.pone.0101607

Brosius J (2003) The contribution of RNAs and retroposition to evolutionary novelties. Genetica 118(2–3):99–116

Brouha B, Schustak J, Badge RM, Lutz-Prigge S, Farley AH, Moran JV, Kazazian HH (2003) Hot L1s account for the bulk of retrotransposition in the human population. Proc Natl Acad Sci U S A 100(9):5280–5285. https://doi.org/10.1073/pnas.0831042100

Bucheton A, Vaury C, Chaboissier MC, Abad P, Pélisson A, Simonelig M (1992) I elements and the Drosophila genome. Genetica 86(1–3):175–190

Bundo M, Toyoshima M, Okada Y, Akamatsu W, Ueda J, Nemoto-Miyauchi T et al (2014) Increased l1 retrotransposition in the neuronal genome in schizophrenia. Neuron 81 (2):306–313. https://doi.org/10.1016/j.neuron.2013.10.053

Campbell IM, Gambin T, Dittwald P, Beck CR, Shuvarikov A, Hixson P et al (2014) Human endogenous retroviral elements promote genome instability via non-allelic homologous recombination. BMC Biol 12:74. https://doi.org/10.1186/s12915-014-0074-4

Cappucci U, Torromino G, Casale AM, Camon J, Capitano F, Berloco M et al (2018) Stress-induced strain and brain region-specific activation of LINE-1 transposons in adult mice. Stress:1–5. https://doi.org/10.1080/10253890.2018.1485647

Carmona LM, Schatz DG (2017) New insights into the evolutionary origins of the recombination-activating gene proteins and V(D)J recombination. FEBS J 284(11):1590–1605. https://doi.org/10.1111/febs.13990

Chu WM, Liu WM, Schmid CW (1995) RNA polymerase III promoter and terminator elements affect Alu RNA expression. Nucleic Acids Res 23(10):1750–1757

Chuong EB, Elde NC, Feschotte C (2017) Regulatory activities of transposable elements: from conflicts to benefits. Nat Rev Genet 18(2):71–86. https://doi.org/10.1038/nrg.2016.139

Cordaux R, Batzer MA (2009) The impact of retrotransposons on human genome evolution. Nat Rev Genet 10(10):691–703. https://doi.org/10.1038/nrg2640

Coufal NG, Garcia-Perez JL, Peng GE, Marchetto MCN, Muotri AR, Mu Y et al (2011) Ataxia telangiectasia mutated (ATM) modulates long interspersed element-1 (L1) retrotransposition in human neural stem cells. Proc Natl Acad Sci U S A 108(51):20382–20387. https://doi.org/10.1073/pnas.1100273108

Crow TJ (1987) Integrated viral genes as potential pathogens in the functional psychoses. J Psychiatr Res 21(4):479–485

Curtin F, Bernard C, Levet S, Perron H, Porchet H, Médina J et al (2018) A new therapeutic approach for type 1 diabetes: rationale for GNbAC1, an anti-HERV-W-Env monoclonal antibody. Diabetes Obes Metab 20(9):2075–2084. https://doi.org/10.1111/dom.13357

D'Agati E, Pitzianti M, Balestrieri E, Matteucci C, Sinibaldi Vallebona P, Pasini A (2016) First evidence of HERV-H transcriptional activity reduction after methylphenidate treatment in a young boy with ADHD. New Microbiol 39(3):237–239

Daskalakis NP, Provost AC, Hunter RG, Guffanti G (2018) Noncoding RNAs: stress, glucocorticoids, and posttraumatic stress disorder. Biol Psychiatry 83(10):849–865. https://doi.org/10.1016/j.biopsych.2018.01.009

de Koning APJ, Gu W, Castoe TA, Batzer MA, Pollock DD (2011) Repetitive elements may comprise over two-thirds of the human genome. PLoS Genet 7(12):e1002384. https://doi.org/10.1371/journal.pgen.1002384

de Souza FSJ, Franchini LF, Rubinstein M (2013) Exaptation of transposable elements into novel Cis-regulatory elements: is the evidence always strong? Mol Biol Evol 30(6):1239–1251. https://doi.org/10.1093/molbev/mst045
DiDomenico BJ, Bugaisky GE, Lindquist S (1982) Heat shock and recovery are mediated by different translational mechanisms. Proc Natl Acad Sci U S A 79(20):6181–6185
Diem O, Schäffner M, Seifarth W, Leib-Mösch C (2012) Influence of antipsychotic drugs on human endogenous retrovirus (HERV) transcription in brain cells. PLoS One 7(1):e30054. https://doi.org/10.1371/journal.pone.0030054
Doyle GA, Crist RC, Karatas ET, Hammond MJ, Ewing AD, Ferraro TN et al (2017) Analysis of LINE-1 elements in DNA from postmortem brains of individuals with schizophrenia. Neuropsychopharmacology 42(13):2602–2611. https://doi.org/10.1038/npp.2017.115
Drongitis D, Rainone S, Piscopo M, Viggiano E, Viggiano A, De Luca B et al (2016) Epigenetics and cortical spreading depression: changes of DNA methylation level at retrotransposon sequences. Mol Biol Rep 43(8):755–760. https://doi.org/10.1007/s11033-016-4000-4
Dupressoir A, Lavialle C, Heidmann T (2012) From ancestral infectious retroviruses to bona fide cellular genes: role of the captured syncytins in placentation. Placenta 33(9):663–671. https://doi.org/10.1016/j.placenta.2012.05.005
Eickbush TH, Furano AV (2002) Fruit flies and humans respond differently to retrotransposons. Curr Opin Genet Dev 12(6):669–674
Eickbush TH, Jamburuthugoda VK (2008) The diversity of retrotransposons and the properties of their reverse transcriptases. Virus Res 134(1–2):221–234. https://doi.org/10.1016/j.virusres.2007.12.010
ENCODE Project Consortium (2012) An integrated encyclopedia of DNA elements in the human genome. Nature 489(7414):57–74. https://doi.org/10.1038/nature11247
Erwin JA, Marchetto MC, Gage FH (2014) Mobile DNA elements in the generation of diversity and complexity in the brain. Nat Rev Neurosci 15(8):497–506. https://doi.org/10.1038/nrn3730
Espinoza CA, Allen TA, Hieb AR, Kugel JF, Goodrich JA (2004) B2 RNA binds directly to RNA polymerase II to repress transcript synthesis. Nat Struct Mol Biol 11(9):822–829. https://doi.org/10.1038/nsmb812
Espinoza CA, Goodrich JA, Kugel JF (2007) Characterization of the structure, function, and mechanism of B2 RNA, an ncRNA repressor of RNA polymerase II transcription. RNA (New York, NY) 13(4):583–596. https://doi.org/10.1261/rna.310307
Ewing AD, Kazazian HH (2011) Whole-genome resequencing allows detection of many rare LINE-1 insertion alleles in humans. Genome Res 21(6):985–990. https://doi.org/10.1101/gr.114777.110
Fedoroff NV (2012) Presidential address. Transposable elements, epigenetics, and genome evolution. Science 338(6108):758–767
Feschotte C, Wessler SR, Zhang X (2002) Miniature inverted-repeat transposable elements and their relationship to established DNA transposons. In: Craig NL, Lambowitz AM, Craigie R, Gellert M (eds) Mobile DNA II. American Society of Microbiology, Washington, pp 1147–1158. https://doi.org/10.1128/9781555817954.ch50
Findly RC, Pederson T (1981) Regulated transcription of the genes for actin and heat-shock proteins in cultured Drosophila cells. J Cell Biol 88(2):323–328
Fornace AJ, Mitchell JB (1986) Induction of B2 RNA polymerase III transcription by heat shock: enrichment for heat shock induced sequences in rodent cells by hybridization subtraction. Nucleic Acids Res 14(14):5793–5811
Frank JA, Feschotte C (2017) Co-option of endogenous viral sequences for host cell function. Curr Opin Virol 25:81–89. https://doi.org/10.1016/j.coviro.2017.07.021
Frank O, Giehl M, Zheng C, Hehlmann R, Leib-Mösch C, Seifarth W (2005) Human endogenous retrovirus expression profiles in samples from brains of patients with schizophrenia and bipolar disorders. J Virol 79(17):10890–10901. https://doi.org/10.1128/JVI.79.17.10890-10901.2005
Frank O, Jones-Brando L, Leib-Mosch C, Yolken R, Seifarth W (2006) Altered transcriptional activity of human endogenous retroviruses in neuroepithelial cells after infection with toxoplasma gondii. J Infect Dis 194(10):1447–1449. https://doi.org/10.1086/508496

Friedli M, Trono D (2015) The developmental control of transposable elements and the evolution of higher species. Annu Rev Cell Dev Biol 31(1):429–451. https://doi.org/10.1146/annurev-cellbio-100814-125514

Gilmour DS, Lis JT (1985) In vivo interactions of RNA polymerase II with genes of Drosophila melanogaster. Mol Cell Biol 5(8):2009–2018

Goodier JL, Kazazian HH (2008) Retrotransposons revisited: the restraint and rehabilitation of parasites. Cell 135(1):23–35. https://doi.org/10.1016/j.cell.2008.09.022

Goodier JL, Ostertag EM, Kazazian HH (2000) Transduction of 3′-flanking sequences is common in L1 retrotransposition. Hum Mol Genet 9(4):653–657

Gould SJ, Vrba ES (1982) Exaptation – a missing term in the science of form. Paleobiology 8 (01):4–15. https://doi.org/10.1017/S0094837300004310

Guffanti G, Gaudi S, Fallon JH, Sobell J, Potkin SG, Pato C, Macciardi F (2014) Transposable elements and psychiatric disorders. Am J Med Genet B Neuropsychiatr Genet 165B (3):201–216. https://doi.org/10.1002/ajmg.b.32225

Guffanti G, Gaudi S, Klengel T, Fallon JH, Mangalam H, Madduri R et al (2016) LINE1 insertions as a genomic risk factor for schizophrenia: preliminary evidence from an affected family. Am J Med Genet B Neuropsychiatr Genet 171(4):534–545. https://doi.org/10.1002/ajmg.b.32437

Guo C, Jeong H-H, Hsieh Y-C, Klein H-U, Bennett DA, De Jager PL et al (2018) Tau activates transposable elements in Alzheimer's disease. Cell Rep 23(10):2874–2880. https://doi.org/10.1016/j.celrep.2018.05.004

Habibi L, Shokrgozar MA, Tabrizi M, Modarressi MH, Akrami SM (2014) Mercury specifically induces LINE-1 activity in a human neuroblastoma cell line. Mutat Res Genet Toxicol Environ Mutagen 759:9–20. https://doi.org/10.1016/j.mrgentox.2013.07.015

Hambor JE, Mennone J, Coon ME, Hanke JH, Kavathas P (1993) Identification and characterization of an Alu-containing, T-cell-specific enhancer located in the last intron of the human CD8 alpha gene. Mol Cell Biol 13(11):7056–7070

Hancks DC, Goodier JL, Mandal PK, Cheung LE, Kazazian HH (2011) Retrotransposition of marked SVA elements by human L1s in cultured cells. Hum Mol Genet 20(17):3386–3400. https://doi.org/10.1093/hmg/ddr245

Harold FM (1986) The vital force: a study of bioenergetics. W.H. Freeman, New York

Houede N, Piazza PV, Pourquier P (2018) LINE-1 as a therapeutic target for castration-resistant prostate cancer. Front Biosci 23:1292–1309

Huang W-J, Liu Z-C, Wei W, Wang G-H, Wu J-G, Zhu F (2006) Human endogenous retroviral pol RNA and protein detected and identified in the blood of individuals with schizophrenia. Schizophr Res 83(2–3):193–199. https://doi.org/10.1016/j.schres.2006.01.007

Huang CRL, Schneider AM, Lu Y, Niranjan T, Shen P, Robinson MA et al (2010) Mobile interspersed repeats are major structural variants in the human genome. Cell 141 (7):1171–1182. https://doi.org/10.1016/j.cell.2010.05.026

Huang W, Li S, Hu Y, Yu H, Luo F, Zhang Q, Zhu F (2011) Implication of the env gene of the human endogenous retrovirus W family in the expression of BDNF and DRD3 and development of recent-onset schizophrenia. Schizophr Bull 37(5):988–1000. https://doi.org/10.1093/schbul/sbp166

Huda A, Tyagi E, Mariño-Ramírez L, Bowen NJ, Jjingo D, Jordan IK (2011) Prediction of transposable element derived enhancers using chromatin modification profiles. PLoS One 6 (11):e27513. https://doi.org/10.1371/journal.pone.0027513

Hunter RG, Murakami G, Dewell S, Seligsohn M, Baker MER, Datson NA et al (2012) Acute stress and hippocampal histone H3 lysine 9 trimethylation, a retrotransposon silencing response. Proc Natl Acad Sci U S A 109(43):17657–17662. https://doi.org/10.1073/pnas.1215810109

Hunter RG, McEwen BS, Pfaff DW (2013) Environmental stress and transposon transcription in the mammalian brain. Mob Genet Elem 3(2):e24555. https://doi.org/10.4161/mge.24555

Hunter RG, Gagnidze K, McEwen BS, Pfaff DW (2015) Stress and the dynamic genome: steroids, epigenetics, and the transposome. Proc Natl Acad Sci U S A 112(22):6828–6833. https://doi.org/10.1073/pnas.1411260111

Jjingo D, Conley AB, Wang J, Mariño-Ramírez L, Lunyak VV, Jordan IK (2014) Mammalian-wide interspersed repeat (MIR)-derived enhancers and the regulation of human gene expression. Mob DNA 5:14. https://doi.org/10.1186/1759-8753-5-14

Karijolich J, Abernathy E, Glaunsinger BA (2015) Infection-induced retrotransposon-derived noncoding RNAs enhance herpesviral gene expression via the NF-κB pathway. PLoS Pathog 11(11):e1005260. https://doi.org/10.1371/journal.ppat.1005260

Karlsson H, Bachmann S, Schröder J, McArthur J, Torrey EF, Yolken RH (2001) Retroviral RNA identified in the cerebrospinal fluids and brains of individuals with schizophrenia. Proc Natl Acad Sci U S A 98(8):4634–4639. https://doi.org/10.1073/pnas.061021998

Karlsson H, Schröder J, Bachmann S, Bottmer C, Yolken RH (2004) HERV-W-related RNA detected in plasma from individuals with recent-onset schizophrenia or schizoaffective disorder. Mol Psychiatry 9(1):12–13. https://doi.org/10.1038/sj.mp.4001439

Kelley D, Rinn J (2012) Transposable elements reveal a stem cell-specific class of long noncoding RNAs. Genome Biol 13(11):R107. https://doi.org/10.1186/gb-2012-13-11-r107

Kerur N, Hirano Y, Tarallo V, Fowler BJ, Bastos-Carvalho A, Yasuma T et al (2013) TLR-independent and P2X7-dependent signaling mediate Alu RNA-induced NLRP3 inflammasome activation in geographic atrophy. Invest Ophthalmol Vis Sci 54 (12):7395–7401. https://doi.org/10.1167/iovs.13-12500

Kesby JP, Eyles DW, McGrath JJ, Scott JG (2018) Dopamine, psychosis and schizophrenia: the widening gap between basic and clinical neuroscience. Transl Psychiatry 8(1):30. https://doi.org/10.1038/s41398-017-0071-9

Kino T, Hurt DE, Ichijo T, Nader N, Chrousos GP (2010) Noncoding RNA gas5 is a growth arrest- and starvation-associated repressor of the glucocorticoid receptor. Sci Signal 3(107):ra8. https://doi.org/10.1126/scisignal.2000568

Kirov G (2015) CNVs in neuropsychiatric disorders. Hum Mol Genet 24(R1):R45–R49. https://doi.org/10.1093/hmg/ddv253

Krug L, Chatterjee N, Borges-Monroy R, Hearn S, Liao W-W, Morrill K et al (2017) Retrotransposon activation contributes to neurodegeneration in a Drosophila TDP-43 model of ALS. PLoS Genet 13(3):e1006635. https://doi.org/10.1371/journal.pgen.1006635

Kugel JF, Goodrich JA (2006) Beating the heat: a translation factor and an RNA mobilize the heat shock transcription factor HSF1. Mol Cell 22(2):153–154. https://doi.org/10.1016/j.molcel.2006.04.003

Kuhn RM, Karolchik D, Zweig AS, Trumbower H, Thomas DJ, Thakkapallayil A et al (2007) The UCSC genome browser database: update 2007. Nucleic Acids Res 35(Database issue): D668–D673. https://doi.org/10.1093/nar/gkl928

Lacar B, Linker SB, Jaeger BN, Krishnaswami S, Barron J, Kelder M, Parylak SL, Paquola ACM, Venepally P, Novotny M, O'Connor C, Fitzpatrick C, Erwin JA, Hsu JY, Husband D, McConnell MJ, Lasken R, Gage FH (2016) Nuclear RNA-seq of single neurons reveals molecular signatures of activation. Nat Commun 7:11022. https://doi.org/10.1038/ncomms11022

Lander ES, Linton LM, Birren B, Nusbaum C, Zody MC, Baldwin J et al (2001) Initial sequencing and analysis of the human genome. Nature 409(6822):860–921. https://doi.org/10.1038/35057062

Lane N, Martin W (2010) The energetics of genome complexity. Nature 467(7318):929–934. https://doi.org/10.1038/nature09486

Lapp HE, Hunter RG (2016) The dynamic genome: transposons and environmental adaptation in the nervous system. Epigenomics 8(2):237–249. https://doi.org/10.2217/epi.15.107

Leboyer M, Tamouza R, Charron D, Faucard R, Perron H (2013) Human endogenous retrovirus type W (HERV-W) in schizophrenia: a new avenue of research at the gene-environment interface. World J Biol Psychiatry 14(2):80–90. https://doi.org/10.3109/15622975.2010.601760

Li T, Spearow J, Rubin CM, Schmid CW (1999) Physiological stresses increase mouse short interspersed element (SINE) RNA expression in vivo. Gene 239(2):367–372

Li W, Jin Y, Prazak L, Hammell M, Dubnau J (2012) Transposable elements in TDP-43-mediated neurodegenerative disorders. PLoS One 7(9):e44099. https://doi.org/10.1371/journal.pone.0044099

Li W, Prazak L, Chatterjee N, Grüninger S, Krug L, Theodorou D, Dubnau J (2013) Activation of transposable elements during aging and neuronal decline in Drosophila. Nat Neurosci 16 (5):529–531. https://doi.org/10.1038/nn.3368

Li W, Lee M-H, Henderson L, Tyagi R, Bachani M, Steiner J et al (2015) Human endogenous retrovirus-K contributes to motor neuron disease. Sci Transl Med 7(307):307ra153. https://doi.org/10.1126/scitranslmed.aac8201

Liu WM, Chu WM, Choudary PV, Schmid CW (1995) Cell stress and translational inhibitors transiently increase the abundance of mammalian SINE transcripts. Nucleic Acids Res 23 (10):1758–1765

Long Y, Wang X, Youmans DT, Cech TR (2017) How do lncRNAs regulate transcription? Sci Adv 3(9):eaao2110. https://doi.org/10.1126/sciadv.aao2110

Lu X, Sachs F, Ramsay L, Jacques P-É, Göke J, Bourque G, Ng H-H (2014) The retrovirus HERVH is a long noncoding RNA required for human embryonic stem cell identity. Nat Struct Mol Biol 21(4):423–425. https://doi.org/10.1038/nsmb.2799

Lunyak VV, Prefontaine GG, Nunez E, Cramer T, Ju B-G, Ohgi KA et al (2007) Developmentally regulated activation of a SINE B2 repeat as a domain boundary in organogenesis. Science 317 (5835):248–251. https://doi.org/10.1126/science.1140871

Lynch VJ, Leclerc RD, May G, Wagner GP (2011) Transposon-mediated rewiring of gene regulatory networks contributed to the evolution of pregnancy in mammals. Nat Genet 43 (11):1154–1159. https://doi.org/10.1038/ng.917

Lynch VJ, Nnamani MC, Kapusta A, Brayer K, Plaza SL, Mazur EC et al (2015) Ancient transposable elements transformed the uterine regulatory landscape and transcriptome during the evolution of mammalian pregnancy. Cell Rep 10(4):551–561. https://doi.org/10.1016/j.celrep.2014.12.052

Mager DL, Stoye JP (2015) Mammalian endogenous retroviruses. Microbiol Spect 3(1):MDNA3-0009-2014. https://doi.org/10.1128/microbiolspec.MDNA3-0009-2014

Makino S, Kaji R, Ando S, Tomizawa M, Yasuno K, Goto S et al (2007) Reduced neuron-specific expression of the TAF1 gene is associated with X-linked dystonia-parkinsonism. Am J Hum Genet 80(3):393–406. https://doi.org/10.1086/512129

Mariño-Ramírez L, Jordan IK (2006) Transposable element derived DNaseI-hypersensitive sites in the human genome. Biol Direct 1:20. https://doi.org/10.1186/1745-6150-1-20

Mattei D, Schweibold R, Wolf SA (2015) Brain in flames – animal models of psychosis: utility and limitations. Neuropsychiatr Dis Treat 11:1313–1329. https://doi.org/10.2147/NDT.S65564

Maxwell PH, Burhans WC, Curcio MJ (2011) Retrotransposition is associated with genome instability during chronological aging. Proc Natl Acad Sci U S A 108(51):20376–20381. https://doi.org/10.1073/pnas.1100271108

Mcclintock B (1956) Intranuclear systems controlling gene action and mutation. Brookhaven Symp Biol 8:58–74

Mili S, Steitz JA (2004) Evidence for reassociation of RNA-binding proteins after cell lysis: implications for the interpretation of immunoprecipitation analyses. RNA 10(11):1692–1694. https://doi.org/10.1261/rna.7151404

Mouse Genome Sequencing Consortium, Waterston RH, Lindblad-Toh K, Birney E, Rogers J, Abril JF et al (2002) Initial sequencing and comparative analysis of the mouse genome. Nature 420(6915):520–562. https://doi.org/10.1038/nature01262

Mukherjee S, Sharma D, Upadhyaya KC (2016) L1 retrotransposons are transcriptionally active in Hippocampus of rat brain. Prague Med Rep 117(1):42–53. https://doi.org/10.14712/23362936.2016.4

Muotri AR, Zhao C, Marchetto MCN, Gage FH (2009) Environmental influence on L1 retrotransposons in the adult hippocampus. Hippocampus 19(10):1002–1007. https://doi.org/10.1002/hipo.20564

Muotri AR, Marchetto MCN, Coufal NG, Oefner R, Yeo G, Nakashima K, Gage FH (2010) L1 retrotransposition in neurons is modulated by MeCP2. Nature 468(7322):443–446. https://doi.org/10.1038/nature09544
Nelson PN, Lever AM, Smith S, Pitman R, Murray P, Perera SA et al (1999) Molecular investigations implicate human endogenous retroviruses as mediators of anti-retroviral antibodies in autoimmune rheumatic disease. Immunol Investig 28(4):277–289
Nitsche A, Stadler PF (2017) Evolutionary clues in lncRNAs. Wiley Interdiscip Rev RNA 8(1). https://doi.org/10.1002/wrna.1376
O'Brien T, Lis JT (1993) Rapid changes in Drosophila transcription after an instantaneous heat shock. Mol Cell Biol 13(6):3456–3463
Ohno S (1972) So much "junk" DNA in our genome. Brookhaven Symp Biol 23:366–370
Okudaira N, Ishizaka Y, Nishio H (2014) Retrotransposition of Long interspersed element 1 induced by methamphetamine or cocaine. J Biol Chem 289(37):25476–25485. https://doi.org/10.1074/jbc.M114.559419
Percharde M, Lin C-J, Yin Y, Guan J, Peixoto GA, Bulut-Karslioglu A et al (2018) A LINE1-nucleolin partnership regulates early development and ESC identity. Cell 174(2):391–405.e19. https://doi.org/10.1016/j.cell.2018.05.043
Perron H, Hamdani N, Faucard R, Lajnef M, Jamain S, Daban-Huard C et al (2012) Molecular characteristics of human endogenous retrovirus type-W in schizophrenia and bipolar disorder. Transl Psychiatry 2:e201. https://doi.org/10.1038/tp.2012.125
Pickeral OK, Makałowski W, Boguski MS, Boeke JD (2000) Frequent human genomic DNA transduction driven by LINE-1 retrotransposition. Genome Res 10(4):411–415
Polavarapu N, Bowen NJ, McDonald JF (2006) Identification, characterization and comparative genomics of chimpanzee endogenous retroviruses. Genome Biol 7(6):R51. https://doi.org/10.1186/gb-2006-7-6-r51
Ponicsan SL, Kugel JF, Goodrich JA (2010) Genomic gems: SINE RNAs regulate mRNA production. Curr Opin Genet Dev 20(2):149–155. https://doi.org/10.1016/j.gde.2010.01.004
Ponomarev I, Wang S, Zhang L, Harris RA, Mayfield RD (2012) Gene coexpression networks in human brain identify epigenetic modifications in alcohol dependence. J Neurosci Off J Soc Neurosci 32(5):1884–1897. https://doi.org/10.1523/JNEUROSCI.3136-11.2012
Ramesh V, Bayam E, Cernilogar FM, Bonapace IM, Schulze M, Riemenschneider MJ et al (2016) Loss of Uhrf1 in neural stem cells leads to activation of retroviral elements and delayed neurodegeneration. Genes Dev 30(19):2199–2212. https://doi.org/10.1101/gad.284992.116
Rana G, Donizetti A, Virelli G, Piscopo M, Viggiano E, De Luca B, Fucci L (2012) Cortical spreading depression differentially affects lysine methylation of H3 histone at neuroprotective genes and retrotransposon sequences. Brain Res 1467:113–119. https://doi.org/10.1016/j.brainres.2012.05.043
Reilly MT, Faulkner GJ, Dubnau J, Ponomarev I, Gage FH (2013) The role of transposable elements in health and diseases of the central nervous system. J Neurosci Off J Soc Neurosci 33(45):17577–17586. https://doi.org/10.1523/JNEUROSCI.3369-13.2013
Ribet D, Harper F, Dupressoir A, Dewannieux M, Pierron G, Heidmann T (2008) An infectious progenitor for the murine IAP retrotransposon: emergence of an intracellular genetic parasite from an ancient retrovirus. Genome Res 18(4):597–609. https://doi.org/10.1101/gr.073486.107
Richardson SR, Doucet AJ, Kopera HC, Moldovan JB, Garcia-Perez JL, Moran JV (2015) The influence of LINE-1 and SINE retrotransposons on mammalian genomes. Microbiol Spect 3(2):MDNA3-0061-2014. https://doi.org/10.1128/microbiolspec.MDNA3-0061-2014
Rusiecki JA, Chen L, Srikantan V, Zhang L, Yan L, Polin ML, Baccarelli A (2012) DNA methylation in repetitive elements and post-traumatic stress disorder: a case-control study of US military service members. Epigenomics 4(1):29–40. https://doi.org/10.2217/epi.11.116
Santangelo AM, de Souza FSJ, Franchini LF, Bumaschny VF, Low MJ, Rubinstein M (2007) Ancient exaptation of a CORE-SINE retroposon into a highly conserved mammalian neuronal enhancer of the proopiomelanocortin gene. PLoS Genet 3(10):1813–1826. https://doi.org/10.1371/journal.pgen.0030166

Sciamanna I, De Luca C, Spadafora C (2016) The reverse transcriptase encoded by LINE-1 retrotransposons in the genesis, progression, and therapy of cancer. Front Chem 4. https://doi.org/10.3389/fchem.2016.00006

Sela N, Mersch B, Hotz-Wagenblatt A, Ast G (2010) Characteristics of transposable element exonization within human and mouse. PLoS One 5(6):e10907. https://doi.org/10.1371/journal.pone.0010907

Senerchia N, Felber F, Parisod C (2015) Genome reorganization in F1 hybrids uncovers the role of retrotransposons in reproductive isolation. Proc Biol Sci 282(1804):20142874. https://doi.org/10.1098/rspb.2014.2874

Shen L, Wu LC, Sanlioglu S, Chen R, Mendoza AR, Dangel AW et al (1994) Structure and genetics of the partially duplicated gene RP located immediately upstream of the complement C4A and the C4B genes in the HLA class III region. Molecular cloning, exon-intron structure, composite retroposon, and breakpoint of gene duplication. J Biol Chem 269(11):8466–8476

Shpyleva S, Melnyk S, Pavliv O, Pogribny I, Jill James S (2018) Overexpression of LINE-1 retrotransposons in autism brain. Mol Neurobiol 55(2):1740–1749. https://doi.org/10.1007/s12035-017-0421-x

Smith MA, Gesell T, Stadler PF, Mattick JS (2013) Widespread purifying selection on RNA structure in mammals. Nucleic Acids Res 41(17):8220–8236. https://doi.org/10.1093/nar/gkt596

Sonea S, Mathieu LG (2000) Prokaryotology: a coherent point of view. Presses de l'Université de Montréal. https://doi.org/10.4000/books.pum.14291

Speek M (2001) Antisense promoter of human L1 retrotransposon drives transcription of adjacent cellular genes. Mol Cell Biol 21(6):1973–1985. https://doi.org/10.1128/MCB.21.6.1973-1985.2001

Startek M, Szafranski P, Gambin T, Campbell IM, Hixson P, Shaw CA et al (2015) Genome-wide analyses of LINE-LINE-mediated nonallelic homologous recombination. Nucleic Acids Res 43 (4):2188–2198. https://doi.org/10.1093/nar/gku1394

Steller H, Pirrotta V (1986) P transposons controlled by the heat shock promoter. Mol Cell Biol 6 (5):1640–1649

Suntsova M, Gogvadze EV, Salozhin S, Gaifullin N, Eroshkin F, Dmitriev SE et al (2013) Human-specific endogenous retroviral insert serves as an enhancer for the schizophrenia-linked gene PRODH. Proc Natl Acad Sci U S A 110(48):19472–19477. https://doi.org/10.1073/pnas.1318172110

Swergold GD (1990) Identification, characterization, and cell specificity of a human LINE-1 promoter. Mol Cell Biol 10(12):6718–6729

Tarallo V, Hirano Y, Gelfand BD, Dridi S, Kerur N, Kim Y et al (2012) DICER1 loss and Alu RNA induce age-related macular degeneration via the NLRP3 inflammasome and MyD88. Cell 149 (4):847–859. https://doi.org/10.1016/j.cell.2012.03.036

Tica J, Lee E, Untergasser A, Meiers S, Garfield DA, Gokcumen O et al (2016) Next-generation sequencing-based detection of germline L1-mediated transductions. BMC Genomics 17:342. https://doi.org/10.1186/s12864-016-2670-x

Ullu E, Tschudi C (1984) Alu sequences are processed 7SL RNA genes. Nature 312 (5990):171–172

Voigt A, Herholz D, Fiesel FC, Kaur K, Müller D, Karsten P et al (2010) TDP-43-mediated neuron loss in vivo requires RNA-binding activity. PLoS One 5(8):e12247. https://doi.org/10.1371/journal.pone.0012247

Wang Z, Kunze R (2015) Transposons in eukaryotes (part A): structures, mechanisms and applications. In: John Wiley & Sons Ltd (ed) eLS. Wiley, Chichester, pp 1–13. https://doi.org/10.1002/9780470015902.a0026264

Wang H, Xing J, Grover D, Hedges DJ, Han K, Walker JA, Batzer MA (2005) SVA elements: a hominid-specific retroposon family. J Mol Biol 354(4):994–1007. https://doi.org/10.1016/j.jmb.2005.09.085

Weber B, Kimhi S, Howard G, Eden A, Lyko F (2010) Demethylation of a LINE-1 antisense promoter in the cMet locus impairs Met signalling through induction of illegitimate transcription. Oncogene 29(43):5775–5784. https://doi.org/10.1038/onc.2010.227

Weis S, Llenos IC, Sabunciyan S, Dulay JR, Isler L, Yolken R, Perron H (2007) Reduced expression of human endogenous retrovirus (HERV)-W GAG protein in the cingulate gyrus and hippocampus in schizophrenia, bipolar disorder, and depression. J Neural Transm (Vienna, Austria: 1996) 114(5):645–655. https://doi.org/10.1007/s00702-006-0599-y

Wheeler BS (2013) Small RNAs, big impact: small RNA pathways in transposon control and their effect on the host stress response. Chromosom Res 21(6–7):587–600. https://doi.org/10.1007/s10577-013-9394-4

Wood JG, Jones BC, Jiang N, Chang C, Hosier S, Wickremesinghe P et al (2016) Chromatin-modifying genetic interventions suppress age-associated transposable element activation and extend life span in Drosophila. Proc Natl Acad Sci U S A 113(40):11277–11282. https://doi.org/10.1073/pnas.1604621113

Yakovchuk P, Goodrich JA, Kugel JF (2009) B2 RNA and Alu RNA repress transcription by disrupting contacts between RNA polymerase II and promoter DNA within assembled complexes. Proc Natl Acad Sci USA 106(14):5569–5574. https://doi.org/10.1073/pnas.0810738106

Yang Z, Boffelli D, Boonmark N, Schwartz K, Lawn R (1998) Apolipoprotein(a) gene enhancer resides within a LINE element. J Biol Chem 273(2):891–897

Yao B, Cheng Y, Wang Z, Li Y, Chen L, Huang L et al (2017) DNA N6-methyladenine is dynamically regulated in the mouse brain following environmental stress. Nat Commun 8 (1):1122. https://doi.org/10.1038/s41467-017-01195-y

Yolken RH, Karlsson H, Yee F, Johnston-Wilson NL, Torrey EF (2000) Endogenous retroviruses and schizophrenia. Brain Res Brain Res Rev 31(2–3):193–199

Multi-, Inter-, and Transgenerational Effects of Drugs of Abuse on Behavior

Fair M. Vassoler, Anika M. Toorie, and Elizabeth M. Byrnes

Contents

Abstract Transgenerational epigenetic inheritance is a burgeoning field that has recently garnered much attention. A growing body of evidence identifies behavioral phenotypes associated with inter-, multi-, and transgenerational studies following a wide variety of parental exposures. This chapter in current topics in behavioral neurogenomics examines the evidence for the presence of behavioral phenotypes and, in particular, the varied and often opposite behavioral responses observed with protocol shifts. Effects following parental exposure to drugs of abuse are used as an example of the wide range of behavioral outcomes and the variability associated with these multiple generation studies. The behavioral phenotypes associated with drug exposure are reviewed in depth.

Keywords Addiction · Behavior · Drugs of abuse · Epigenetic · Transgenerational

F. M. Vassoler (✉), A. M. Toorie, and E. M. Byrnes
Department of Biomedical Sciences, Cummings School of Veterinary Medicine, Tufts University, North Grafton, MA, USA
e-mail: fair.vassoler@tufts.edu

© Springer Nature Switzerland AG 2019

Curr Topics Behav Neurosci (2019) 42: 247–258
DOI 10.1007/7854_2019_106
Published Online: 9 August 2019

1 Introduction

Historically, the prevailing hypothesis has been that phenotype is determined largely by genotype, with the acknowledgement that some aspects of an individual's phenotype may be further influenced by environmental factors. Thus, both nature (genetics) and nurture (environmental experiences) play a role in phenotypic variations observed in both human and animal populations. A growing body of evidence, however, suggests that phenotypic variation may also be shaped by parental experiences occurring prior to conception, which are then transmitted to offspring in the absence of alterations in DNA sequence. This phenomenon has been termed *epigenetic inheritance*. Because the term epigenetics is used mechanistically to describe a multitude of specific genomic events, it is worth devoting attention to a precise and careful definition that will be used in the current chapter.

The term epigenetics was first coined by Conrad Hal Waddington in 1942 to describe the relationship between genotype and the embryonic environment, which together determine cellular differentiation, and ultimately, phenotype. Importantly, most definitions of epigenetics include some factor of heritability, whether it be across cell divisions or across generations. Perhaps one of the most well-cited definitions is that of Adrian Bird, who in 2007 wrote, "The following could be a unifying definition of epigenetic events: the structural adaptation of chromosomal regions so as to register, signal or perpetuate altered activity states." This definition encompasses chromosomal marks and includes both stable changes maintained across multiple cell generations and changes inherited in offspring. For the purposes of this chapter, we will utilize a relatively conservative definition that defines epigenetics as *the perpetuation of specific gene activation states resulting in phenotypic variations that persist across generations in the absence of changes to the DNA sequence*.

Within the burgeoning field of epigenetic inheritance, another important distinction must be recognized, that of inter-, multi-, and transgenerational phenomena. This definition relates to the nature of the exposure (direct versus indirect) experienced by a particular generation (Daxinger and Whitelaw 2012; Ferguson-Smith and Patti 2011; Heard and Martienssen 2014; Lim and Brunet 2013; Nagy and Turecki 2015). The terms inter- and multigenerational refer to studies that involve an environmental stimulus applied to the parental generation (F0 or P0) and the resulting phenotypic variations observed in the offspring. Intergenerational studies examine first-generation offspring (F1) following a parental manipulation; however, because the germ cells that produce the F1 generation are present during the manipulation, F1 animals are considered to have experienced a direct exposure. Such studies are deemed intergenerational, i.e., they span two generations with direct exposure. A multigenerational design often spans more than two generations but still suggests direct exposure across all generations examined. For example, during in utero manipulations, the developing fetus is directly exposed to the environmental manipulation (F1 direct exposure). The second generation (F2), however, also has a direct exposure as the developing fetus already contains the germ cells that will become the F2 animals. This is a classic example of a multigenerational study (spans multiple generations all with direct exposure). Traditionally, to be considered a transgenerational study, at least one generation is examined that never received any direct exposure. A clear definition, proposed by Skinner in 2011, states that transgenerational epigenetic inheritance is "germline-mediated inheritance of epigenetic information between generations in the absence of direct environmental influences that leads to phenotypic variation" (Skinner 2011). Thus, this is a distinct phenomenon because it implies that there is some stable change in the epigenome of the animal that is inherited beyond any generation that experienced direct exposure.

There is a growing body of evidence that shows a wide range of behavioral, molecular, and physiological phenotypes in inter-, multi-, and transgenerational studies. Interestingly, the phenotypes often vary widely even within similar exposure protocols. It is the goal of this chapter to detail inter-, multi-, and transgenerational behavioral endpoints following parental exposures. Given the breadth of current findings, we will utilize the behavioral endpoints following parental exposure to drugs of abuse as an example of the varying and complex behavioral outcomes in order to highlight some of the challenges facing researchers. In fact, skepticism remains in the field due to difficulties with conducting these complex experiments (Martos et al. 2015; Nagy and Turecki 2015). Moreover, variation observed between different models (rat versus mouse models) also serves to demonstrate the wide variability in outcomes with these studies and underscore the uniqueness of this phenomenon in different species. Both prenatal studies and exposure-specific reviews have recently arisen, so the goal of this chapter will be to define appropriate next steps for this burgeoning field.

For reviews of prenatal studies, see Behnke and Eyler (1993), Bock et al. (2015), Graignic-Philippe et al. (2014), Rohan et al. (2011), Sithisarn et al. (2012), and Veru et al. (2014).

For a review of behavioral phenotypes following different exposures, see Babenko et al. (2015), Chastain and Sarkar (2017), Park et al. (2017), Skinner (2014), Vassoler et al. (2014), and Yuan et al. (2016).

2 Behavioral Outcomes Following Parental Exposure to Drugs of Abuse

One of the most studied phenomena in the field remains that of consequences of exposure to drugs of abuse on subsequent generations. This is particularly compelling because of the high rates of recreational use, prescriptions, misuse, and abuse of drugs by humans. In terms of behavioral effects that have been tested, they can be broadly divided into four categories: effects on locomotor activity, affective behaviors, learning and memory/attention, and effects on reward-related behaviors.

2.1 Changes in Basal Activity Level Following F0 Drug Exposure

The monitoring of locomotor activity in rodents is one of the most common methods to track and screen for generalized changes in behavior. Locomotor activity studied under specific environmental conditions can differentiate behavioral outcomes. Specific categories are often investigated, including measures of general hypo- or hyperactivity, anxiety-like behavior, motivated exploration, and highly repetitive sequences of locomotor behavior (i.e., stereotypy). The focus on locomotor behavior in the context of drug exposure also demonstrates translational value, with evidence from human studies suggesting an association between exposure to drugs of abuse and alteration in locomotor activity in children. For example, there is a correlation between parental alcoholism and offspring childhood hyperactivity (Goodwin et al. 1975). Therefore, offspring locomotor behavior is often measured when examining the phenotypic differences in multi-, inter-, and transgenerational studies.

A number of studies in rodents have examined the effects of parental alcohol consumption on locomotor activity in offspring. In one such study, male rats were intubated twice daily with alcohol for 9 weeks and then bred with untreated females. Controls included pair-fed and vehicle-intubated males, as well as non-treated ad-lib-fed males. Locomotor activity in subsequent offspring was tested at 21, 42, and 90 days of age. Increased locomotor activity across all ages was associated with paternal alcohol consumption, and these effects were mediated, at least in part, by changes within the cholinergic system, which suggests alterations that could affect many aspects of behavior including learning and memory as well as attention (Abel 1993, 1994). Another study, however, observed decreased activity associated with paternal alcohol consumption in rats (Abel 1989b). Indeed, it appears that effects on locomotor activity are dependent upon both the duration of drinking prior to insemination and the age of the offspring at the time of testing (Abel 1989a). In addition to basal locomotor activity, psychostimulant-induced hyperlocomotion has also been examined following paternal alcohol consumption. Dose-dependent effects on amphetamine-induced locomotor behavior in the offspring of alcohol-exposed males have been observed, with increased locomotor activity observed following high doses and blunted locomotor activity following low doses (Abel 1993). Moreover, if the paternal alcohol consumption is voluntary (e.g., in the drinking water), there is a dose-dependent decrease in locomotor activity on post-natal days (PND) 20 and 24 (Abel and Lee 1988) accompanied by a dose-related decreased in testosterone at PND 55 (Abel 1989b; Abel and Lee 1988). It should be noted that these studies on voluntary intake were conducted in mice, while all of the other aforementioned studies were performed in rats. Overall, however, these studies suggest that divergent effects on locomotor behavior may be dependent on the duration of exposure, on the age at testing, and perhaps on whether alcohol exposure is forced or voluntary.

Alcohol is not the only drug that affects offspring locomotor behavior. Paternal cocaine exposure also results in hyperlocomotion in the offspring. For example, cocaine administered to male Long Evans rats for a minimum of 72 days results in offspring hyperactivity (Abel 1989b). This effect, however, was not observed when examined using a mouse model (Killinger et al. 2012), suggesting that species and perhaps strain can influence such intergenerational effects. Indeed, strain differences in behavioral response to drugs of abuse have been demonstrated (Cabib et al. 2000), and therefore it would not be surprising if strain interacted with environment to produce different generational effects. Interestingly, prenatal nicotine exposure in female mice causes hyperlocomotion in F2 and F3 offspring of the maternal but not paternal lineage (Zhu et al. 2014). Taken together, studies performed to date demonstrate both intergenerational and transgenerational effects on locomotor activity following parental drug exposure. The specific direction and magnitude of the effect are dependent on numerous factors, including the sex of the parent, the age of the offspring at testing, the type and duration of the exposure regimen, the drug, as well as the species/strain examined. Such wide variations are not unexpected when one considers that at its essence epigenetic inheritance involves interplay between genotype, embryological milieu, and environmental factors throughout development.

2.2 *Changes in Affective Behaviors Following F0 Drug Exposure*

Substance use disorders and mood disturbances are highly correlated with significant comorbidity. For example, compared with the general population, individuals with a substance use disorder are more than twice as likely to suffer from a mood and/or anxiety disorder; and the reverse is also true (Geller 1988; Hasin and Kilcoyne 2012; Hasin and Grant 1987; Pasche 2012; Wu and Blazer 2014). Such findings suggest significant interaction between the neural mechanisms underlying substance abuse and affective disorders. Based on this interrelatedness, measures of anxiety- and depressive-like behaviors in rodents are often examined following F0 drug exposure. These studies have revealed significant epigenetic effects as a function of parental drug exposure.

With regard to alcohol, paternal alcohol consumption leads to decreased immobility in the forced swim test in the offspring, which has been interpreted as reduced depressive-like behavior (Abel 1991a). However, such findings may be confounded by the generalized alterations in locomotor behavior detailed previously. Moreover, rat offspring of alcohol-consuming males also demonstrate decreased grooming, which may suggest a blunted response to environmental stimuli, which is often associated with a depressive-like phenotype (Abel 1991b). Similar to locomotor effects, rats and mice seem to have opposite effects in the forced swim test following paternal alcohol consumption. Mice offspring sired by alcohol-consuming males had increased immobility on the FST, which was reversed by imipramine (Abel and Bilitzke 1990). Of note, this same study included rats and replicated the previous forced swim test finding of decreased immobility, providing additional support for significant species differences on the intergenerational effects of paternal alcohol (Abel and Bilitzke 1990).

Even very limited exposures to alcohol prior to mating have been shown to affect subsequent generations. Thus, males administered acute alcohol just prior to mating sired pups that had delays in meeting developmental milestones in comparison to controls. They showed delayed surface righting, clinging, tail-pull reflex, rotation, linear movement, and climbing an inclined surface (Meek et al. 2007). As adults, the alcohol-sired group showed blunted risk assessment behavior, worse attention, and decreased anxiety-like behavior as measured with defensive probe burying (Meek et al. 2007). They were also more aggressive in social interactions (Meek et al. 2007). Together, the results indicate that acute alcohol administered just prior to mating causes offspring with delayed physical development, increased aggression, and decreased fearfulness (Meek et al. 2007). The mechanisms involved in the transmission of these effects remain unknown.

Opioid exposure in the F0 generation has also been shown to affect mood-related behaviors in subsequent generations. In one model, adolescent females are administered morphine or saline for 10 days during adolescence and then bred with naïve males after a 3–4-week drug-free period. Differences in their offspring are apparent even during early postnatal development, with F1 offspring demonstrating decreased maternal separation-evoked ultrasonic vocalizations (Bodi et al. 2016). During adulthood, F1 rats show increased time spent in the center of an open field indicating reduced anxiety-like behavior, although this effect was only observed in females during diestrus (Byrnes et al. 2011). In contrast, there was increased anxiety-like behavior as measured in the elevated plus maze (less time on the open arms) and decreased novel environment exploration in the offspring of adolescently morphine-exposed females compared to saline controls (Byrnes 2005). In addition, adolescent F1 males demonstrate decreased rough and tumble play behavior, while F1 females show increased play behaviors (Johnson et al. 2011). This effect seems to suggest convergence of gender-typical social behaviors, which could have significant implications over time.

Paternal cocaine exposure in a mouse model did not affect anxiety-like behavior in the offspring yet increased immobility in the tail suspension test. As tail suspension is used as a measure of depressive-like behavior, such findings suggest an increase in depressive-like behaviors (Killinger et al. 2012). However, another study reported the opposite outcome, i.e., male offspring of cocaine-exposed males demonstrated increased anxiety-like behavior with no change in depressive-like behavior (Fischer et al. 2017). This second data set matches results from rats, using a model in which males were allowed to self-administer cocaine for an extended period during adulthood prior to siring offspring. In that study, there was increased anxiety-like behavior in the male offspring as measured by novelty-induced hyperphagia paradigm and defensive probe burying task (White et al. 2016). Although there was no effect on depressive-like behavior as measured by forced swim task (White et al. 2016), once again, there seem to be divergent effects depending on the animal model, exposure regimen, and age at exposure and testing. These discrepancies actually serve to reconcile the divergent literature showing a wide range of behavioral phenotypes in multi-, inter-, and transgenerational models.

2.3 Changes in Learning and Memory and Attention Following F0 Drug Exposure

Substance use disorders are sometimes thought of as a dysfunction of the learning and memory system. Drug taking causes a pathological subversion of normal brain learning such that the motivation to seek and take a drug as well as stimuli associated with drug intake leads to compulsive use (Robbins and Everitt 2002). In addition, there is some evidence that drug use itself disrupts normal learning, memory, and attention processes (Taubenfeld et al. 2010; Torregrossa et al. 2011). Thus, these same processes are examined in the offspring in order to determine if there is any transmission of these acquired deficiencies. Indeed, multiple drug models have observed deficits in learning, memory, and/or attention in subsequent generations.

Male rats intubated with alcohol produce offspring that have delays in learning a passive avoidance task during early adolescence (Abel 1994). In rats that were allowed to freely consume alcohol (as opposed to oral gavage), female offspring showed deficits in learning a two-way shock avoidance task, although there were no differences in passive avoidance or spontaneous alternation (Abel and Tan 1988). Extended paternal alcohol consumption that ceased 2 weeks prior to conception produced offspring that did not differ on developmental milestones or locomotor activity nor on object exploration and recognition tasks (Wozniak et al. 1991). However, male offspring did demonstrate impaired acquisition of a spatial learning discrimination task utilizing either a radial arm maze or a T-maze (Wozniak et al. 1991). In contrast, in mice, high doses of paternal alcohol produced offspring requiring fewer trials to learn a passive avoidance task which was associated with longer latencies to reach a choice point in a T-maze (Abel and Lee 1988). Paternal preconception ethanol in mice also produces F1 offspring with attention deficits such as inattention and increased impulsivity (Kim et al. 2014).

Paternal cocaine exposure also results in F1 offspring with memory deficits. Extended adult cocaine self-administration during adulthood causes deficits in novel place recognition task in male offspring but not novel object recognition (Wimmer et al. 2017). Similarly, with paternal cocaine exposure in mice, the offspring have deficits in sustained visuospatial attention and spatial working memory utilizing the 5-arm maze paradigm (He et al. 2006). There was also a decreased percentage of correct trials at the shortest stimulus duration in male F1 as well as deficits in working memory (He et al. 2006). The deficit was worse in females appearing at longer stimulus durations and shorter intertrial intervals (He et al. 2006).

2.4 *Altered Responsivity to Drugs of Abuse Following F0 Drug Exposure*

Substance use disorders run in families, yet a genetic component that completely explains the heredity has yet to be determined. There have been many genome-wide association studies that have searched for genetic components to explain inheritance patterns of addiction, and the lack of any definitive gene or set of genes has been termed "missing heritability." It has been proposed that transgenerational epigenetic inheritance may play a role in this missing heritability. It is not surprising then to find that considerable work on F0 exposure to drugs of abuse has focused on alterations in the reward system and offspring susceptibility to addiction-like disorders.

Based on clinical observations, it was widely predicted that increased sensitivity to drugs of abuse would be observed in future generations. There is some evidence to support this hypothesis. For example, in rats, paternal alcohol exposure caused increased locomotor response to amphetamine in F1 offspring (Abel 1993). This could be interpreted as a cross-sensitization phenotype of the F1 animals. Cross-sensitization is a phenomenon whereby sensitization to a particular stimulus (such as a drug of abuse) is generalized to a related stimulus (a different drug of abuse), which results in the amplification of a response to both the original and the related stimulus. Therefore, the presence of an increased response to amphetamine may indicate an increase in the addiction liability in those animals. Some of the most compelling evidence for increased susceptibility to addiction-like disorders in offspring arises from adolescent cannabinoid exposure studies. For example, it was shown that adolescent cannabinoid exposure to the synthetic cannabinoid WIN-55,212 induces an enhanced sensitivity to morphine-induced conditioned place preference in male F1 offspring (Byrnes et al. 2012). In addition, female offspring demonstrate an enhanced expression of morphine-induced locomotor sensitization (Vassoler et al. 2013a). It was later shown that adolescent THC exposure (the psychoactive cannabinoid found in marijuana) in both parents increases the effort that offspring will work toward heroin self-administration (Szutorisz et al. 2014). Cocaine exposure in females prior to pregnancy enhances the acute locomotor response to cocaine in adult male offspring (Sasaki et al. 2014). Finally, intergenerational effects on drug sensitivity have also been shown with adolescent morphine exposure. Female adolescent morphine exposure causes a more rapid induction of morphine-induced locomotor sensitization in female offspring and a significant enhancement of sensitization in male offspring (Byrnes 2005). Moreover, increased sensitivity to the analgesic effects of morphine and more rapid development of tolerance have also been observed in these F1 offspring (Byrnes et al. 2011). Lastly, there were increases in the preference for lower doses of morphine as measured by conditioned place preference (CPP; Vassoler et al. 2016).

Yet, for each of these classes of drugs and exposure regimens, there has been an equally large (if not larger) body of evidence that suggests a protective effect in the offspring where exposures in parental generations seem to be preparing offspring for similar environments. This hypothesis, termed the "fetal origins of adult disease

(FOAD) hypothesis," has been gaining supporting evidence from multiple fields (Calkins and Devaskar 2011). Thus, paternal alcohol exposure reduces ethanol preference and consumption in adult male offspring yet increases sensitivity to the anxiolytic and motor-enhancing effects of ethanol (Finegersh and Homanics 2014). Adolescent female cannabinoid exposure diminishes the reward-facilitating effects of low doses of THC as well as amphetamine in adult male offspring as measured using intracranial self-stimulation (Pitsilis et al. 2017). Female adolescent morphine exposure attenuates quinpirole-induced locomotor sensitization in F1 and F2 male offspring, demonstrating transgenerational reductions in sensitivity to the dopaminergic system (Byrnes et al. 2013). Indeed, when examining self-administration, there are a decrease in self-administration acquisition across all doses tested in females and a particularly striking blunting of morphine-induced reinstatement in males that extends to the F1 and F2 generations, another transgenerational effect (Vassoler et al. 2017). In one study where fathers were allowed to binge-eat high-sucrose foods, cocaine- and sucrose-seeking behaviors were blunted in the offspring (Le et al. 2017a). The offspring also showed decreases in reward seeking and increases in anxiety-like behavior (Le et al. 2017a).

Further evidence in support of the FOAD hypothesis stems from paternal cocaine work. In mice, paternal cocaine injections for 75 days produced female offspring with blunted cocaine preference (Fischer et al. 2017). Additionally, females and male F1 s exhibited increased sensitivity to the psychomotor effects of cocaine and amphetamine (Fischer et al. 2017). This point reconciles some of the disparate data as it appears that reward and locomotor response are not necessarily correlated. Therefore, although animals may demonstrate increased locomotor response to particular drugs of abuse, their reward-related response may be blunted. F1 animals sired by cocaine-exposed males also consumed more sucrose, demonstrating specific alterations in reward pathways (Fischer et al. 2017).

In a study where male rats were trained to self-administer cocaine for 60 days during adulthood, 24 h prior to mating with naïve females, it was found that male F1 offspring demonstrated decreased acquisition of cocaine self-administration and reduced effort in a progressive ratio schedule (Vassoler et al. 2013b). Intriguingly, this study was then replicated by an independent group, and it was further discovered that if the sires are separated into addiction-like groups and non-addiction-like groups, the F1 and F2 offspring of high-responding sires had increased incentive response to cocaine (Le et al. 2017b). Thus, when all animals are examined together, there is a general trend toward decreased self-administration in the offspring, but when the animals are separated based on response, the highly drug-motivated sires transmit this phenotype to their offspring and grand offspring (Le et al. 2017b). This is particularly interesting when the human substance use patterns are taken into consideration. The majority of the population can consume rewarding substances recreationally and not develop a substance use disorder. The minority that does develop a substance use disorder may have additional genetic and epigenetic changes that prime them for the disease.

Taken together these reviewed studies demonstrate the breadth and variability of behavioral phenotypes in multi-, inter-, and transgenerational studies. Moreover, it is clear that relatively minor exposures result in long-term and transgenerational behavioral phenotypes but also fairly minor modifications to protocols result in divergent and sometimes opposite responses. While some skepticism exists about the mechanism of action of transfer between generations, it is clear that environmental exposures can interact with genetics to alter behavior in subsequent generations.

References

Abel EL (1989a) Paternal alcohol consumption: effects of age of testing and duration of paternal drinking in mice. Teratology 40:467–474

Abel EL (1989b) Paternal behavioral mutagenesis. Neurotoxicology 10:335–345

Abel EL (1991a) Alcohol consumption does not affect fathers but does affect their offspring in the forced swimming test. Pharmacol Toxicol 68:68–69

Abel EL (1991b) Paternal alcohol consumption affects grooming response in rat offspring. Alcohol 8:21–23

Abel EL (1993) Paternal alcohol exposure and hyperactivity in rat offspring: effects of amphetamine. Neurotoxicol Teratol 15:445–449

Abel EL (1994) Effects of physostigmine on male offspring sired by alcohol-treated fathers. Alcohol Clin Exp Res 18:648–652

Abel EL, Bilitzke P (1990) Paternal alcohol exposure: paradoxical effect in mice and rats. Psychopharmacology 100:159–164

Abel EL, Lee JA (1988) Paternal alcohol exposure affects offspring behavior but not body or organ weights in mice. Alcohol Clin Exp Res 12:349–355

Abel EL, Tan SE (1988) Effects of paternal alcohol consumption on pregnancy outcome in rats. Neurotoxicol Teratol 10:187–192

Babenko O, Kovalchuk I, Metz GA (2015) Stress-induced perinatal and transgenerational epigenetic programming of brain development and mental health. Neurosci Biobehav Rev 48:70–91

Behnke M, Eyler FD (1993) The consequences of prenatal substance use for the developing fetus, newborn, and young child. Int J Addict 28:1341–1391

Bock J, Wainstock T, Braun K, Segal M (2015) Stress in utero: prenatal programming of brain plasticity and cognition. Biol Psychiatry 78:315–326

Bodi CM, Vassoler FM, Byrnes EM (2016) Adolescent experience affects postnatal ultrasonic vocalizations and gene expression in future offspring. Dev Psychobiol 58:714–723

Byrnes EM (2005) Transgenerational consequences of adolescent morphine exposure in female rats: effects on anxiety-like behaviors and morphine sensitization in adult offspring. Psychopharmacology 182:537–544

Byrnes JJ, Babb JA, Scanlan VF, Byrnes EM (2011) Adolescent opioid exposure in female rats: transgenerational effects on morphine analgesia and anxiety-like behavior in adult offspring. Behav Brain Res 218:200–205

Byrnes JJ, Johnson NL, Schenk ME, Byrnes EM (2012) Cannabinoid exposure in adolescent female rats induces transgenerational effects on morphine conditioned place preference in male offspring. J Psychopharmacol 26:1348–1354

Byrnes JJ, Johnson NL, Carini LM, Byrnes EM (2013) Multigenerational effects of adolescent morphine exposure on dopamine D2 receptor function. Psychopharmacology 227:263–272

Cabib S, Orsini C, Le Moal M, Piazza PV (2000) Abolition and reversal of strain differences in behavioral responses to drugs of abuse after a brief experience. Science 289:463–465

Calkins K, Devaskar SU (2011) Fetal origins of adult disease. Curr Probl Pediatr Adolesc Health Care 41:158–176

Chastain LG, Sarkar DK (2017) Alcohol effects on the epigenome in the germline: role in the inheritance of alcohol-related pathology. Alcohol 60:53–66

Daxinger L, Whitelaw E (2012) Understanding transgenerational epigenetic inheritance via the gametes in mammals. Nat Rev Genet 13:153–162

Ferguson-Smith AC, Patti ME (2011) You are what your dad ate. Cell Metab 13:115–117

Finegersh A, Homanics GE (2014) Paternal alcohol exposure reduces alcohol drinking and increases behavioral sensitivity to alcohol selectively in male offspring. PLoS One 9:e99078

Fischer DK, Rice RC, Martinez Rivera A, Donohoe M, Rajadhyaksha AM (2017) Altered reward sensitivity in female offspring of cocaine-exposed fathers. Behav Brain Res 332:23–31

Geller B (1988) Pharmacotherapy of concomitant psychiatric disorders in adolescent substance abusers. NIDA Res Monogr 77:94–112

Goodwin DW, Schulsinger F, Hermansen L, Guze SB, Winokur G (1975) Alcoholism and the hyperactive child syndrome. J Nerv Ment Dis 160:349–353

Graignic-Philippe R, Dayan J, Chokron S, Jacquet AY, Tordjman S (2014) Effects of prenatal stress on fetal and child development: a critical literature review. Neurosci Biobehav Rev 43:137–162

Hasin DS, Grant BF (1987) Psychiatric diagnosis of patients with substance abuse problems: a comparison of two procedures, the DIS and the SADS-L. Alcoholism, drug abuse/dependence, anxiety disorders and antisocial personality disorder. J Psychiatr Res 21:7–22

Hasin D, Kilcoyne B (2012) Comorbidity of psychiatric and substance use disorders in the United States: current issues and findings from the NESARC. Curr Opin Psychiatry 25:165–171

He F, Lidow IA, Lidow MS (2006) Consequences of paternal cocaine exposure in mice. Neurotoxicol Teratol 28:198–209

Heard E, Martienssen RA (2014) Transgenerational epigenetic inheritance: myths and mechanisms. Cell 157:95–109

Johnson NL, Carini L, Schenk ME, Stewart M, Byrnes EM (2011) Adolescent opiate exposure in the female rat induces subtle alterations in maternal care and transgenerational effects on play behavior. Front Psych 2:29

Killinger CE, Robinson S, Stanwood GD (2012) Subtle biobehavioral effects produced by paternal cocaine exposure. Synapse 66:902–908

Kim P, Choi CS, Park JH, Joo SH, Kim SY, Ko HM, Kim KC, Jeon SJ, Park SH, Han SH, Ryu JH, Cheong JH, Han JY, Ko KN, Shin CY (2014) Chronic exposure to ethanol of male mice before mating produces attention deficit hyperactivity disorder-like phenotype along with epigenetic dysregulation of dopamine transporter expression in mouse offspring. J Neurosci Res 92:658–670

Le Q, Li Y, Hou W, Yan B, Yu X, Song H, Wang F, Ma L (2017a) Binge-like sucrose self-administration experience inhibits cocaine and sucrose seeking behavior in offspring. Front Behav Neurosci 11:184

Le Q, Yan B, Yu X, Li Y, Song H, Zhu H, Hou W, Ma D, Wu F, Zhou Y, Ma L (2017b) Drug-seeking motivation level in male rats determines offspring susceptibility or resistance to cocaine-seeking behaviour. Nat Commun 8:15527

Lim JP, Brunet A (2013) Bridging the transgenerational gap with epigenetic memory. Trends Genet 29:176–186

Martos SN, Tang WY, Wang Z (2015) Elusive inheritance: transgenerational effects and epigenetic inheritance in human environmental disease. Prog Biophys Mol Biol 118:44–54

Meek LR, Myren K, Sturm J, Burau D (2007) Acute paternal alcohol use affects offspring development and adult behavior. Physiol Behav 91:154–160

Nagy C, Turecki G (2015) Transgenerational epigenetic inheritance: an open discussion. Epigenomics 7:781–790

Park JH, Kim SH, Lee MS, Kim MS (2017) Epigenetic modification by dietary factors: implications in metabolic syndrome. Mol Asp Med 54:58–70

Pasche S (2012) Exploring the comorbidity of anxiety and substance use disorders. Curr Psychiatry Rep 14:176–181

Pitsilis G, Spyridakos D, Nomikos GG, Panagis G (2017) Adolescent female cannabinoid exposure diminishes the reward-facilitating effects of delta(9)-tetrahydrocannabinol and d-amphetamine in the adult male offspring. Front Pharmacol 8:225

Robbins TW, Everitt BJ (2002) Limbic-striatal memory systems and drug addiction. Neurobiol Learn Mem 78:625–636
Rohan AJ, Monk C, Marder K, Reame N (2011) Prenatal toxicology screening for substance abuse in research: codes and consequences. Subst Abus 32:159–164
Sasaki A, Constantinof A, Pan P, Kupferschmidt DA, McGowan PO, Erb S (2014) Cocaine exposure prior to pregnancy alters the psychomotor response to cocaine and transcriptional regulation of the dopamine D1 receptor in adult male offspring. Behav Brain Res 265:163–170
Sithisarn T, Granger DT, Bada HS (2012) Consequences of prenatal substance use. Int J Adolesc Med Health 24:105–112
Skinner MK (2011) Environmental epigenetic transgenerational inheritance and somatic epigenetic mitotic stability. Epigenetics 6:838–842
Skinner MK (2014) Endocrine disruptor induction of epigenetic transgenerational inheritance of disease. Mol Cell Endocrinol 398:4–12
Szutorisz H, DiNieri JA, Sweet E, Egervari G, Michaelides M, Carter JM, Ren Y, Miller ML, Blitzer RD, Hurd YL (2014) Parental THC exposure leads to compulsive heroin-seeking and altered striatal synaptic plasticity in the subsequent generation. Neuropsychopharmacology 39:1315–1323
Taubenfeld SM, Muravieva EV, Garcia-Osta A, Alberini CM (2010) Disrupting the memory of places induced by drugs of abuse weakens motivational withdrawal in a context-dependent manner. Proc Natl Acad Sci U S A 107:12345–12350
Torregrossa MM, Corlett PR, Taylor JR (2011) Aberrant learning and memory in addiction. Neurobiol Learn Mem 96:609–623
Vassoler FM, Johnson NL, Byrnes EM (2013a) Female adolescent exposure to cannabinoids causes transgenerational effects on morphine sensitization in female offspring in the absence of in utero exposure. J Psychopharmacol 27:1015–1022
Vassoler FM, White SL, Schmidt HD, Sadri-Vakili G, Pierce RC (2013b) Epigenetic inheritance of a cocaine-resistance phenotype. Nat Neurosci 16:42–47
Vassoler FM, Byrnes EM, Pierce RC (2014) The impact of exposure to addictive drugs on future generations: physiological and behavioral effects. Neuropharmacology 76(Pt B):269–275
Vassoler FM, Wright SJ, Byrnes EM (2016) Exposure to opiates in female adolescents alters mu opiate receptor expression and increases the rewarding effects of morphine in future offspring. Neuropharmacology 103:112–121
Vassoler FM, Oliver DJ, Wyse C, Blau A, Shtutman M, Turner JR, Byrnes EM (2017) Transgenerational attenuation of opioid self-administration as a consequence of adolescent morphine exposure. Neuropharmacology 113:271–280
Veru F, Laplante DP, Luheshi G, King S (2014) Prenatal maternal stress exposure and immune function in the offspring. Stress 17:133–148
White SL, Vassoler FM, Schmidt HD, Pierce RC, Wimmer ME (2016) Enhanced anxiety in the male offspring of sires that self-administered cocaine. Addict Biol 21:802–810
Wimmer ME, Briand LA, Fant B, Guercio LA, Arreola AC, Schmidt HD, Sidoli S, Han Y, Garcia BA, Pierce RC (2017) Paternal cocaine taking elicits epigenetic remodeling and memory deficits in male progeny. Mol Psychiatry 22:1641–1650
Wozniak DF, Cicero TJ, Kettinger L 3rd, Meyer ER (1991) Paternal alcohol consumption in the rat impairs spatial learning performance in male offspring. Psychopharmacology 105:289–302
Wu LT, Blazer DG (2014) Substance use disorders and psychiatric comorbidity in mid and later life: a review. Int J Epidemiol 43:304–317
Yuan TF, Li A, Sun X, Ouyang H, Campos C, Rocha NBF, Arias-Carrion O, Machado S, Hou G, So KF (2016) Transgenerational inheritance of paternal neurobehavioral phenotypes: stress, addiction, ageing and metabolism. Mol Neurobiol 53:6367–6376
Zhu J, Lee KP, Spencer TJ, Biederman J, Bhide PG (2014) Transgenerational transmission of hyperactivity in a mouse model of ADHD. J Neurosci 34:2768–2773

Correction to: Role of MicroRNAs in Anxiety and Anxiety-Related Disorders

Conor P. Murphy and Nicolas Singewald

Correction to: Chapter "Role of MicroRNAs in Anxiety and Anxiety-Related Disorders" in: C. P. Murphy and N. Singewald, Curr Topics Behav Neurosci, DOI 10.1007/7854_2019_109

The chapter 'Role of MicroRNAs in Anxiety and Anxiety-Related Disorders' has now been made available open access under a CC BY 4.0 license.

The updated online version of this chapter can be found at DOI 10.1007/7854_2019_109

© Springer Nature Switzerland AG 2019

Curr Topics Behav Neurosci (2019) 42: 259–260
https://doi.org/10.1007/7854_2019_118
Published Online: 17 November 2019

The manufacturer's authorised representative in the EU is Springer Nature Customer Service Centre GmbH, Europaplatz 3, 69115 Heidelberg, Germany. If you have any concerns regarding our products, please contact ProductSafety@springernature.com

Printed and bound by CPI Group (UK) Ltd, Croydon, CR0 4YY
25/09/2026
02222892-0008